T0309795

Geophysical Monograph Series

Including
IUGG Volumes
Maurice Ewing Volumes
Mineral Physics Volumes

Geophysical Monograph Series

164 Archean Geodynamics and Environments *Keith Benn, Jean-Claude Mareschal, and Kent C. Condie (Eds.)*
165 Solar Eruptions and Energetic Particles *Natchimuthukonar Gopalswamy, Richard Mewaldt, and Jarmo Torsti (Eds.)*
166 Back-Arc Spreading Systems: Geological, Biological, Chemical, and Physical Interactions *David M. Christie, Charles Fisher, Sang-Mook Lee, and Sharon Givens (Eds.)*
167 Recurrent Magnetic Storms: Corotating Solar Wind Streams *Bruce Tsurutani, Robert McPherron, Walter Gonzalez, Gang Lu, José H. A. Sobral, and Natchimuthukonar Gopalswamy (Eds.)*
168 Earth's Deep Water Cycle *Steven D. Jacobsen and Suzan van der Lee (Eds.)*
169 Magnetospheric ULF Waves: Synthesis and New Directions *Kazue Takahashi, Peter J. Chi, Richard E. Denton, and Robert L. Lysak (Eds.)*
170 Earthquakes: Radiated Energy and the Physics of Faulting *Rachel Abercrombie, Art McGarr, Hiroo Kanamori, and Giulio Di Toro (Eds.)*
171 Subsurface Hydrology: Data Integration for Properties and Processes *David W. Hyndman, Frederick D. Day-Lewis, and Kamini Singha (Eds.)*
172 Volcanism and Subduction: The Kamchatka Region *John Eichelberger, Evgenii Gordeev, Minoru Kasahara, Pavel Izbekov, and Johnathan Lees (Eds.)*
173 Ocean Circulation: Mechanisms and Impacts—Past and Future Changes of Meridional Overturning *Andreas Schmittner, John C. H. Chiang, and Sidney R. Hemming (Eds.)*
174 Post-Perovskite: The Last Mantle Phase Transition *Kei Hirose, John Brodholt, Thorne Lay, and David Yuen (Eds.)*
175 A Continental Plate Boundary: Tectonics at South Island, New Zealand *David Okaya, Tim Stem, and Fred Davey (Eds.)*
176 Exploring Venus as a Terrestrial Planet *Larry W. Esposito, Ellen R. Stofan, and Thomas E. Cravens (Eds.)*
177 Ocean Modeling in an Eddying Regime *Matthew Hecht and Hiroyasu Hasumi (Eds.)*
178 Magma to Microbe: Modeling Hydrothermal Processes at Oceanic Spreading Centers *Robert P. Lowell, Jeffrey S. Seewald, Anna Metaxas, and Michael R. Perfit (Eds.)*
179 Active Tectonics and Seismic Potential of Alaska *Jeffrey T. Freymueller, Peter J. Haeussler, Robert L. Wesson, and Göran Ekström (Eds.)*
180 Arctic Sea Ice Decline: Observations, Projections, Mechanisms, and Implications *Eric T. DeWeaver, Cecilia M. Bitz, and L.-Bruno Tremblay (Eds.)*
181 Midlatitude Ionospheric Dynamics and Disturbances *Paul M. Kintner, Jr., Anthea J. Coster, Tim Fuller-Rowell, Anthony J. Mannucci, Michael Mendillo, and Roderick Heelis (Eds.)*
182 The Stromboli Volcano: An Integrated Study of the 2002–2003 Eruption *Sonia Calvari, Salvatore Inguaggiato, Giuseppe Puglisi, Maurizio Ripepe, and Mauro Rosi (Eds.)*
183 Carbon Sequestration and Its Role in the Global Carbon Cycle *Brian J. McPherson and Eric T. Sundquist (Eds.)*
184 Carbon Cycling in Northern Peatlands *Andrew J. Baird, Lisa R. Belyea, Xavier Comas, A. S. Reeve, and Lee D. Slater (Eds.)*
185 Indian Ocean Biogeochemical Processes and Ecological Variability *Jerry D. Wiggert, Raleigh R. Hood, S. Wajih A. Naqvi, Kenneth H. Brink, and Sharon L. Smith (Eds.)*
186 Amazonia and Global Change *Michael Keller, Mercedes Bustamante, John Gash, and Pedro Silva Dias(Eds.)*
187 Surface Ocean–Lower Atmosphere Processes *Corinne Le Quèrè and Eric S. Saltzman (Eds.)*
188 Diversity of Hydrothermal Systems on Slow Spreading Ocean Ridges *Peter A. Rona, Colin W. Devey, Jérôme Dyment, and Bramley J. Murton (Eds.)*
189 Climate Dynamics: Why Does Climate Vary? *De-Zheng Sun and Frank Bryan (Eds.)*
190 The Stratosphere: Dynamics, Transport, and Chemistry *L. M. Polvani, A. H. Sobel, and D. W. Waugh (Eds.)*
191 Rainfall: State of the Science *Firat Y. Testik and Mekonnen Gebremichael (Eds.)*
192 Antarctic Subglacial Aquatic Environments *Martin J. Siegert, Mahlon C. Kennicut II, and Robert A. Bindschadler*
193 Abrupt Climate Change: Mechanisms, Patterns, and Impacts *Harunur Rashid, Leonid Polyak, and Ellen Mosley-Thompson (Eds.)*
194 Stream Restoration in Dynamic Fluvial Systems: Scientific Approaches, Analyses, and Tools *Andrew Simon, Sean J. Bennett, and Janine M. Castro (Eds.)*
195 Monitoring and Modeling the *Deepwater Horizon* Oil Spill: A Record-Breaking Enterprise *Yonggang Liu, Amy MacFadyen, Zhen-Gang Ji, and Robert H. Weisberg (Eds.)*
196 Extreme Events and Natural Hazards: The Complexity Perspective *A. Surjalal Sharma, Armin Bunde, Vijay P. Dimri, and Daniel N. Baker (Eds.)*
197 Auroral Phenomenology and Magnetospheric Processes: Earth and Other Planets *Andreas Keiling, Eric Donovan, Fran Bagenal, and Tomas Karlsson (Eds.)*
198 Climates, Landscapes, and Civilizations *Liviu Giosan, Dorian Q. Fuller, Kathleen Nicoll, Rowan K. Flad, and Peter D. Clift (Eds.)*
199 Dynamics of the Earth's Radiation Belts and Inner Magnetosphere *Danny Summers, Ian R. Mann, Daniel N. Baker, Michael Schulz (Eds.)*
200 Lagrangian Modeling of the Atmosphere *John Lin (Ed.)*
201 Modeling the Ionosphere-Thermosphere *Jospeh D. Huba, Robert W. Schunk, and George V Khazanov (Eds.)*

Geophysical Monograph 202

The Mediterranean Sea

Temporal Variability and Spatial Patterns

Gian Luca Eusebi Borzelli
Miroslav Gačić
Piero Lionello
Paola Malanotte-Rizzoli
Editors

This work is a co-publication of the American Geophysical Union and John Wiley & Sons, Inc.

WILEY

This Work is a co-publication between the American Geophysical Union and John Wiley & Sons, Inc.

Published by John Wiley & Sons, Inc., Hoboken, New Jersey
Published simultaneously in Canada

Library of Congress Cataloging-in-Publication Data

Borzelli, Gianluca Eusebi.
The Mediterranean sea : temporal variability and spatial patterns / Gianluca Eusebi Borzelli. – First edition.
pages cm. – (Geophysical monograph series)
Includes index.
ISBN 978-1-118-84734-3 (hardback)
1. Mediterranean Sea. 2. Chemical oceanography. 3. Biogeochemical cycles. 4. Ocean circulation.
5. Ocean currents. I. Title.
GC651B57 2014
551.46′138–dc23

2013042748

Front cover image: Time averages of Sea Level Anomalies (SLA) obtained from altimeter data. Red-Yellow tones indicate high values, Blue tones indicate low values. Top panel: Average SLA in the period 1993–1997. Bottom panel: Average SLA in the period 1997–2001. The figure represents the dynamical transition occurred in the central Mediterranean in 1997. Altimeter data represented in the cover were produced by Ssalto/Duacs and distributed by Aviso with support of CNES. Ssalto/Duacs, Aviso and CNES staffs are gratefully acknowledged for the services provided. Map: Courtesy of the Gian Luca Eusebi Borzelli, volume editor. Map of the region showing temporal and spatial variations in sea circulation patterns.
Sea: Carpe89. Beautiful landscape sea in Sardinia.

Cover design by Modern Alchemy LLC

Printed in Singapore

10 9 8 7 6 5 4 3 2 1

This book is a tribute to our colleagues Allan Robinson and Volfango Rupolo.

CONTENTS

Color plate section is located between pages 84 and 85.

CONTRIBUTORS

Ali Aydoğdu
Institute of Marine Sciences, METU, Erdemli, Mersin, Turkey

Manuel Bensi
Istituto Nazionale di Oceanografia e di Geofisica Sperimentale (OGS), Sgonico (Trieste), Italy

Mireno Borghini
Istituto di Scienze Marine—ISMAR, Consiglio Nazionale delle Ricerche (CNR), Lerici (SP), Italy

Harry Bryden
National Oceanography Centre Southampton, University of Southampton, Empress Dock, Southampton,
United Kingdom; Istituto di Scienze Marine—ISMAR, Consiglio Nazionale delle Ricerche (CNR),
Venice, Italy

Vanessa Cardin
Istituto Nazionale di Oceanografia e di Geofisica Sperimentale (OGS), Sgonico (Trieste), Italy

Sotiria Georgiou
University of Athens, Athens, Greece

Isaac Gertman
Israel Oceanographic and Limnological Research, Haifa, Israel

Dagmar Hainbucher
Institut für Meereskunde, ZMAW, Universität Hamburg, Hamburg, Germany

Artur Hecht
Israel Oceanographic and Limnological Research, Haifa, Israel

R. Iacono
ENEA—C. R. Casaccia, Roma, Italy

Birgit Klein
Bundesanstalt für Seeschifffahrt und Hydrographie, Hamburg, Germany

Gerasimos Korres
Hellenic Center for Marine Research, Institute of Oceanography, Anavyssos, Greece

George Krokos
Hellenic Center for Marine Research, Institute of Oceanography,
Anavyssos, Greece

Alex Lascaratos
University of Athens, Athens, Greece

Mohammed Abdul Latif
Institute of Marine Sciences, METU, Erdemli, Mersin, Turkey

L. Liberti
Istituto Superiore per la Protezione e la Ricerca Ambientale, Rome, Italy

Anneta Mantziafou
University of Athens, Athens, Greece

S. Marullo
ENEA—C. R. Frascati, Frascati, Italy

E. Napolitano
ENEA—C. R. Casaccia, Roma, Italy

Emin Özsoy
Institute of Marine Sciences, METU, Erdemli, Mersin, Turkey

Vassilis Papadopoulos
Hellenic Centre for Marine Research, Patras, Greece

Ananda Pascual
IMEDEA (CSIC-UIB), Esporles, Mallorca, Spain

L. Pratt
Woods Hole Oceanographic Institution, Woods Hole, Massachusetts, USA

Wolfgang Roether
Institut für Umweltphysik, Univ. Bremen, Bremen, Germany

Angelo Rubino
Dipartimento di Scienze Ambientali, Informatica e Statistica, Universita' Ca' Foscari di Venezia, Venezia, Italy

Simón Ruiz
IMEDEA (CSIC-UIB), Esporles, Mallorca, Spain

J. C. Sánchez-Garrido
Grupo de Oceanografía Física. Dpto. Física Aplicada II, Campus de Teatinos, University of Malaga, Malaga, Spain

Gianmaria Sannino
Climate and Impact Modeling Lab—Energy and Environment Modeling Unit—ENEA, CR Casaccia, Rome, Italy

Katrin Schroeder
Istituto di Scienze Marine—ISMAR, Consiglio Nazionale delle Ricerche (CNR), Venice, Italy

Sarantis Sofianos
University of Athens, Athens, Greece

Samuel Somot
MeteoFrance, Toulouse, France

Stefania Sparnocchia
Istituto di Scienze Marine—ISMAR, Consiglio Nazionale delle Ricerche (CNR), Trieste, Italy

Alexander Theocharis
Hellenic Center for Marine Research, Institute of Oceanography, Anavyssos, Greece

Ersin Tutsak
Institute of Marine Sciences, METU, Erdemli, Mersin, Turkey

Dimitris Velaoras
Hellenic Center for Marine Research, Institute of Oceanography, Anavyssos, Greece

Anna Vetrano
Istituto di Scienze Marine—ISMAR, Consiglio Nazionale delle Ricerche (CNR), Lerici (SP), Italy

Enrique Vidal-Vijande
IMEDEA (CSIC-UIB), Esporles, Mallorca, Spain

PREFACE

This book covers important issues of the Mediterranean dynamics. In this relatively small body of water, fundamental ocean processes, such as surface wind forcing, buoyancy fluxes, lateral mass exchange, and deep convection, take place analogously to the world ocean, but over shorter spatial and temporal scales, simplifying the logistics necessary for monitoring. This makes the Mediterranean a laboratory for processes characterizing the global ocean and its climate.

The 1980s represent a crossroad in the study of the Mediterranean Sea. Indeed, during the second half of the 1980s, several large-scale, long-term international experiments were conducted and provided a wealth of oceanographic information. Four large international programs—the Gibraltar Experiment, the Physical Oceanography of the Eastern Mediterranean, which in 1990 evolved into the fully interdisciplinary program named POEM-Biology and Chemistry (POEM-BC), the Western Mediterranean Circulation Experiment, and PRIMO—defined the major characteristics of the Mediterranean Sea. The picture of the Mediterranean variability emerging from this fieldwork was complex and showed that multiple interacting time and spatial scales (basin, subbasin, and mesoscale), representing a wide variety of physical processes, characterize the Mediterranean dynamics.

The aforementioned programs ended by the second half of the 1990s. Since then, although valuable studies were carried out, uncoordinated research efforts, driven mainly by national interests, provided fragmented and sporadic results.

To establish the state of the art of the research in the Mediterranean and allow interested scientists to interact, the Space Academy Foundation, a nonprofit organization to promote space-science, and technology, CIESM, and the OGS (Istituto Nazionale di Oceanografia e di Geofisica Sperimentale), under the aegis of the Alta Presidenza della Repubblica Italiana, organized a fully interdisciplinary meeting, which was held in Rome on 7–8 November 2011 at the Accademia Nazionale dei Lincei. This book is the outcome of this effort and it is meant to be an important and original contribution to the knowledge of the phenomena that regulate the oceanography of the Mediterranean Sea.

Gian Luca Eusebi Borzelli, *CERSE (Center for Remote Sensing of the Earth), Rome, Italy; OGS (Istituto Nazionale di Oceanografia e di Geofisica Sperimenta), Italy*

Miroslav Gačić, *National Institute of Oceanography and Experimental Geophysics, Italy*

Piero Lionello, *University of Salento, Italy*

Paola Malanotte-Rizzoli, *Massachusetts Institute of Technology, USA*

1

Introduction to *The Mediterranean Sea: Temporal Variability and Spatial Patterns*

Gian Luca Eusebi Borzelli[1], Paola Malanotte-Rizzoli[2], Miroslav Gačić[3], and Piero Lionello[4]

This book stems from a workshop held in Rome in November 2011 at Accademia Nazionale dei Lincei to mark the twenty-fifth anniversary of the POEM (Physical Oceanography of the Eastern Mediterranean) program. The objectives of the workshop, however, were more ambitious than a memorial. First, the workshop was meant to provide a synopsis of the state of the art of the present knowledge of the Mediterranean Sea circulation. Second, it aimed at offering the opportunity to scientists working in different areas of the sea, both in the western and eastern basins, to meet and share ideas, fostering pan-Mediterranean collaborations.

The members of the POEM program gratefully acknowledge the crucial support provided over the years by the Intergovernmental Oceanographic Commission (IOC/UNESCO Ocean Sciences) and the Mediterranean Science Commission (CIESM).

This book collects eight original research articles describing new results in the study of the Mediterranean Sea physical properties. Until the beginning of the 1980s, the Mediterranean was considered of marginal importance being characterized by specific, regional phenomena with limited interest for global processes. The second half of the 1980s represents a crossroad in the study of this basin. Four large international programs—the Gibraltar Experiment [*Kinder and Bryden*, 1987], the Physical Oceanography of the Eastern Mediterranean [*Malanotte-Rizzoli and Robinson*, 1988], which in 1990 evolved in to the fully interdisciplinary program named POEM-Biology and Chemistry (POEM-BC), the Western Mediterranean Circulation Experiment [*WMCE Consortium*, 1989], and PRIMO [*EUROMODEL Group*, 1995]—defined the major characteristics of the Mediterranean Sea. The picture of its variability emerging from these studies was complex and it showed that multiple interacting time and spatial scales (basin, subbasin, and mesoscale), representing a wide variety of physical processes, characterize the Mediterranean dynamics.

This new observational and theoretical knowledge established that the Mediterranean is a laboratory basin, where the processes characterizing the global ocean and its climate can be investigated. In fact, all major forcing mechanisms (such as surface wind forcing, buoyancy fluxes, lateral mass exchange, and deep convection) determining the global oceanic circulation are present in the Mediterranean Sea. Deep and intermediate water masses are formed in different areas and drive the Mediterranean thermohaline cells, which show important analogies with the global ocean conveyor belt. However, the Mediterranean Sea presents important advantages as temporal and spatial scales are shorter than in the global ocean, simplifying the logistics necessary for monitoring the circulation.

The aforementioned programs ended by the second half of the 1990s revealing a number of important features and opened a series of scientific questions. These can be summarized as follows:

1. The Eastern Mediterranean Transient (EMT)
   The main source of dense water driving the eastern Mediterranean deep convection cell, normally localized in the Adriatic Sea, by the end of the 1980s, shifted to

[1] *CERSE (Center for Remote Sensing of the Earth), Rome, Italy; OGS (Istituto Nazionale di Oceanografia e di Geofisica Sperimenta), Sgonico (TS), Italy*

[2] *Massachusetts Institute of Technology, Cambridge, MA—USA*

[3] *Istituto Nazionale di Oceanografia e di Geofisica Sperimentale—OGS, Trieste, Italy*

[4] *University of Salento, Dipartimento di Scienze e Tecnologie Biologiche ed Ambientali, Lecce, Italy*

The Mediterranean Sea: Temporal Variability and Spatial Patterns, Geophysical Monograph 202. First Edition.
Edited by Gian Luca Eusebi Borzelli, Miroslav Gačić, Piero Lionello, and Paola Malanotte-Rizzoli.

the Aegean and determined changes in properties of water masses in the deep layers of the eastern [*Roether et al.*, 1996] and western Mediterranean [*Schroeder et al.*, 2006; *Gačić et al.*, 2013]. Is this effect, which determines a nonsteady picture of the entire Mediterranean thermohaline circulation, a sporadic event or a recurrent feature of the circulation?

2. The Ionian upper-layer circulation reversals
   Experimental data collected during POEM surveys indicate that, by the second half of the 1980s, the Ionian upper-layer circulation reversed from cyclonic to anticyclonic [*Malanotte-Rizzoli et al.*, 1997]. The reversal was ascribed to wind forcing, which, in the eastern Mediterranean is characterized by a prevailing anticyclonic pattern [*Pinardi et al.*, 1997; *Demirov and Pinardi*, 2002; *Molcard et al.*, 2002]. In 1997 another inversion of the Ionian near-surface circulation, from anticyclonic to cyclonic, took place in presence of an anticyclonic wind pattern. This indicates that this inversion is sustained by redistribution of water masses in the Ionian abyss [*Eusebi Borzelli et al.*, 2009; *Gačić et al.*, 2010; *Gačić et al.*, 2011] and not by the wind field pattern. The question then arises: is this reversal a consequence of the redistribution of water masses in the Ionian abyss or does it trigger the shift of the eastern Mediterranean deep water formation site from the Adriatic to the Aegean and vice-versa?
3. The Mediterranean Sea salinity increase
   Lacombe et al. [1985] examined historical hydrographic observations in the western Mediterranean Sea and concluded that there had been no measurable change in deep-water salinity up to 1969. Since 1969, western Mediterranean waters below 200 m depth have become progressively saltier. The increase in salinity occurs in both Levantine Intermediate Water and Western Mediterranean Deep Water and amounts to 0.07‰ over 40 years when averaged below 200 m depth. In terms of net water mass balance, such salinity increase can be related to an increase in evaporation or a decrease in precipitation or runoff larger than 10 cm/year. Can we distinguish the role of gradual changes and singular events in causing the salinity increase? Do the changes in the salinity penetrate downward from the surface due to uniform local evaporation, laterally through advection of salty intermediate water, or upward from the bottom after injection of new salty deep waters?
4. Functioning of the Gibraltar Strait ("The Gibraltar valve")
   The Mediterranean basin scale circulation is broadly described in terms of a surface flow from the Atlantic Ocean entering through the Strait of Gibraltar and proceeding to the eastern basin, and a return flow of intermediate water, originating in the Levantine basin, proceeding toward Gibraltar, and finally exiting into the Atlantic (e.g., *Tsimplis et al.* 2006 and *Schroeder et al.* 2006 for a review). This basin scale open cell is mainly driven by thermohaline forcing: an east-west density gradient, associated with enhanced heat and moisture fluxes in the Levantine sea, drives the eastward flow of surface Atlantic water. In the Levantine basin, the ocean releases buoyancy to the atmosphere through heat loss and an evaporation/precipitation deficit. The buoyancy loss reduces the stability of the water column, with loss of potential energy, which is compensated by a buoyancy gain associated with the inflow of the fresh surface Atlantic water. For this open cell, the forcing of the Mediterranean basin-scale circulation is due to the inflows through the Gibraltar and Sicily straits. The narrow and shallow sill at Gibraltar passage, however, imposes an upper bound to the flow rate of Atlantic water at this strait. How do the orography of the Gibraltar strait and variations of the Atlantic water inflow determine variations in the western Mediterranean circulation pattern?

Since the end of POEM and WMCE, although valuable studies were carried out aiming to respond to the above issues, uncoordinated research efforts, driven mainly by national interests, provided fragmented and sporadic results.

To establish the state of the art of the research in the Mediterranean and allow interested scientists to interact, the Space Academy Foundation, a nonprofit organization to promote space science and technology, CIESM, and the OGS (Istituto Nazionale di Oceanografia e di Geofisica Sperimentale), under the aegis of the Alta Presidenza della Repubblica Italiana, organized a fully interdisciplinary meeting, which was held in Rome November 7–8, 2011. More than 35 scientists convened from different countries (France, Germany, Greece, Italy, Spain, Turkey, UK, and United States) summarized the current thinking about the Mediterranean, exposed new research ideas, and agreed to propose to AGU a collection of original papers inspired by to the workshop presentations.

This book is the outcome of this common effort and is meant to be an important and original contribution to the knowledge of the phenomena that regulate the oceanography of this basin. Furthermore, this book is a valuable tool for those not directly involved in Mediterranean studies who want to use the Mediterranean as a basin for processes of interest for the global ocean and climate.

The studies in this volume can be regarded individually or as parts of an integrated dissertation on spatial patterns and temporal variability of the Mediterranean Sea. Each one has its own conclusion and is written in such a way that a general conclusion to the entire volume is not needed. Overall, these studies indicate directions for future research and show that, though progress has been made over the last 10 years, coordinated efforts are still necessary to understand the variability of the Mediterranean Sea circulation.

REFERENCES

Demirov, E., and N. Pinardi (2002), Simulation of the Mediterranean Sea circulation from 1979 to 1993, Part I, The interannual variability, *J. Mar. Syst.*, 33–34, 23–50.

EUROMODEL Group (1995), Progress from 1989 to 1992 in understanding the circulation of the western Mediterranean Sea, *Ocean. Acta*, *18*, 2, 255–271.

Eusebi Borzelli, G. L., M. Gačić, V. Cardin, and G. Civitarese (2009), Eastern Mediterranean Transient and reversal of the Ionian Sea circulation, *Geophys. Res. Lett.*, *36*, 15, doi:10.1029/2009GL039261.

Gačić M., G. Civitarese, G. L. Eusebi Borzelli, V. Kovačević, P.-M. Poulain, A. Theocharis, M. Menna, A. Catucci, and N. Zarokanellos (2011), On the relationship between the decadal oscillations of the northern Ionian Sea and the salinity distributions in the eastern Mediterranean, *J. Geophys. Res.*, *116*, doi: 10.1029/2011JC007280.

Gačić, M., G. L. Eusebi Borzelli, G. Civitarese, V. Cardin, and S. Yari (2010), Can internal processes sustain reversals of the ocean upper circulation? The Ionian Sea example, *Geophys. Res. Lett.*, *37*, L09608, doi:10.1029/2010GL043216.

Gačić M., K. Schroeder, G. Civitarese, S. Consoli, A. Vetrano, and G. L. Eusebi Borzelli (2013), Salinity in the Sicily Channel corroborates the role of the Adriatic–Ionian Bimodal Oscillating System (BiOS) in shaping the decadal variability of the Mediterranean overturning circulation, *Ocean Sci.*, *9*, doi:10.5194/os-9-83-2013.

Kinder, T. H., and H. L. Bryden (1987), The 1985–1986 Gibraltar Experiment: Data collection and preliminary results, *EOS*, *68*, 786–787.

Lacombe, H., P. Tchernia, and L. Gamberoni (1985), Variable bottom water in the Western Mediterranean basin, *Progr. Ocean.*, *14*, 319–338.

Malanotte-Rizzoli, P., and A. R. Robinson (1988), POEM: Physical oceanography of the eastern Mediterranean, *EOS*, *69*, 194–203.

Malanotte-Rizzoli, P., B. B. Manca, M. Ribera d'Alcalà, A. Theocharis, A. Bergamasco, D. Bregant, G. Budillon, G. Civitarese, D. Georgopoulos, A. Michelato, E. Sansone, P. Scarazzato, and E. Souvermezoglou (1997), A synthesis of the Ionian Sea hydrography, circulation and water mass pathways during POEM-Phase I, *Prog. Oceanogr.*, *39*, 153–204.

Molcard, A., N. Pinardi, M. Iskandarani, and D. B. Haidvogel (2002), Wind driven general circulation of the Mediterranean Sea simulated with a Spectral Element Ocean Model, *Dyn. Atm. Oceans*, *35*, 97–130.

Pinardi N., G. Korres, A. Lascaratos, V. Rousenov, and E. Stanev (1997), Numerical simulation of the interannual variability of the Mediterranean sea upper ocean circulation, *Geophys. Res. Lett.*, *24*, 4, 425–428.

Roether, W., B. B. Manca, B. Klein, D. Bregant, D. Georgopoulos, V. Beitzel, V. Kovačević, and A. Lucchetta (1996), Recent changes in Eastern Mediterranean deep waters, *Science*, *271*(5247), DOI: 10.1126/science.271.5247.333.

Schroeder, K., G. P. Gasparini, M. Tangherlini, and M. Astraldi (2006), Deep and intermediate water in the western Mediterranean under the influence of the Eastern Mediterranean Transient, *Geophys. Res. Lett.*, *33*, L21607, doi:10.1029/2006GL027121.

Tsimplis M., V. Zervakis, S. Josey, E. Peneva, M. V. Struglia, E. Stanev, P. Lionello, V. Artale, A. Theocharis, E. Tragou, and J. Rennell (2006), Variability of the Mediterranean Sea level and oceanic circulation and their relation to climate patterns, in P. Lionello, P. Malanotte-Rizzoli, R. Boscolo (eds.), *Mediterranean Climate Variability*, Amsterdam: Elsevier (NETHERLANDS), 227–282.

WMCE Consortium (1989), Western Mediterranean Circulation Experiment: A preliminary review of results, *EOS, Trans. Amer. Geophys.* Union, *70*, 746.

2

Spatiotemporal Variability of the Surface Circulation in the Western Mediterranean: A Comparative Study Using Altimetry and Modeling

Ananda Pascual[1], Enrique Vidal-Vijande[1], Simón Ruiz[1], Samuel Somot[2], and Vassilis Papadopoulos[3]

2.1. INTRODUCTION

The progress in oceanographic research and the increase of available measurements over the past half century have greatly improved our knowledge about the ocean variability and highlighted its complexity and ubiquity over a wide range of space and timescales. However, observational datasets still remain too short, too superficial, or too dispersed in time and space to allow detailed studies of many of the physical processes governing the ocean variability. In order to advance our understanding, it is crucial to complement observational data with ocean numerical modeling studies.

The Mediterranean Sea is a very interesting basin for ocean modeling as many of the oceanic processes found throughout the world's oceans can be studied in a reduced scale [*Robinson et al.*, 2001; *Malanotte-Rizzoli and the Pan-Med Group*, 2012]. In particular, dynamic processes such as intermediate and deep convection are generally difficult to simulate because of their dependence on intense episodic atmospheric events [*Herrmann and Somot*, 2008] as well as on small mesoscale baroclinic instabilities [*Herrmann et al.*, 2008].

Several modeling studies have focused on the Mediterranean Sea during the last two decades. In the nineties, *Roussenov et al.* [1995] described the seasonal characteristics of the Mediterranean Sea's general circulation as simulated by a primitive equation general circulation model (MOM) at a 1/4° spatial resolution. The model was forced with climatological monthly mean atmospheric parameters. *Herbaut et al.* [1997] performed an 18-year run at a 1/8° resolution in the western Mediterranean, using a perpetual atmospheric forcing and the LODYC model. They investigated the influences of atmospheric forcing on the circulation and they found that the major characteristics of the circulation were well reproduced and that several features are were linked to the wind stress action.

In the last decade, *Beckers et al.* [2002] presented an intercomparison between different simulations, carried out in the frame of the MEDMEX project. They showed that the seasonal cycle of the sea surface temperature was represented similarly by the different models. *Demirov and Pinardi* [2002] analyzed the interannual variability of the Mediterranean circulation from 1979 to 1993 with a 1/8° resolution OGCM (MOM). They identified significant changes in the circulation of the eastern basin, which were in good agreement with available observational data.

More recent works have used 1/8° and 1/16° models such as DieCAST [*Fernández et al.*, 2005], OPAMED8 [*Somot et al.*, 2006], MFS [*Tonani et al.*, 2008], NEMOMED8 [*Sevault et al.*, 2009], and MED16 [*Béranger et al.*, 2010] to name a few. The DieCAST model was run for assessing the circulation and transport variability in the Mediterranean Sea, forced by climatological monthly mean winds and relaxation toward monthly climatological surface temperature and salinity. The OPAMED8 1/8° simulation was a scenario of the Mediterranean Sea under climate change IPCC-A2 conditions and run with an atmosphere regional climate model (ARPEGE) over the 1960–2009 period using a hierarchy of three different models. The MFS simulation is based on the OPA code [*Madec et al.*, 1998] with a 1/16°

[1] *IMEDEA (CSIC-UIB), Esporles, Mallorca, Spain*
[2] *MeteoFrance, Toulouse, France*
[3] *Hellenic Centre for Marine Research, Patras, Greece*

The Mediterranean Sea: Temporal Variability and Spatial Patterns, Geophysical Monograph 202. First Edition.
Edited by Gian Luca Eusebi Borzelli, Miroslav Gačić, Piero Lionello, and Paola Malanotte-Rizzoli.

spatial resolution. It is presently used for operational daily forecasts over the Mediterranean Sea and has been available since 1997. The NEMOMED8 1/8° simulation is a 51-year-long hindcast (now covering 1961–2011) carried out with an eddy permitting model driven by realistic interannual high resolution air sea fluxes, which has also been used in the recent studies by *Beuvier et al.* [2010] and *Herrmann et al.* [2010] that focused on the Eastern Mediterranean Transient and the 2004–2005 northwestern Mediterranean convection events, respectively. Regarding the MED16 simulation, it is a modeling experiment to assess the performance of atmospheric forcing resolution on winter ocean convection in the Mediterranean Sea by running two 4-year simulations, 1998–2002 (with 11 years spin-up), one using ERA40 and one using ECMWF analyzed surface fields, which have twice the resolution of ERA40.

Additionally, some recent studies such as *Shaeffer et al.* [2011] have used short, very high resolution models (~1 km resolution) to study specific areas and processes in the Mediterranean (such as eddies in the Gulf of Lions). From a different perspective, *Vidal-Vijande et al.* [2011] and *Vidal-Vijande et al.* [2012] have also analyzed the performance of three global hindcasts (ORCA-70, ORCA-85, and GLORYS) in the whole Mediterranean Sea, yet particularly focusing on the western Mediterranean subbasin.

However, in order to improve our knowledge of the functioning of the ocean system, it is necessary to perform accurate and robust intercomparisons between model simulations and observations, covering a wide range of spatiotemporal scales. In this context, satellite altimetry has demonstrated prominent capabilities to characterize surface ocean variability as a result of the continuous sea surface topography measurements provided by several altimeter missions (TOPEX/Poseidon, ERS-1, ERS-2, Geosat Follow-On, Jason-1, Envisat, and OSTM/Jason-2).

The objective of this study is to analyze the interannual, seasonal, and large mesoscale variability of the WMED surface circulation through the combined use of one of the simulations previously mentioned (NEMOMED8) and satellite altimetry observations. It is noteworthy that this is the first work that intercompares an eddy-permitting simulation (forced with realistic high-resolution atmospheric fields) with altimetry data covering a period of 15 years. We focus on the mean geostrophic circulation derived from the simulation and compare it to the mean currents derived from altimetry measurements. To assess the ocean variability, we analyze the Eddy kinetic energy (EKE) from both model and altimetry data. The main patterns of the WMED circulation are examined by using standard EOF analysis in three subbasins of the western Mediterranean: the Alborán Basin, the Algerian Basin, and the northwestern Mediterranean.

The paper is organized as follows: section 2 presents the numerical simulations, the altimetry records, and the methods; section 3 gives results on the mean surface circulation, EKE, and EOF analysis; and section 4 provides a summary and brief discussion.

2.2. DATA AND METHODS

2.2.1. Data

2.2.1.1. The NEMOMED8 numerical simulation The NEMOMED8 (NM8 hereinafter) simulation [*Beuvier et al.*, 2010; *Herrmann et al.*, 2010] is a 1/8° Mediterranean configuration of the NEMO numerical framework. It is an updated version of the OPAMED8 model by *Somot et al.* [2006], which is a regional configuration of the global OPA ocean model [*Madec et al.*, 1998].

NM8's resolution of 1/8° provides a mesh size in the 9–12-km range from the north to the south of the domain, covering the entire Mediterranean Sea and an Atlantic buffer zone near the Strait of Gibraltar. NM8 has 43 vertical Z-levels with an inhomogeneous distribution (from $\Delta Z = 6$ m at the surface to $\Delta Z = 200$ m at the bottom with 25 levels in the first 1000 m). Bathymetry is based on the ETOPO $5' \times 5'$ database and the deepest level has a variable height in order to adapt to the real bathymetry.

Sea surface height evolution is parameterized by a filtered free-surface [*Roullet and Madec*, 2000]. With this parameterization, the volume of the Mediterranean Sea is not conserved due to the loss of water induced by evaporation. In order to conserve volume, at each time step, the evaporated water over the whole basin is redistributed in the Atlantic buffer zone west of 7.5° W.

NM8 is forced by the 50 km resolution ARPERA atmospheric forcing [*Herrmann and Somot*, 2008; *Tsimplis et al.*, 2008], which is a dynamic downscaling of the ERA40 reanalysis (125 km resolution) from ECMWF [*Simmons and Gibson*, 2000] by the ARPEGE-Climate regional climate model developed at CNRM [*Déqué and Piedelievre*, 1995]. For the period 1958–2001, ARPEGE-Climate is driven by the ERA40 reanalysis, with a spatial resolution of 125 x 125 km^2, and from 2002 to 2006, it is driven by ECMWF fields with spatial resolution of 55 × 55 km^2. For consistency, ECMWF fields resolution has been downgraded to ERA40 resolution.

The forcing fields for NM8 are daily momentum, freshwater, and heat fluxes. The heat flux is adjusted to the model SST by a surface relaxation toward the SST used in the ERA40 reanalysis. This term is considered as a first order coupling between the SST of the ocean model and the atmosphere heat flux. It ensures the consistency between the surface heat flux coming from

ARPERA and the SST computed by the ocean model. The relaxation coefficient is −40 W/m2/K, as defined in *Barnier et al.* [1995]. It is equivalent to an 8-day restoring timescale. With this value, the relaxation is not too strong and the ocean model is able to create and maintain small-scale structures. We are aware that this relaxation term may have a slight influence on our results, damping long-lasting eddies. No salinity restoring is used at the surface and river runoff freshwater flux is explicitly added to complete the surface water budget. The Black Sea is not included in NM8, instead, its freshwater input into the Mediterranean is considered as an interannual river influx to the Aegean Sea. All river inputs are based on monthly climatologies including interannual variability (all details about the river forcing are found in *Beuvier et al.*, 2010).

The exchanges with the Atlantic Ocean are performed through a buffer zone from 11° W to 7.5° W. Temperature and salinity in this area are relaxed to an interannual 3-D T-S field coming from the combination of the *Reynaud et al.* [1998] seasonal climatology and of the interannual global reanalysis by *Daget et al.* [2009]. The relaxation is weaker closer to Gibraltar and stronger away from the strait. Initial conditions for the Mediterranean Sea are given by the MEDATLAS-II climatology *MEDAR-Group* [2002] and by *Reynaud et al.* [1998] for the Atlantic.

The NM8 simulation covers the 1961–2011 period and has been validated by *Beuvier et al.* [2010] over the 1961–2007 period, who showed a general agreement between the simulation and the observations. The interannual variations of the heat content were well reproduced while those of the salt content showed more discrepancies. The main bias was found at the intermediate layer, which was slightly warmer and saltier but less compared to other simulations [e.g., *Vidal-Vijande et al.*, 2011; *Vidal-Vijande et al.*, 2012]. On the other hand, the water budget was in good balance: the net water transport at the Gibraltar Strait compensated the water loss over the Mediterranean surface. Furthermore, NM8 proved to produce a reasonable basin scale circulation (in the entire Mediterranean), but since the main goal of *Beuvier et al.* [2010] was to study the Eastern Mediterranean Transient (EMT), very little attention was devoted to the detailed analysis of circulation variability in the western Mediterranean, which is the objective of this article.

2.2.1.2. Satellite altimetry Satellite altimetry provides realistic high-resolution sea surface height (SSH) observations making it a useful tool in the validation of sea level and surface circulation numerical models. However, altimeter measurements include Earth's geoid, which varies tens of meters across the ocean and needs to be subtracted from altimetry data to obtain a usable product. Since the exact shape of the geoid is largely unknown (the best data available come from the GRACE and GOCE satellite measurements that have a resolution of approximately 200 km, which is only sufficient for large-scale studies), the typical solution to overcome this limitation is to subtract a temporal mean of SSH at each location, leading only to the variable part of sea surface height, known as sea level anomaly (SLA).

Altimetry data are available as a global product at 1/4° resolution, but in this study we use the higher-resolution product specific for the Mediterranean Sea, available through the AVISO FTP (http://www.aviso.oceanobs.com). These data consist of gridded, delayed-time sea level anomaly fields specific for the Mediterranean Sea merging several altimeter missions (TOPEX/Poseidon, ERS-1, ERS-2, Jason-1, Envisat, and OSTM/Jason-2) provided weekly with a resolution of 1/8°. The reference product has been selected in order to have a homogeneous time series with the same configuration of satellite missions included in the objective analysis scheme. In the following, we present a brief summary of the methodology used in AVISO to build the Mediterranean gridded fields (a complete description is available in *Pujol and Larnicol*, 2005). First, a homogeneous and intercalibrated SSH data set is obtained by performing a global crossover adjustment [*Le Traon and Ogor*, 1998]. Second, these data are geophysical corrected (tides, wet and dry troposphere, ionosphere, wind, and pressure response). Along-track data are resampled every 7 km using cubic splines and the SLA is computed by removing a 7-year mean SSH corresponding to the 1993–1999 period. Measurement noise is then reduced by applying a 35-km median filter and a Lanczos low-pass filter with a cut-off wavelength of 42 km. Finally, the data are subsampled every second point.

The mapping method to produce gridded fields of SLA from along-track data is based on a suboptimal objective analysis, which takes into account the long wavelength correlated errors [*Le Traon and Ogor*, 1998] remaining from the orbit errors and the atmospheric effects. The analysis is performed using space and time correlation functions with 100-km and 10-day correlation radius (zero crossing of the correlation function). These parameters were obtained from along-track correlation statistics [*Pujol and Larnicol*, 2005]. Note that, given the small Rossby radius of deformation in the Mediterranean Sea (15–20 km according to *Robinson et al.*, 2001), the typical mesoscale features are characterized by scales of the order of 10–100 km [*Malanotte-Rizzoli and the Pan-Med group*, 2012]. Thus, present altimetry gridded fields only resolve large mesoscale structures in this basin.

An important issue to be analyzed in the models described in this work is the capability to reproduce the main features of mean circulation. However, as mentioned above, the SLA altimetry fields include only the variable part of the total sea level signal. Thus, the mean circulation has to be investigated from alternative sources. Here, we consider the synthetic mean dynamic topography (MDT) produced by *Rio et al.* [2007], which uses an average over 1993–1999 of SSH outputs from MFS simulation [*Tonani et al.*, 2008] as a first guess. This is then improved in a second step by combining drift buoy velocities and altimetry using an empirical method to obtain local estimates of the mean geostrophic circulation that are merged with the first guess through an inverse technique. The synthetic MDT is mapped onto a 1/8° grid and added to the SLA, leading to absolute dynamic topography (ADT).

2.2.2. Methods

2.2.2.1. Eddy Kinetic Energy

In this study, we use EKE as a measure of the degree of variability to identify regions with highly variable phenomena such as eddies, current meanders, fronts, or filaments. It is calculated by making the geostrophic assumption:

$$EKE = \frac{1}{2}\left[U'^2_g + V'^2_g\right]$$

$$U'_g = -\frac{g}{f}\frac{\partial \eta'}{\partial y}$$

$$V'_g = \frac{g}{f}\frac{\partial \eta'}{\partial x}$$

where η' represents the surface dynamic topography anomaly provided either by altimetry or by a model; U'_g and V'_g denote the zonal and meridional geostrophic velocity anomalies; f is the Coriolis parameter, g is the acceleration of gravity, and the derivatives $\frac{\partial \eta'}{\partial x}$ and $\frac{\partial \eta'}{\partial y}$ are computed by finite differences where x and y are the distances in longitude and latitude, respectively. The variables used for the EKE computation are ADT for altimetry, and SSH from NM8 simulation. Anomalies are calculated by removing the temporal mean (1993–2007) at each grid point.

2.2.2.2. Empirical Orthogonal Functions (EOFs) EOFs are a convenient statistical tool to identify and quantify the spatial and temporal variability. They consistently reduce the dimensions of large datasets to few significant orthogonal modes of variability, and to estimate the amount of variance associated with each mode in percentage terms [*Emery and Thomson*, 1997; *Bjornsson and Venegas*, 1997; *Beckers et al.*, 2002; *Molcard et al.*, 2002]. In this study, we use EOFs to decompose the surface ocean circulation signal into its major modes of spatial variability $F_k(x,y)$ and their associated temporal components $a_k(t)$.

Thus, the signal $O(x, y, t)$ is decomposed using the following formulation:

$$O(x,y,t) = \overline{O(x,y,t)} + \sum_{k=1}^{n} F_k(x,y)a_k(t)$$

where $\overline{O(x,y,t)}$ denotes the temporal mean, $|a_k| = 1$, and $|F_k| \neq 1$.

As for the EKE, the variables analyzed are ADT for altimetry and SSH for NM8 simulation. The temporal mean (1993–2007) is removed at each grid point and then the modes are computed.

The EOFs analysis is performed dividing the WMED into subregions to better account for the variability in each area. For the interpretation of the results, we adopt a seasonal approach: Winter (January–March), Spring (April–June), Summer (July–September), Autumn (October–December).

The first two EOF modes have been considered for the analysis as they have been found to correspond to signals associated with physical processes and explain more than 70% of the total variance.

2.3. RESULTS

2.3.1. Mean Surface Circulation

In the following sections, we will focus on the dynamic characteristics of the WMED and in particular on the capability of NM8 simulation to reproduce them by comparing to altimetry. Figure 2.1 presents the WMED mean circulation derived from ADT averaged over the period of study. In the Alborán Basin, remote sensing data reproduce correctly the quasi-permanent gyre at its western part, called the West Alborán Gyre [WAG, *Videz et al.*, 1998; *Baldacci et al.*, 2001; *Flexas et al.*, 2006], with associated geostrophic velocities of about 1 m/s, in agreement with characteristic values found in previous studies [*Gomis et al.*, 2001]. The Atlantic water entering through the Strait of Gibraltar forms a surface powerful inflow known as the Atlantic jet [*Tintoré et al.*, 1988], which is also visible in Figure 2.1. This jet meanders and forms an intermittent eastern Alborán gyre [EAG, *Tintoré et al.*, 1988]. In the eastern Alborán Basin, the altimetry provides the well-known Almería-Orán front [*Allen et al.*, 2001], which marks the start of the Algerian Current (AC), close to 0° longitude. The relatively narrow AC (30–50 km) and the magnitude of the mean velocity (0.35–0.45 cm/s) given by the altimetry are well fitted with the values given by other studies [*Astraldi et al.*, 1999; *Millot*, 1999; *Morán et al.*, 2001].

The altimetry-based mean circulation shows that the AC flows eastward with its width and separation from the

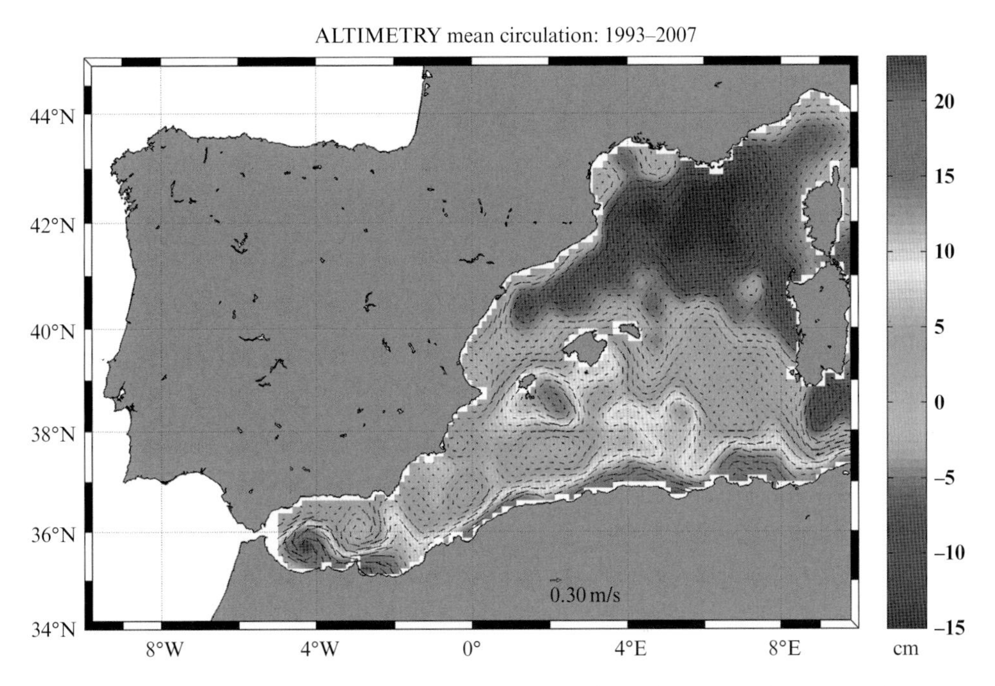

Figure 2.1 Altimetry mean geostrophic circulation in the WMED derived from Absolute Dynamic Topography for the 1993–2007 period. For color detail, please see color plate section.

coast to vary considerably. Due to baroclinic instabilities, the AC regularly forms meanders that can eventually be detached from the current and become both cyclonic and anticyclonic coastal eddies as already observed in *Puillat et al.* [2002] and also in previous altimetric works [*Iudicone et al.*, 1998; *Larnicol et al.*, 2002; *Olita et al.*, 2011]. Some of the stronger eddies, which are mostly anticyclonic, may become "open sea eddies" such as the eddy observed around 5° E (Figure 2.1). Several of these detached eddies propagate westward or northward reaching the Balearic Islands and the Liguro-Provencal Basin [*Millot*, 1999; *Ruiz et al.*, 2002]. As shown in Figure 2.1, the AC and its associated eddy-generating capacity dominate the circulation of the Algerian Basin.

In the northern part of the WMED, altimetry data give a mean surface circulation controlled by the Northern Current, which flows along the coasts of Italy, France, and Spain following the bathymetry as described by *Millot* [1999]. The Gulf of Lions is a site of intense mesoscale activity and winter deep-water formation [*MEDOC Group*, 1970]. Due to the action of the Northern Current, the circulation here is generally cyclonic. The Northern Current flows southward along the Iberian coast and splits into two branches, one recirculating into the Balearic Current [*Pascual and Gomis* 2003; *Ruiz et al.*, 2009] and the other flowing south through the Ibiza Channel [*Pinot et al.*, 2002]. The mean northward inflow observed by altimetry in the Ibiza Channel (Figure 2.1) has also been documented by *Pinot et al.* [2002], who provided a detailed description of the mean seasonal variability of the surface circulation.

The mean circulation patterns derived from altimetry, while assumed to be mostly correct, can comprise errors associated to the MDT. A clear example of this is the anticyclonic eddy found southeast of Ibiza. This is probably an artifact of the MDT and may not be a predominant circulation feature.

Figure 2.2 presents the WMED mean circulation derived from NM8 SSH. At first glance, the overall circulation displays a smoother field with less intricate mesoscale structures compared to altimetry. In the Alborán Basin, the WAG is not present, only the EAG, which in this case is more intense than in altimetry. The Almeróa-Orán Front that sits between the EAG and an intense cyclonic gyre to the northeast is strong. The AC that flows east is not so clearly defined, with its outer edge extending farther out into open sea. The center of the Algerian Basin is dominated by anticyclonic gyres, in good agreement with altimetry observations.

In the northern part of the WMED, the Northern Current and the intense cyclonic circulation of the Gulf of Lions are clearly visible (Figure 2.2). The Northern Current has most of its recirculation occurring farther north than in the altimetry-derived circulation, with a weaker signal reaching west of Mallorca and recirculating into the Balearic Current. Also visible is the flow through the Ibiza Channel, completing the circulation loop of the Mediterranean.

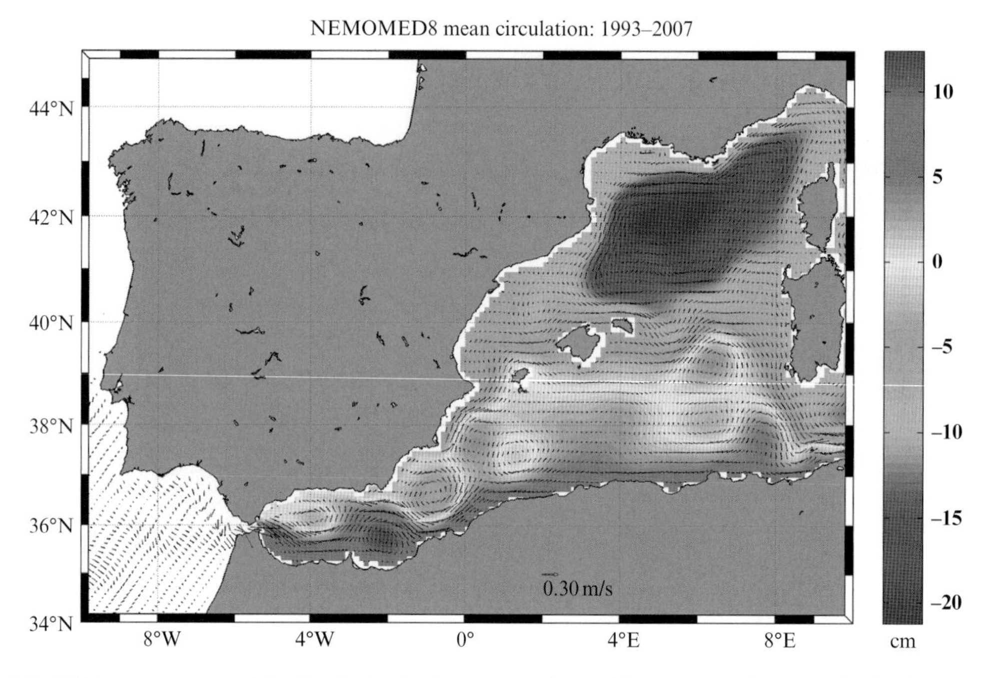

Figure 2.2 NM8 mean geostrophic circulation in the WMED derived from Sea Surface Height for the 1993–2007 period. For color detail, please see color plate section.

2.3.2. Eddy Kinetic Energy

Figure 2.3 shows the mean geostrophic eddy kinetic energy (EKE) calculated from altimetry and averaged for the 1993–2007 period. There is a clear north–south gradient in EKE levels, with values higher than $250\,cm^2/s^2$ corresponding to the Alborán Basin (WAG and EAG) and the path of the Algerian Current and associated gyres (between 100 and $250\,cm^2/s^2$). In contrast, the northern WMED has values lower than $50\,cm^2/s^2$. Another area of high EKE is found southwest of Sardinia, which is characterized by the regular formation of anticyclonic eddies that propagate toward both the east and north. These general findings agree with the description presented by previous authors [*Iudicone et al.*, 1998; *Pujol and Larnicol*, 2005; *Pascual et al.*, 2007], who have already employed altimetry data in the Mediterranean Sea to study EKE variability. The main innovations of the present study are the extension of the time series up to 15 years (which gives more statistical robustness to the results) and that the EKE map is subsequently compared to the equivalent model EKE map.

More in detail, in the northwestern part of the Mediterranean Sea, a weaker signal corresponding to the 1998 anticyclonic eddy north of Mallorca [*Pascual et al.*, 2002] is also visible. Although this 1998 eddy was an episodic feature lasting only for a few months, it was so strong that it remains visible even when a 15-year mean is considered. As pointed by *Pujol and Larnicol* [2005], singular instances of high EKE signals can skew the overall mean, making difficult the interpretation of the mean EKE state of the ocean. It is therefore useful to examine the temporal EKE evolution. Figure 2.4 shows the temporal evolution EKE for the three regions of the WMED considered in this study. The Alborán Basin is the region where higher EKE is found (maximum values of $450–500\,cm^2/s^2$), with intense interannual variability but few distinct peaks. This denotes an area where EKE is regularly high, and the mean spatial EKE signal is probably representative of the mean state. In the Algerian Basin, the mean EKE shows a 2.5-year peak between the end of 1996 and the beginning of 1999. According to *Pujol and Larnicol* [2005], this is due to the presence of two anticyclonic eddies detected in 1997. They were also studied by *Puillat et al.* [2002] and had a lifetime of about 2–3 years, therefore considerably influencing the EKE variability. In the North Balearic Sea, the clear spike related to the 1998 anticyclonic eddy [*Pascual et al.*, 2002] is also affecting the overall mean state of the region.

Figure 2.5 shows the mean EKE calculated from NM8 for the 1993–2007 period. In comparison with the altimetry, NM8 has a similar overall distribution of EKE with the clear north–south gradient and maximum values in the Alborán Basin and Algerian Basin. In this case, the strongest signal is seen in the EAG (higher than $300\,cm^2/s^2$), which is very clearly defined. The WAG is also visible, with a maximum at the eastern edge. Higher EKE is also found at the western part of the Algerian Current (between 200 and $250\,cm^2/s^2$) decreasing to values below $200\,cm^2/s^2$ in the southwest of Sardinia. What is present in

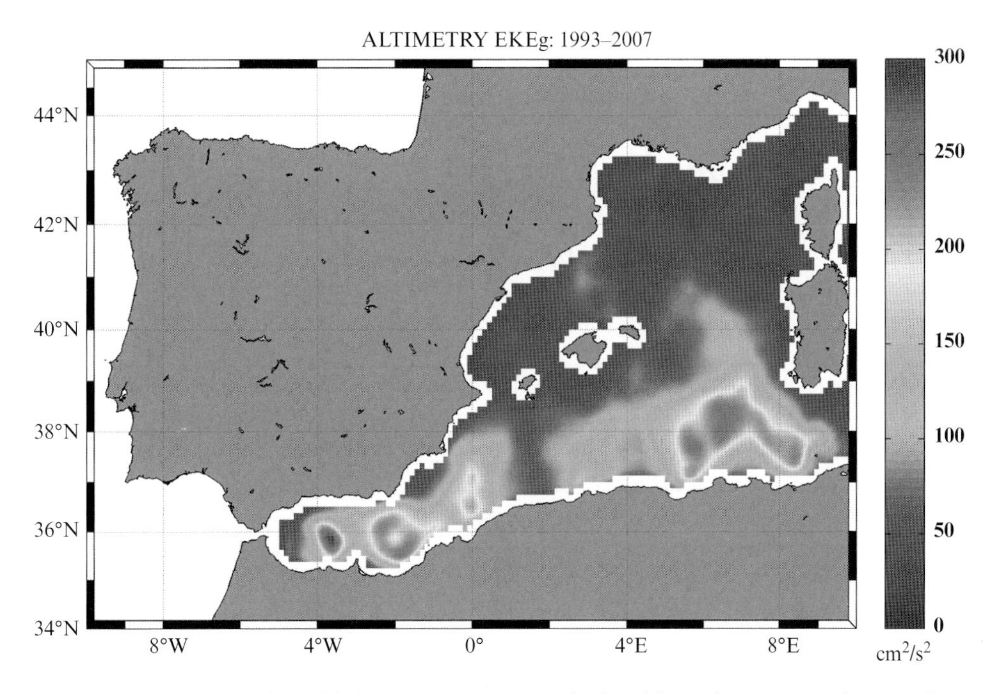

Figure 2.3 WMED mean geostrophic eddy kinetic energy (EKE) calculated from altimetry over the period 1993–2007. For color detail, please see color plate section.

NM8 but not clearly visible in altimetry is the high EKE values at the Ibiza Channel and to a lower extent at the Mallorca Channel, which are perhaps linked to the northward flow through these channels and its continuation to the Balearic Current (Figures 2.1 and 2.2).

The temporal evolution of NM8 EKE (Figure 2.6) also shows the occurrence of singular events with the potential to influence the mean spatial EKE patterns. In the Alborán Basin, a very intense peak at the end of 1993 with a value of ~700 cm^2/s^2, which likely represents a very intense EAG, is certainly affecting the distribution of EKE values in the spatial means. In the Alborán Basin, there are seasonal maxima but no discernible outlying peaks. In the northwestern Mediterranean region, a few higher EKE peaks are likely related to increased flow through the Ibiza Strait.

2.3.3. Surface Circulation Variability Using EOFs

2.3.3.1. Alborán Basin The first altimetry EOF mode (Figure 2.7, upper panels) is characterized mainly by the EAG variability and less by the WAG variability. The temporal amplitude is maximum in September-November and minimum in February-March, which corresponds to the steric contribution of the seasonal signal, as already pointed out in *Cazenave et al.*[2002]. The first EOF accounts for 70.6% of the total variance. These results are in agreement with the detailed analysis published recently by *Renault et al.* [2012].

NM8 shows similar pattern and amplitude to altimetry (Figure 2.7, lower panels), without the presence of the WAG, and a positive anomaly on the eastern boundary of the domain. The explained variance (55.2%) is slightly lower than altimetry's first mode, but still remains far leading.

The second EOF mode of the altimetry (Figure 2.8) exhibits only interannual variability with a remarkable peak around August 2001. The spatial pattern reveals a well-defined dipole between the EAG and the cyclonic eddy east of the EAG, with a very intense Almeróa-Orán front. This mode explains 8.5% of the total variance.

The spatial pattern of NM8 is extremely similar to altimetry, with almost identical gradients, but with different temporal variability and higher variance (18%). This probably reflects the limited capability of NM8 to reproduce some interannual signals, which is expected in a free run without data assimilation.

2.3.3.2. Algerian Basin The first EOF of altimetry (Figure 2.9) explains the 67.9% of the total variance and is dominated by a well-defined dipole between an anticyclonic eddy to the south of Sardinia and a cyclonic large mesoscale feature to the southeast of the Balearic Islands. This pattern is maximum in September-November and minimum in February-March, indicating, as in the Alborán Basin, the seasonality of the steric signal. The Algerian Current is hardly seen in this first EOF mode. NM8 shows very similar spatial pattern and amplitude to

ALBORAN ALTI EKEg (SSH) 1993–2007

550 500 450 400 350 300 250 200 150

1993 1994 1995 1996 1997 1998 1999 2000 2001 2002 2003 2004 2005 2006 2007

Time

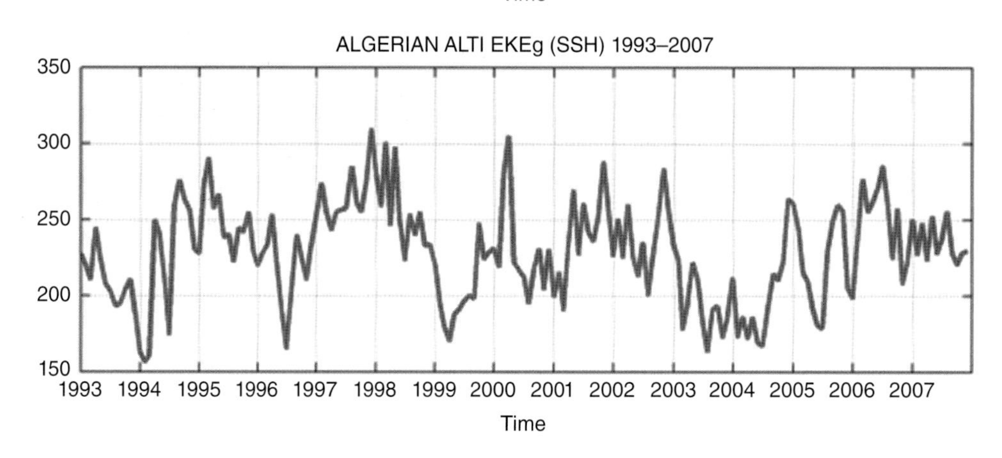

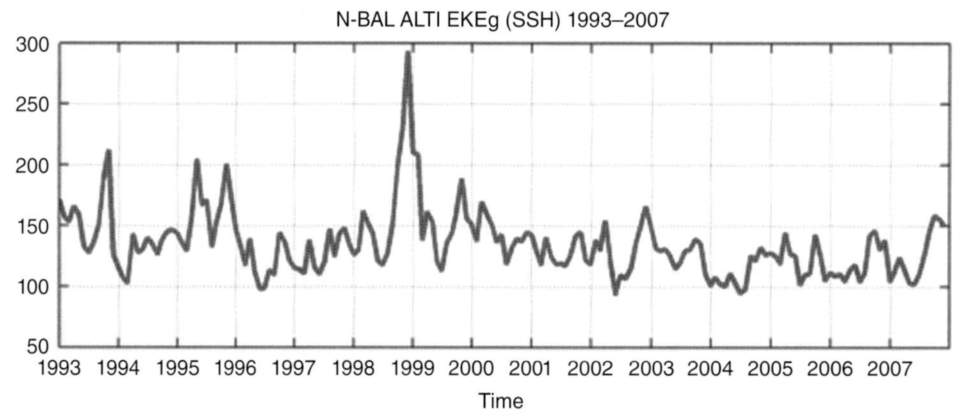

Figure 2.4 Altimetry EKE time series for the Alborán Basin (top), Algerian Basin (middle), and North Balearic Sea (bottom).

altimetry with the dipole mentioned above. In this case, the gradient between both cyclonic and anticyclonic features is steeper. The explained variance (58.9%) is also of a similar magnitude to that of the altimetry. In this case, there is a clear evidence of the AC.

The second EOF mode of both altimetry and NM8 (Figure 2.10) shows another strong dipole to the southwest of Sardinia. Both of them explain similar variance (5% and 4.4%, respectively). For altimetry, the signal could be attributed to isolated events in 1997, 2003, and 2006, whereas in NM8 they are more frequent. In NM8, there are also a number of weaker (in amplitude) structures along the North African coastline associated with the AC variability.

2.3.3.3. Northwestern Mediterranean Basin (NWMED)
The NWMED domain used for this analysis covers part of the Ligurian Basin, the Gulf of Lions, and the Balearic Sea. The first EOF mode of altimetry (Figure 2.11) explains 81.5% of the total variance. The

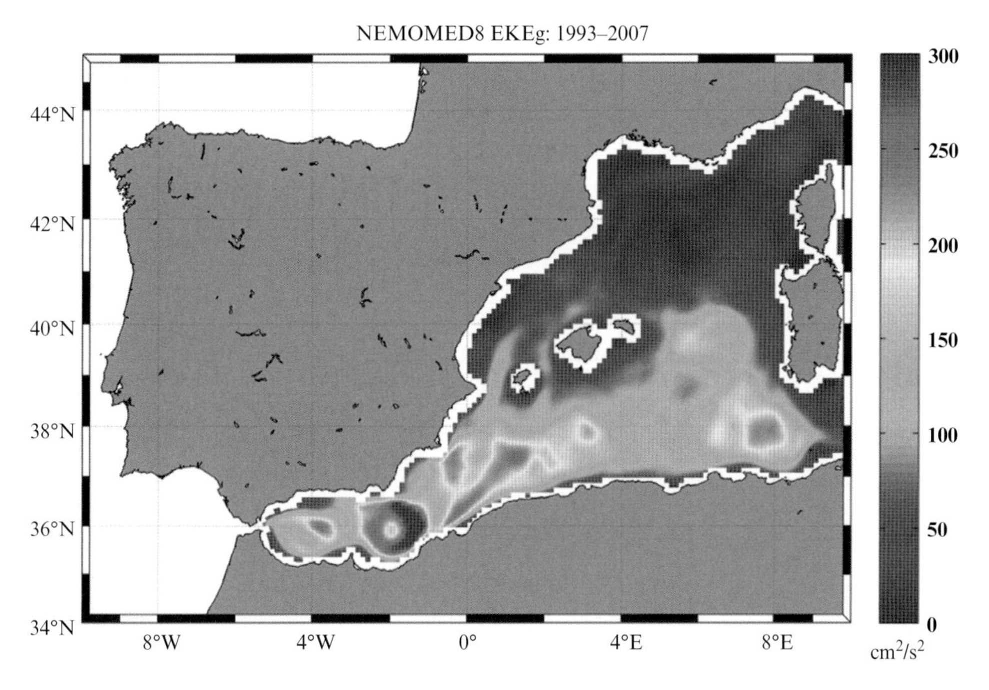

Figure 2.5 WMED mean geostrophic eddy kinetic energy (EKE) calculated from NM8 over the period 1993–2007. For color detail, please see color plate section.

temporal amplitude is again maximum in beginning of autumn and minimum by the end of winter, denoting again the seasonal cycle of the steric signal. The spatial pattern is associated with a strong north–south gradient centered at approximately 41.5°N where, usually, the North Balearic front is found [*Lípez-Garcóa et al.*, 1994]. This front is caused by the wind-channeling effect of the Pyrenees, especially in the summer, where the more stable weather conditions in the northern part of the NWMED are disturbed by this effect. NM8 first mode also accounts for a similar fraction of explained variance (76.5%) compared to altimetry and displays a comparable general pattern, albeit with lower intensity and less well defined. The second EOF mode for both datasets (Figure 2.12) is dominated by discrete subbasin scale features in the northern Balearic Sea with an associated explained variance of 3.1% for altimetry and 6.1% for the model. In altimetry, the main signal is an anticyclonic eddy that occurred north of Mallorca in 1998 and has been studied in depth by *Pascual et al.* [2002]. NM8 shows a weaker signal at that location; instead, a stronger eddy is found north of Ibiza.

2.4. SUMMARY AND DISCUSSION

The performance of an eddy-permitting simulation (forced with realistic high-resolution atmospheric fields) has been assessed by means of a comparison with satellite altimetry measurements in the western Mediterranean Sea (WMED). This is the first time, to our knowledge, that such an approach has been followed covering a period of 15 years with the aim of characterizing the interannual, seasonal, and large mesoscale surface variability. This study represents a new contribution toward a better understanding of ocean variability by evaluating the capabilities and limitations of present numerical simulations and remote sensing observations.

First, mean surface circulation provided by the model has been compared to the mean currents calculated from observations. Second, we have computed the eddy kinetic energy (EKE) from both model and remote sensing to assess the surface variability. Finally, the principal main patterns of the WMED circulation have been examined based on standard EOF analysis.

The simulation (NM8) displayed a correct overall mean circulation pattern but produced a smoother field than altimetry. It exhibited discrepancies with respect to altimetry in specific areas of the WMED, summarized as follows. NM8 failed to reproduce the West Alborán Gyre (WAG) and gave a less-defined Algerian Current (AC). It reproduced reasonably the Northern Current along the Italian and French coast but the recirculation that happens in the Balearic Sea is weaker, with most of it happening farther north.

Both datasets (altimetry and NM8) agree that the Alborán Basin and the Algerian Basin are the regions of the maximum EKE. Altimetry and NM8 have similar levels of EKE and reproduce the WAG and EAG

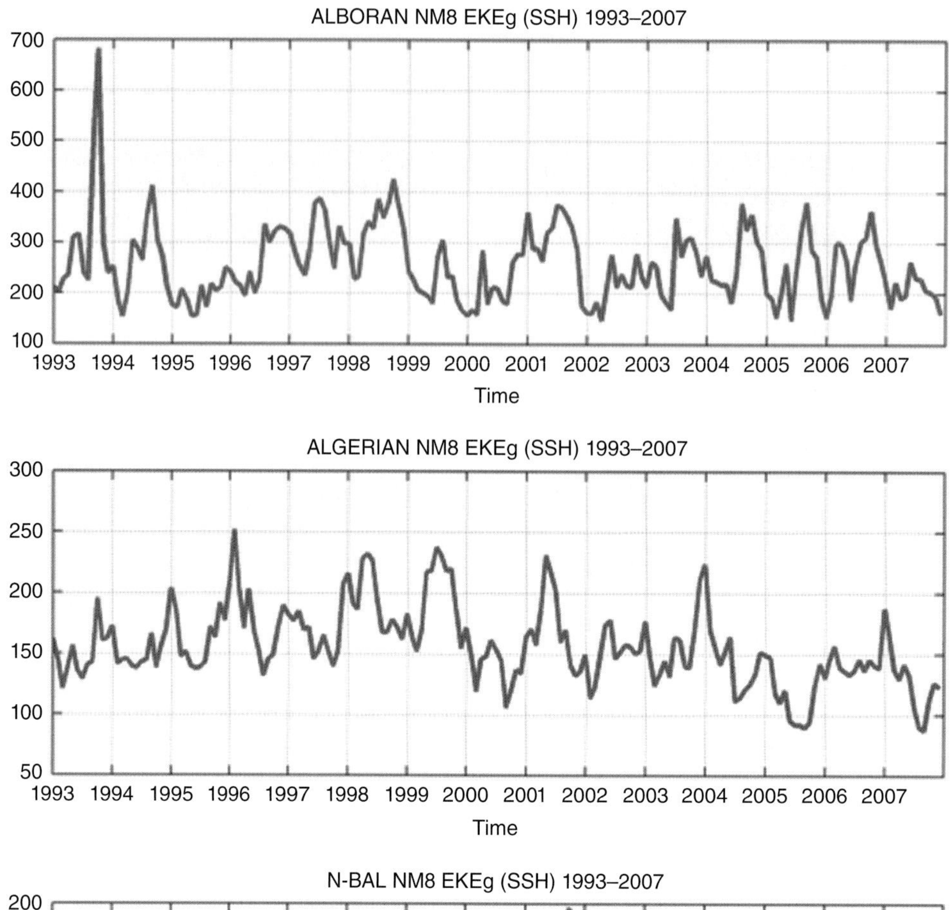

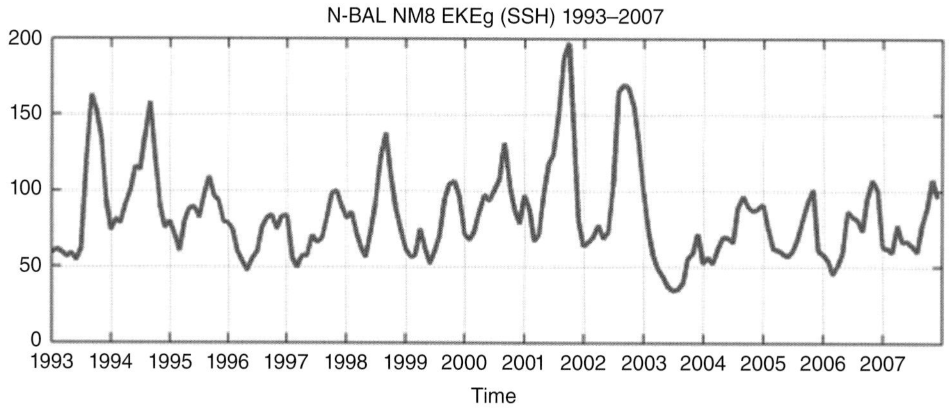

Figure 2.6 NM8 EKE time series for the Alborán Basin (top), Algerian Basin (middle) and North Balearic Sea (bottom).

variability as well as the high EKE regions along the path of the AC.

The first EOF mode was clearly dominant explaining between 55% and 81% of the total variance and corresponding to the seasonal steric cycle. The second mode had a much lower explained variance (3%–18%) but represented evident interannual and large mesoscale signals that are very relevant in the Mediterranean Sea. The third mode (not shown) has only a slightly lower level of explained variance compared to the second one, but the physical interpretation was less conclusive. In the first two EOF modes, NM8 showed an exceptional performance when compared to altimetry in all regions, with very similar spatial patterns and temporal phase. However, it is noteworthy that their interannual variability did not match, which is expected since the simulation does not assimilate data and it would be very difficult for it to reproduce actual interannual variability.

In summary, NM8 appears as a mature and solid simulation, consistently producing good results when

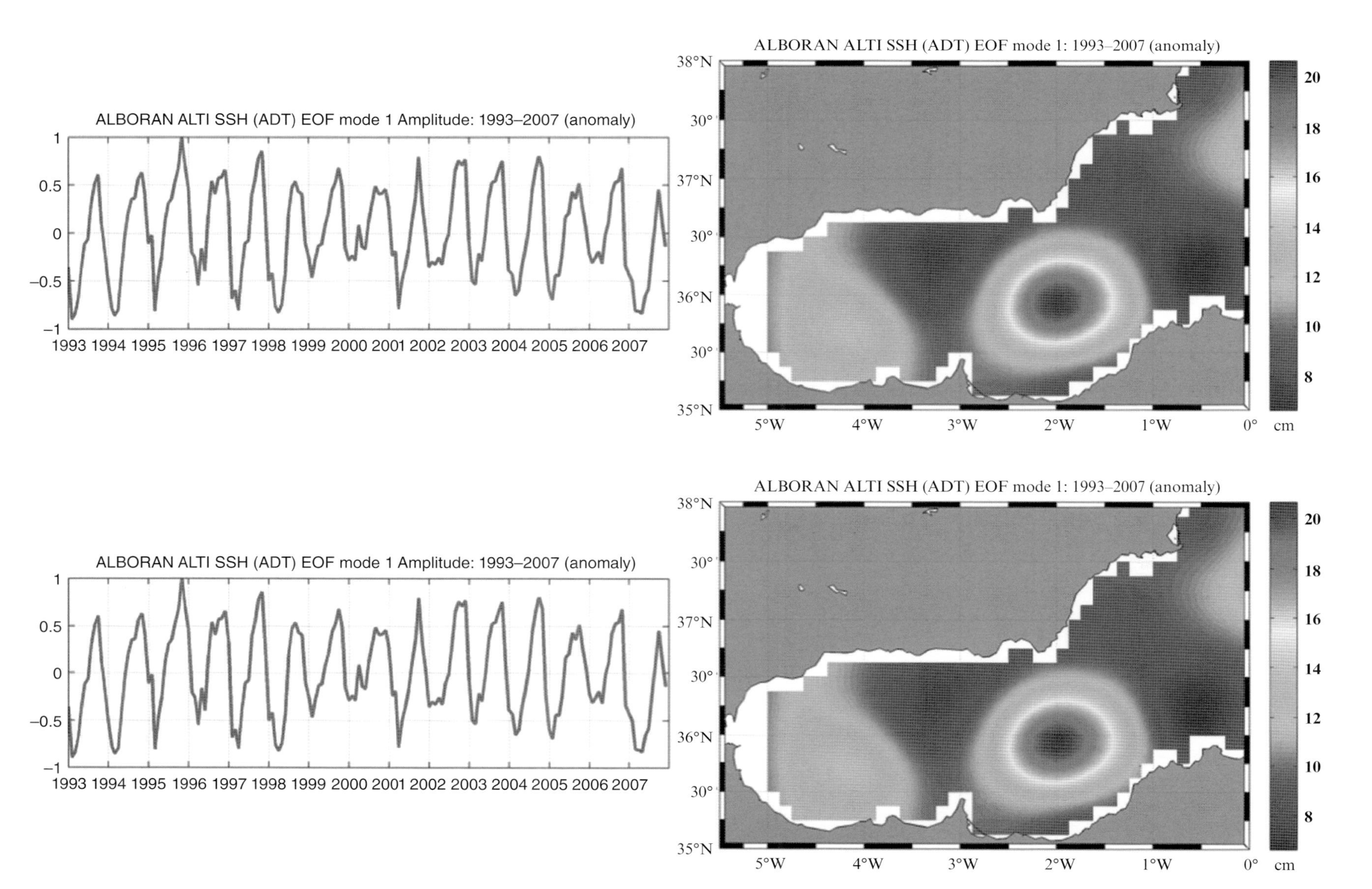

Figure 2.7 First EOF for the Alborán Sea with amplitude (left) and pattern (right); altimetry ADT (top), NM8 SSH (bottom). For color detail, please see color plate section.

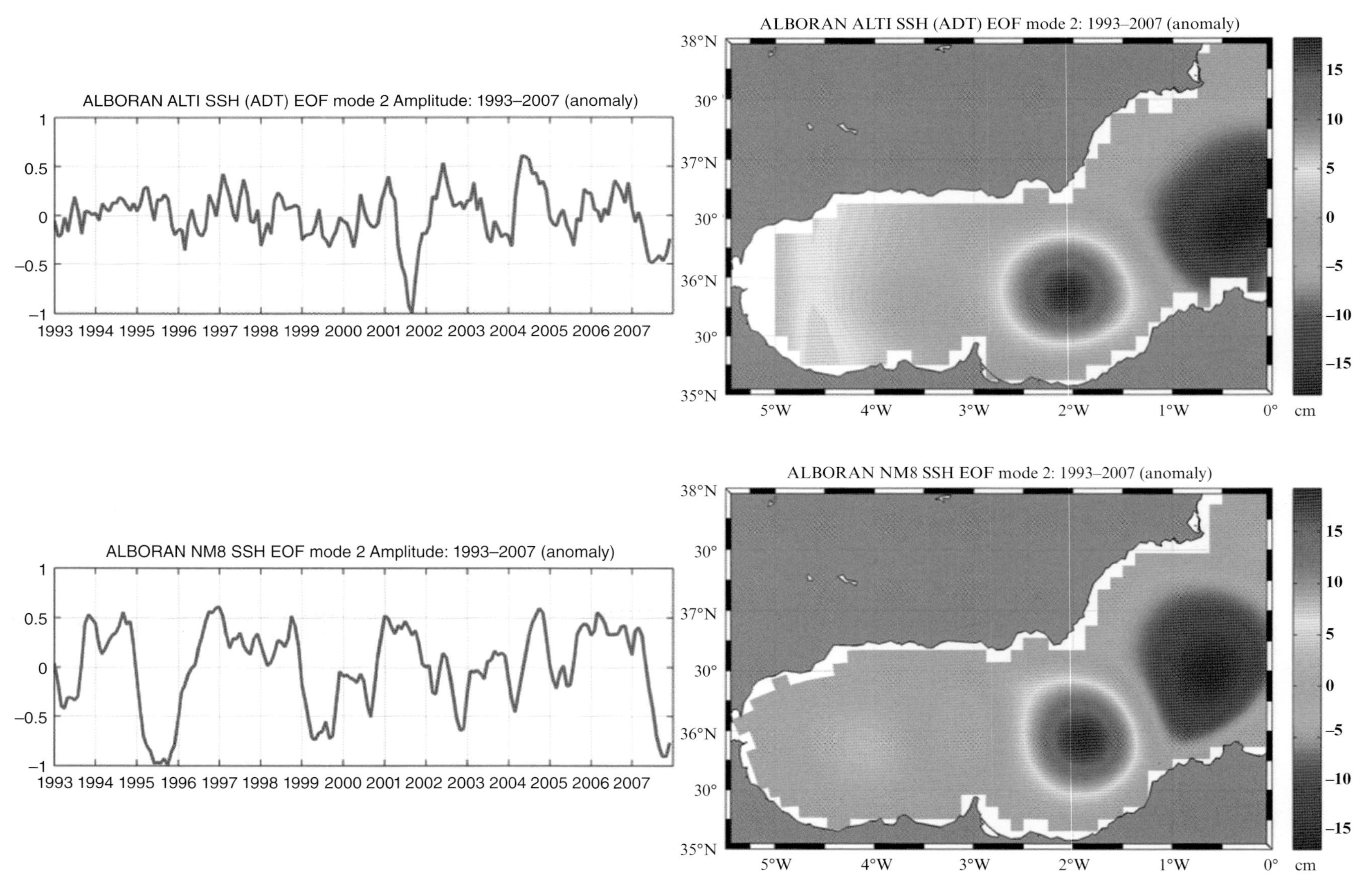

Figure 2.8 Second EOF for the Alborán Sea with amplitude (left) and pattern (right); altimetry ADT (top), NM8 SSH (bottom). For color detail, please see color plate section.

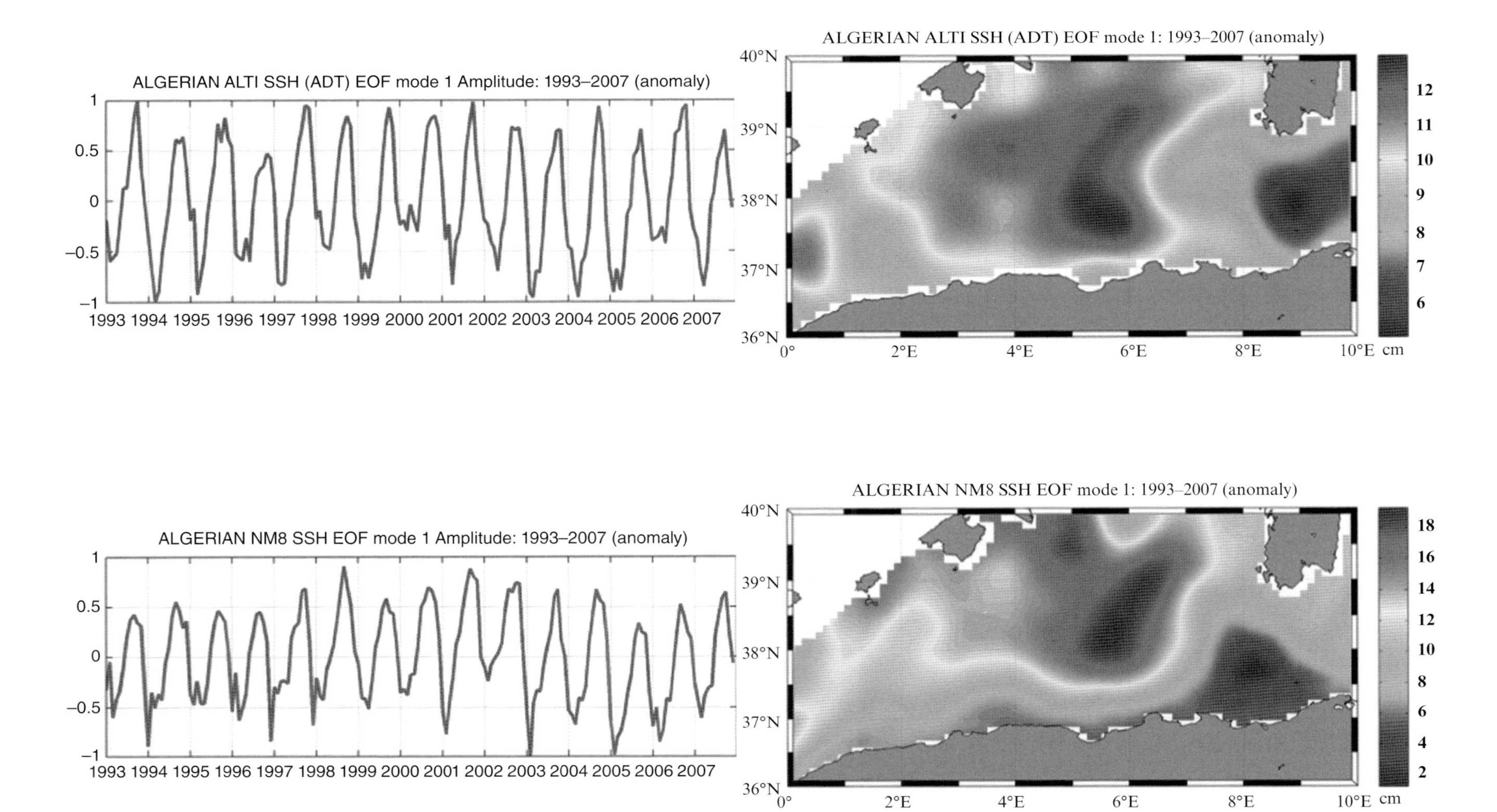

Figure 2.9 First EOF for the Algerian Basin with amplitude (left) and pattern (right); altimetry ADT (top), NM8 SSH (bottom). For color detail, please see color plate section.

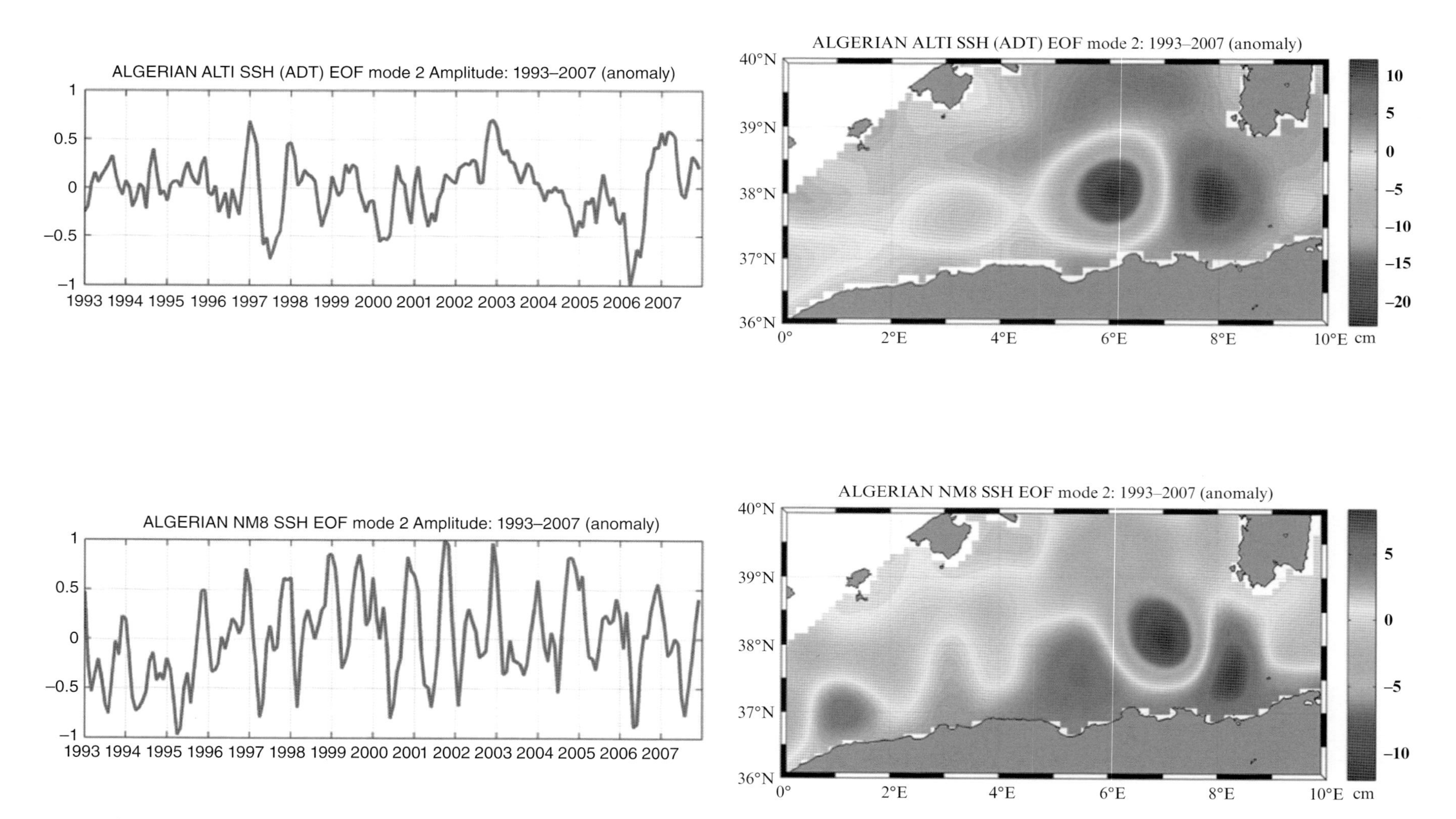

Figure 2.10 Second EOF for the Algerian Basin with amplitude (left) and pattern (right); altimetry ADT (top), NM8 SSH (bottom). For color detail, please see color plate section.

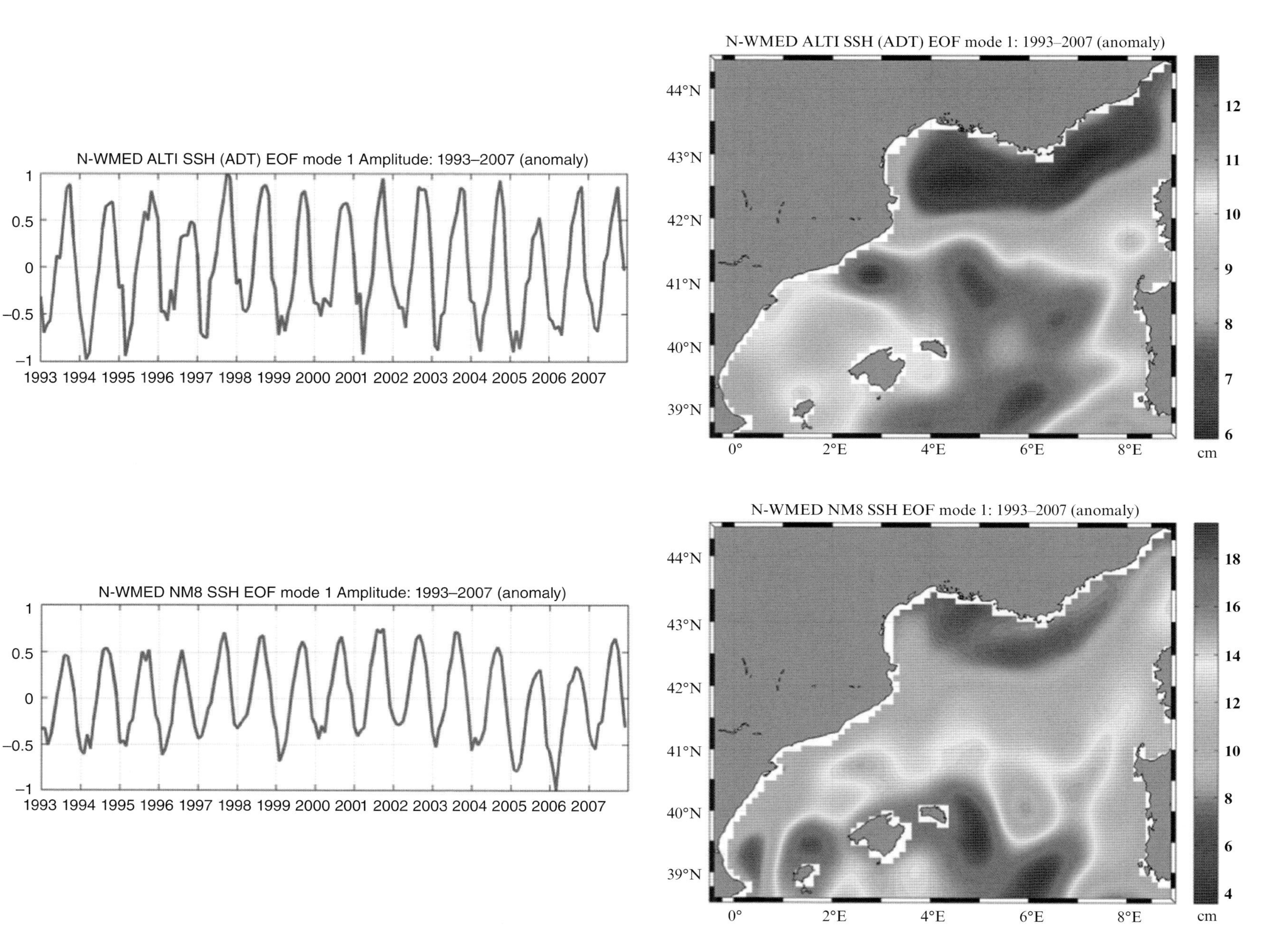

Figure 2.11 First EOF for the NWMED with amplitude (left) and pattern (right); altimetry ADT (top), NM8 SSH (bottom). For color detail, please see color plate section.

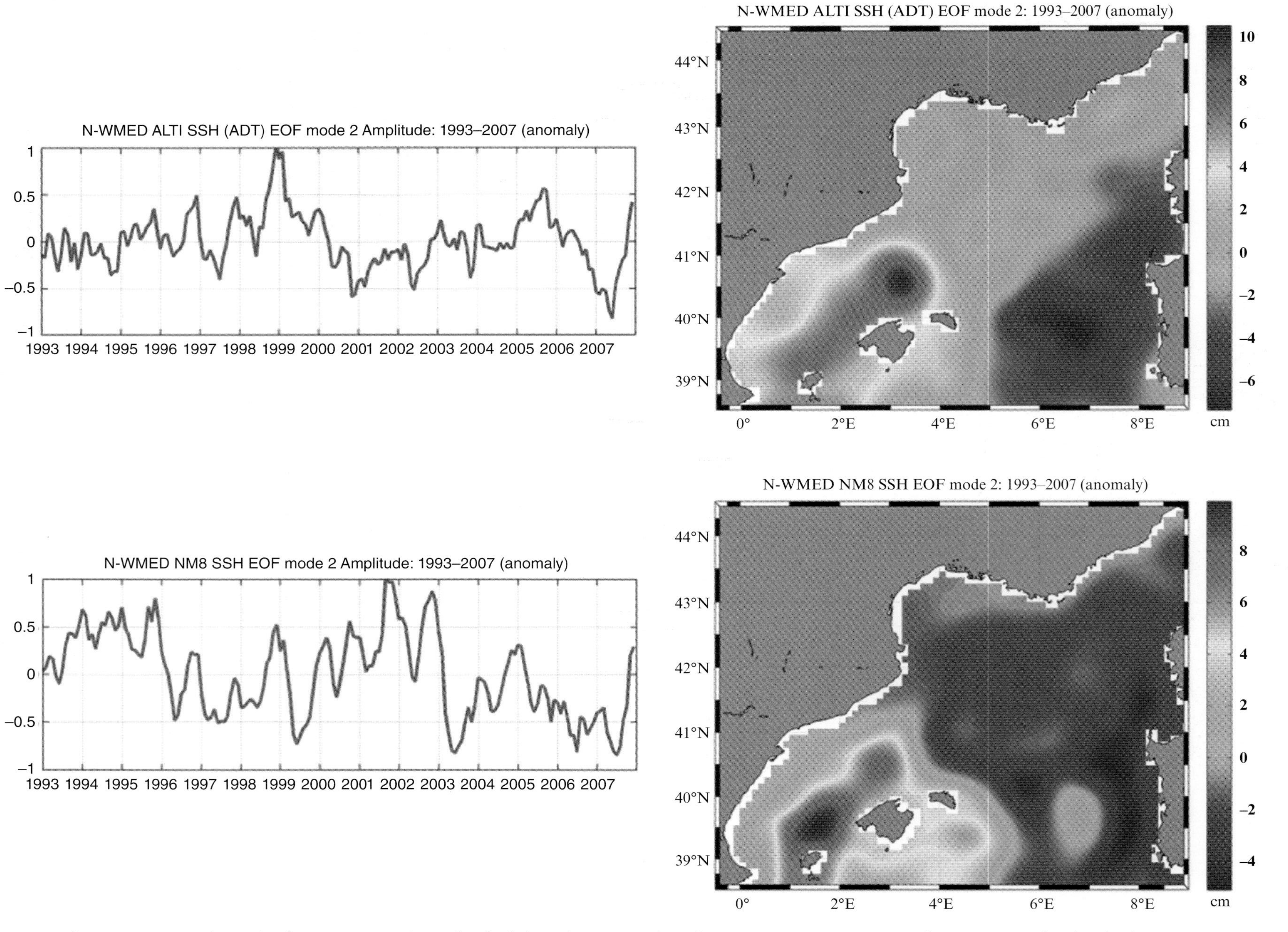

Figure 2.12 Second EOF for the NWMED with amplitude (left) and pattern (right); altimetry ADT (top), NM8 SSH (bottom). For color detail, please see color plate section.

validated against observational datasets. However, despite these encouraging results, some key circulation features such as the WAG or other large mesoscale signals related to interannual variability are not reproduced properly.

Forthcoming simulations based on recent modeling developments will probably allow us to answer some of these issues soon, either using the Mediterranean Sea eddy-permitting (2–3 km, NEMOMED36) ocean model with higher vertical resolution over a multiyear period [*Beuvier et al.*, 2008], higher spatial resolution atmospheric forcing at 10 km [*Herrmann et al.*, 2011], or fully interactive atmosphere–ocean regional climate models laterally driven by reanalysis [*Herrmann et al.*, 2011].

On the other hand, altimetric measurements present their own limitations. The data used in this study were based on the combination of two altimeter missions, which do not fully resolve the typical scale of mesoscale structures. Indeed, the mesoscale in the Mediterranean has a typical length scale of around 10–100 km [*Malanotte-Rizzoli and the Pan-Med group*, 2012], that is, up to a few times the first internal Rossby radius of deformation (15–20 km according to *Robinson et al.* [2001] and *Pinardi and Masetti* [2000]) and a temporal scale of the order of 5–10 days, which is much shorter and much faster than the global ocean mesoscale (50–500 km and 10–100 days according to *Morrow and Le Traon* [2012]).

In this respect, *Pascual et al.* [2007] merged four altimeter missions to improve the description of surface mesoscale activity in the Mediterranean Sea. Mean eddy kinetic energy was estimated from different altimeter configurations, providing evidence that the classical combination of two missions failed to reproduce several intense signals, as was revealed through a comparison with independent data. Conversely, when a third mission was included in the analysis, these features were retrieved. The fourth mission was more moderately effective but contributed to the detection of certain structures more accurately. The study concluded that at least three, but preferably four, altimeter missions are required to effectively monitor the Mediterranean surface mesoscale variability.

Despite these advances, the resulting altimetry maps are still spatially smooth and are missing an important part of the small-scale signals, as evidenced by previous *in situ* experiments [*Nencioli et al.*, 2011; *Pascual et al.*, 2010]. Thus, it is still necessary to complement altimetry data with alternative remote and in-situ sensors to fully characterize the three-dimensional circulation covering a wide range of spatiotemporal scales [*Pascual et al.*, 2012].

In waiting for the new SWOT altimeter mission [*Fu and Ferrari*, 2008], which will be more suited for fine-scale studies, alternative methods [e.g., *Escudier et al.*, 2012] aiming at the generation of remotely sensed high-resolution sea surface topography fields, should be explored and implemented. In parallel, an integrated approach combining high-resolution numerical models and observations would be proven more valuable for improving our knowledge on the spatial and temporal variability of the Mediterranean ocean circulation.

Acknowledgments. The authors would like to thank Meteo-France for running and distributing the ARPERA forcing (Michel Déqué) and the NEMOMED8 simulation (Florence Sevault). The altimeter products were produced by SSALTO-DUACS and distributed by AVISO with support from CNES. We thank CSIC for the financial support of E. Vidal-Vijande's Ph.D. Funding from the European Union (MyOcean2 EU FP7 project), Conselleria d'Educacií, Cultura i Universitat from the Local Government of the Balearic Islands, and FEDER funds (CAIB-51/2011) is acknowledged. This study is a contribution to the international HyMeX program.

REFERENCES

Allen, J. T., D. A. Smeed, J. Tintoreé, and S. Ruiz (2001), Mesoscale subduction at the Almeria-Oran front, Part 1: Agesotrophic flow, *J. Mar. Syst.*, *30*, 263–285, doi:10.1016/S0924-7963(01)00062-8.

Astraldi, M., et al. (1999), The role of straits and channels in understanding the characteristics of Mediterranean circulation, *Prog. Oceanogr.*, *44* (1–3), 65–108.

Baldacci, A., G. Corsini, R. Grasso, G. Manzella, J. T. Allen, P. Cipollini, T. H. Guymer, and H. M. Snaith (2001), A study of the Alborán Basin mesoscale system by means of empirical orthogonal function decomposition of satellite data, *J. Mar. Syst.*, *29* (1–4), 293–311.

Barnier, B., L. Siefridt, and P. Marchesiello (1995), Thermal forcing for a global ocean circulation model using a three year climatology of ECMWF analyses, *J. Mar. Syst.*, *6* (4): 363–380

Beckers, J. M., et al. (2002), Model intercomparison in the Mediterranean: MEDMEX simulations of the seasonal cycle, *J. Mar. Syst.*, *33–34*, 215–251.

Béranger, K., et al. (2010), Impact of the spatial distribution of the atmospheric forcing on water mass formation in the Mediterranean Sea, *J. Geophys. Res.*, *115* (C12041), 1–22.

Beuvier, J., et al. (2010), Modeling the Mediterranean Sea interannual variability during 1961–2000: Focus on the eastern Mediterranean transient, *J. Geophys. Res.*, *115* (C8), C08 017.

Beuvier, J., F. Sevault, and S. Somot (2008), Modélisation de la variabilité interannuelle de la mer Méditerranée sur la période 1960–2000 a l'aide de NEMOMED8, *Note Cent.*, 105, Groupe de Meteorol. de Grande Echelle et Clim., Cent. Natl. de Rech. Meteorol., Toulouse, France.

Bjornsson, H., and S. A. Venegas (1997), *A Manual for EOF and SVD Analyses of Climate Data*, McGill University, CCGCR Report No. 97–1, Montreal, Quebec.

Cazenave, A., P. Bonnefond, F. Merciera, K. Dominha, and V. Toumazou (2002), Sea level variations in the Mediterranean Sea and Black Sea from satellite altimetry and tide gauges, *Global Planet. Change*, *34*(1–2), 59–86, doi:10.1016/S0921-8181(02)00106-6.

Daget, N., A. T. Weaver, and M. A. Balmaseda (2009), Ensemble estimation of background-error variances in a three-dimensional variational data assimilation system for the global ocean, *Q. J. R. Meteorol. Soc.*, *135*, 1071–1094.

Demirov, E., and N. Pinardi (2002), Simulation of the Mediterranean sea circulation from 1979 to 1993: Part i., the interannual variability, *J. Mar. Syst.*, *33–34*, 23–50.

Déqué M., and J.-P. Piedelievre (1995), High-resolution climate simulation over Europe, *Climate Dynamics*, *11*, 321–339.

Emery, W. J., and R. E. Thomson (1997), *Data Analysis and Methods in Physical Oceanograph*, Elsevier, New York, USA.

Escudier, R., J. Bouffard, A. Pascual, P.-M. Poulain, and M.-I. Pujol (2012), Improvement of coastal and mesoscale observation from space: Application to the northwestern Mediterranean Sea, Submitted to *Geophysical Research Letters*.

Fernández, V., D. E. Dietrich, R. L. Haney, and J. Tintoré (2005), Mesoscale, seasonal and interannual variability in the Mediterranean Sea using a numerical ocean model, *Prog. Oceanogr.*, *66*, 321–340.

Flexas, M., D. Gomis, S. Ruiz, A. Pascual, and P. Leín (2006), In situ and satellite observations of the eastward migration of the western Alborán Basin gyre, *Prog. Oceanogr.*, *70* (2–4), 486–509.

Fu, L.-L., and R. Ferrari (2008), Observing oceanic submesoscale processes from space, *Eos Trans. AGU*, *89*(48), 488, doi:10.1029/2008EO480003.

Gomis, D., S. Ruiz, and M. A. Pedder (2001), Diagnostic analysis of the 3D ageostrophic circulation from a multivariate spatial interpolation of CTD and ADCP data, *Deep Sea Res., Part I*, *48*, 269–295, doi:10.1016/ S0967-0637(00)00060-1.

Herbaut, C., F. Martel, and M. Crépon (1997), A sensitivity study of the general circulation of the western Mediterranean Sea, Part II: The response to atmospheric forcing, *Journal of Physical Oceanography*, *27*(10), 2126–2145.

Herrmann, M., F. Sevault, J. Beuvier, and S. Somot (2010), What induced the exceptional 2005 convection event in the northwestern Mediterranean basin? Answers from a modeling study, *J. Geophys. Res., C: Oceans*, *115*(12).

Herrmann, M., S. Somot, F. Sevault, C. Estournel, and M. Déqué (2008), Modeling the deep convection in the northwestern Mediterranean Sea using an eddy-permitting and an eddy-resolving model: Case study of the 1986–87 winter, *J. Geophys. Res.*, *113*, C04011, doi:10.1029/2006JC003991.

Herrmann, M., S. Somot, S. Calmanti, C. Dubois, and F. Sevault (2011), Representation of daily wind speed spatial and temporal variability and intense wind events over the Mediterranean Sea using dynamical downscaling: Impact of the regional climate model configuration, *Nat. Hazards Earth Syst. Sci.*, *11*, 1983–2001, doi:10.5194/nhess-11-1983-2011.

Herrmann, M. J., and S. Somot (2008), Relevance of era40 dynamical downscaling for modeling deep convection in the Mediterranean Sea, *Geophys. Res. Lett.*, *35* (4).

Iudicone, D., R. Santoleri, S. Marullo, and P. Gerosa (1998), Sea level variability and surface eddy statistics in the Mediterranean Sea from TOPEX/POSEIDON data, *J. Geophys. Res., C: Oceans*, *103*(C2), 2995–3011.

Larnicol, G., N. Ayoub, and P. Y. Le Traon (2002), Major changes in Mediterranean Sea level variability from 7 years of TOPEX/Poseidon and ERS-1/2 data, *J. Mar. Syst.*, *33*, 63–89, doi:10.1016/S0924-7963(02)00053-2.

Le Traon, P.-Y., and F. Ogor (1998), ERS-1/2 orbit improvement using TOPEX/POSEIDON: The 2 cm challenge, *J. Geophys. Res.*, *103*, 8045–8057.

Lípez García, M., C. Millot, J. Font, and E. Garcóa-Ladona (1994), Surface circulation variability in the Balearic basin, *J. Geophys. Res.*, *99* (C2), 3285–3296.

Madec, G., P. Delecluse, M. Imbard, and C. Levy (1998), OPA 8.1, Ocean general circulation model, reference manual, Université P. et M. Curie, B102 T15-E5, 4 place Jussieu, Paris cedex 5, IPSL/LODYC, France, Note du Pôle de mod élisation.

Malanotte-Rizzoli, P., and the Pan-Med Group (2012), Physical forcing and physical/biochemical variability of the Mediterranean Sea: A review of unresolved issues and directions of future research, Report of the Workshop "Variability of the Eastern and Western Mediterranean Circulation and Thermohaline Properties: Similarities and Differences," Rome, November 7–9, 2011, 48 pp.

MEDAR-Group (2002), Mediterranean and black sea database of temperature, salinity and biochemical parameters and climatological atlas [4 cd-roms].

MEDOC Group (1970), Observation of formation of deep water in the Mediterranean Sea, 1969, *Nature*, *227*, 1037–1040.

Millot, C. (1999), Circulation in the western Mediterranean Sea, *J. Mar. Syst.*, *20*, 423–442.

Molcard, A., N. Pinardi, M. Iskandarani, and D. Haidvogel (2002), Wind driven general circulation of the Mediterranean Sea simulated with a spectral element, *Dynamics of Atmospheres and Oceans*, *35*, 97–130.

Morán, X. A. G, I. Taupier-Letage, E. Vázquez-Domónguez, S. Ruiz, L. Arin, P. Raimbault, and M. Estrada (2001), Physical-biological coupling in the Algerian Basin (SW Mediterranean): Influence of mesoscale instabilities on the biomass and production of phytoplankton and bacterioplankton, *Deep Sea Research Part I: Oceanographic Research Papers*, *48*, 2, 405–437, doi:10.1016/S0967-0637(00)00042-X.

Morrow, R., and P.-Y. Le Traon (2012), Recent advances in observing mesoscale ocean dynamics with satellite altimetry, *Advances in Space Research*, *50*(8), 1062–1076.

Nencioli, F., F. d'Ovidio, A. M. Doglioli, and A. A. Petrenko (2011), Surface coastal circulation patterns by in-situ detection of Lagrangian coherent structures, *Geophys. Res. Lett.*, *38*, L17604, doi:10.1029/2011GL048815.

Olita, A., A. Ribotti, R. Sorgente, L. Fazioli, and A. Perilli (2011), Sla- chlorophyll-a variability and covariability in the Algero-Provencal Basin (1997–2007) through combined use of EOF and wavelet analysis of satellite data, *Ocean Dynamics*, *61* (1), 89–102.

Pascual, A., and D. Gomis (2003), Use of surface data to estimate geostrophic transport, *Journal of Atmospheric and Oceanic Technology*, *20*(6), 912–926.

Pascual, A., B. Buongiorno Nardelli, G. Larnicol, M. Emelianov, and D. Gomis (2002), A case of an intense anticyclonic eddy in the Balearic Sea (western Mediterranean), *J. Geophys. Res. C*, 107.

Pascual, A., J. Bouffard, S. Ruiz, B. Nardelli, E. Vidal-Vijande,; R. Escudier, J. M. Sayol, and A. Orfila (2012), Recent improvements on mesoscale characterization in the western Mediterranean Sea: Synergy between satellite altimetry and other observational approaches, *Scientia Marina*, under review.

Pascual, A., M.-I. Pujol, G. Larnicol, P.-Y. Le Traon, and M.–H. Rio (2007), Mesoscale mapping capabilities of multi-satellite altimeter missions: First results with real data in the Mediterranean Sea, *J. Mar. Syst.*, *65* (1–4), 190–211.

Pascual, A., S. Ruiz, and J. Tintoré (2010), Combining new and conventional sensors to study the Balearic current, *Sea Technology*, *51*, 7, 32–36.

Pinardi, N., and E. Masetti (2000), Variability of the large scale general circulation of the Mediterranean Sea from observations and modelling: A review, *Palaeoecology*, *158*, 153–173.

Pinot, J. M., J. L. Lípez-Jurado, and M. Riera (2002), The canales experiment (1996–1998): Interannual, seasonal, and mesoscale variability of the circulation in the Balearic channels, *Prog. Oceanogr.*, *55* (3–4), 335–370.

Puillat, I., I. Taupier-Letage, and C. Millot (2002), Algerian eddies lifetime can near 3 years, *J. Mar. Syst.*, *31* (4), 245–259.

Pujol, M.-I., and G. Larnicol (2005), Mediterranean Sea eddy kinetic energy variability from 11 years of altimetric data, *J. Mar. Syst.*, *58*, 121–142.

Renault, L., T. Oguz, A. Pascual, G. Vizoso, and J. Tintore (2012), Surface circulation in the Alborán Sea (western Mediterranean) inferred from remotely sensed data, *J. Geophys. Res.*, *117*, C08009, doi:10.1029/2011JC007659.

Reynaud, T., H. Legrand, and B. Barnier (1998), A new analysis of hydrographic data in the Atlantic and its application to an inverse modeling study, *International WOCE Newsletter*, *32*, 29–31.

Rio, M.–H., P.-M. Poulain, A. Pascual, E. Mauri, G. Larnicol, and R. Santoleri (2007), A mean dynamic topography of the Mediterranean Sea computed from altimetric data, in-situ measurements and a general circulation model, *J. Mar. Syst.*, *65* (1–4), 484–508.

Robinson, A., W. Leslie, A. Theocharis, and A. Lascaratos (2001), Mediterranean Sea circulation, 1689–1706, in *Encyclopedia of Ocean Sciences*, Academic Press Ltd., London.

Roullet, G., and G. Madec (2000), Salt conservation, free surface, and varying levels: A new formulation for ocean general circulation models, *J. Geophys. Res.*, *105*, 23 927–23 942.

Roussenov, V., E. Stanev, V. Artale, and N. Pinardi (1995), A seasonal model of the Mediterranean Sea general circulation, *J. Geophys. Res.*, *100*(C7), 13,515–13,538, doi:10.1029/95JC00233.

Ruiz, S., A. Pascual, B. Garau, Y. Faugere, A. Alvarez, and J. Tintoré (2009), Mesoscale dynamics of the Balearic front, integrating glider, ship and satellite data, *J. Mar. Syst.*

Ruiz, S., J. Font, M. Emelianov, J. Isern-Fontanet, C. Millot, J. Salas, and I. Taupier-Letage (2002), Deep structure of an open sea eddy in the Algerian Basin, *J. Mar. Syst.*, *33–34*, 179–195.

Sevault, F., S. Somot, and J. Beuvier (2009), A regional version of the NEMO Ocean Engine on the Mediterranean Sea: NEMOMED8 user's guide, Note de centre GMGEC, CNRM.

Shaeffer, A., P. Garreau, A. Molcard, P. Fraunié, and Y. Seity (2011), Influence of high-resolution wind forcing on hydrodynamic modeling of the gulf of lions, *Ocean Dynamics*, *61*, 1823–1844.

Simmons, A. J., and J. K. Gibson (2000), The ERA-40 Project Plan, ERA-40 Project Report Series No. 1. Available from: http://www.ecmwf.int/publications/library/ecpublications/_pdf/ERA40_PRS_1.pdf

Somot, S., F. Sevault, and M. Déqué (2006), Transient climate change scenario simulation of the Mediterranean Sea for the twenty-first century using a high-resolution ocean circulation model, *Climate Dynamics*, *27* (7), 851–879.

Somot, S., F. Sevault, M. Déqué, M. Crépon (2008), 21st century climate change scenario for the Mediterranean using a coupled Atmosphere–ocean Regional Climate Model, *Global and Planetary Change*, *63*(2–3), 112–126, doi:10.1016/j.gloplacha.2007.10.003.

Tintoré, J., P. La Violette, I. Blade, and A. Cruzado (1988), A study of an intense density front in the eastern Alborén Basin: The Almetóa-Oran front, *J. Phys. Oceanogr.*, *18* (10), 1384–1397.

Tonani, M., N. Pinardi, S. Dobricic, I. Pujol, and C. Fratianni (2008), A high- resolution free-surface model of the Mediterranean Sea, *Ocean Science*, *4* (1), 1–14.

Tsimplis, M. N., M. Marcos, S. Somot, and B. Barnier (2008), Sea level forcing in the Mediterranean Sea between 1960 and 2000, *Global Planet. Change*, *63* (4), 325–332.

Vidal-Vijande, E., A. Pascual, B. Barnier, J.-M. Molines, and J. Tintoré (2011), Analysis of a 44-year hindcast for the Mediterranean Sea: Comparison with altimetry and in-situ observations, *Scientia Marina*, *75* (1).

Vidal-Vijande, E., A. Pascual, B. Barnier, J.-M. Molines, and J. Tintoré (2012), Multiparametric analysis and validation in the western Mediterranean of three global OGCM hindcasts, *Scientia Marina*, *76S1*, 147–164.

Videz, A., J. M. Pinot, and R. L. Haney (1998), On the upper layer circulation in the Alborán Basin. *J. Geophys. Res. C*, *103*, 21 653–21 666.

3

Exchange Flow through the Strait of Gibraltar as Simulated by a σ-Coordinate Hydrostatic Model and a z-Coordinate Nonhydrostatic Model

Gianmaria Sannino[1], J. C. Sánchez Garrido[2], L. Liberti[3], and L. Pratt[4]

3.1. INTRODUCTION

The Mediterranean Sea is a semi-enclosed basin displaying an active thermohaline circulation (MTHC) that is sustained by the atmospheric forcing and controlled by the narrow and shallow Strait of Gibraltar (hereinafter SoG). The atmospheric forcing drives the Mediterranean basin toward a negative budget of water and heat. Over the basin, evaporation exceeds the sum of precipitation and river discharge, while a net heat flux is transferred to the overlying atmosphere through the sea surface. These fluxes are balanced by the exchange flow that takes place in Gibraltar. Within the SoG, the MTHC takes the form of a two-way exchange: an upper layer of fresh and relatively warm Atlantic water spreads in the Mediterranean basin, and a lower layer of colder and saltier Mediterranean water sinks as a tongue in the North Atlantic at intermediate depths. The interaction between the intense tidal forcing [*Candela et al.*, 1990] and the complex geometry of the SoG (Figure 3.1a) influences the two-way exchange via hydraulic control [*Bryden and Stommel*, 1984]. The exchange is subject to vigorous mixing and entrainment [*Wesson and Gregg*, 1994] as well as intermittent hydraulic controls over the main sills and in its narrowest sections [*Sannino et al.*, 2007; *Sannino et al.*, 2009a]. The simultaneous presence in the SoG of at least two cross sections in which the exchange is controlled drives the strait dynamics toward the so-called maximal regime [*Bryden and Stommel*, 1984; *Armi and Farmer*, 1988]. If the exchange is subject to only one hydraulic control, the regime is called submaximal. The two regimes have different implications for property fluxes, response time, and other physical characteristics of the coupled circulation in the SoG and Mediterranean Sea. The maximal regime can be expected to have larger heat, salt, and mass fluxes and to respond more slowly to changes in stratification and thermohaline forcing within the Mediterranean Sea and the North Atlantic Ocean [*Sannino et al.*, 2009a].

As first recognized by *Bray et al.* [1995], the strong entrainment and mixing present in the Strait of Gibraltar lead to the formation of a thick interfacial layer where density and velocity change gradually in the vertical direction. They also argued that the classical two-layer approach used to describe the two-way exchange was insufficient to account for the flow regime in the SoG. They found that a three-layer system, which includes an active interface layer, best represents the exchange through the SoG. The presence of a thick interfacial layer complicates the estimation of the hydraulic state of the flow exchange using the two-layer hydraulic theory. Such difficulty has been recently overcome by *Sannino et al.* [2007] who analyzed for the fist time the hydraulic regime of the exchange flow applying a three-layer hydraulic theory. Doing so they considered the thick interfacial layer as an active participant of the hydraulic regime. The hydraulic studies conducted by *Sannino et al.* [2007] were based on the analysis of numerical simulations. The simulations were carried out using a σ-coordinate model

[1] *Climate and Impact Modeling Lab—Energy and Environment Modeling Unit—ENEA, CR Casaccia, Rome, Italy*

[2] *Grupo de Oceanografía Física. Dpto. Física Aplicada II, Campus de Teatinos, University of Malaga, Malaga, Spain*

[3] *Istituo Superiore per la Protezione e la Ricerca Ambientale, Rome, Italy*

[4] *Woods Hole Oceanographic Institution, Woods Hole, Massachusetts, USA*

The Mediterranean Sea: Temporal Variability and Spatial Patterns, Geophysical Monograph 202. First Edition.
Edited by Gian Luca Eusebi Borzelli, Miroslav Gačić, Piero Lionello, and Paola Malanotte-Rizzoli.

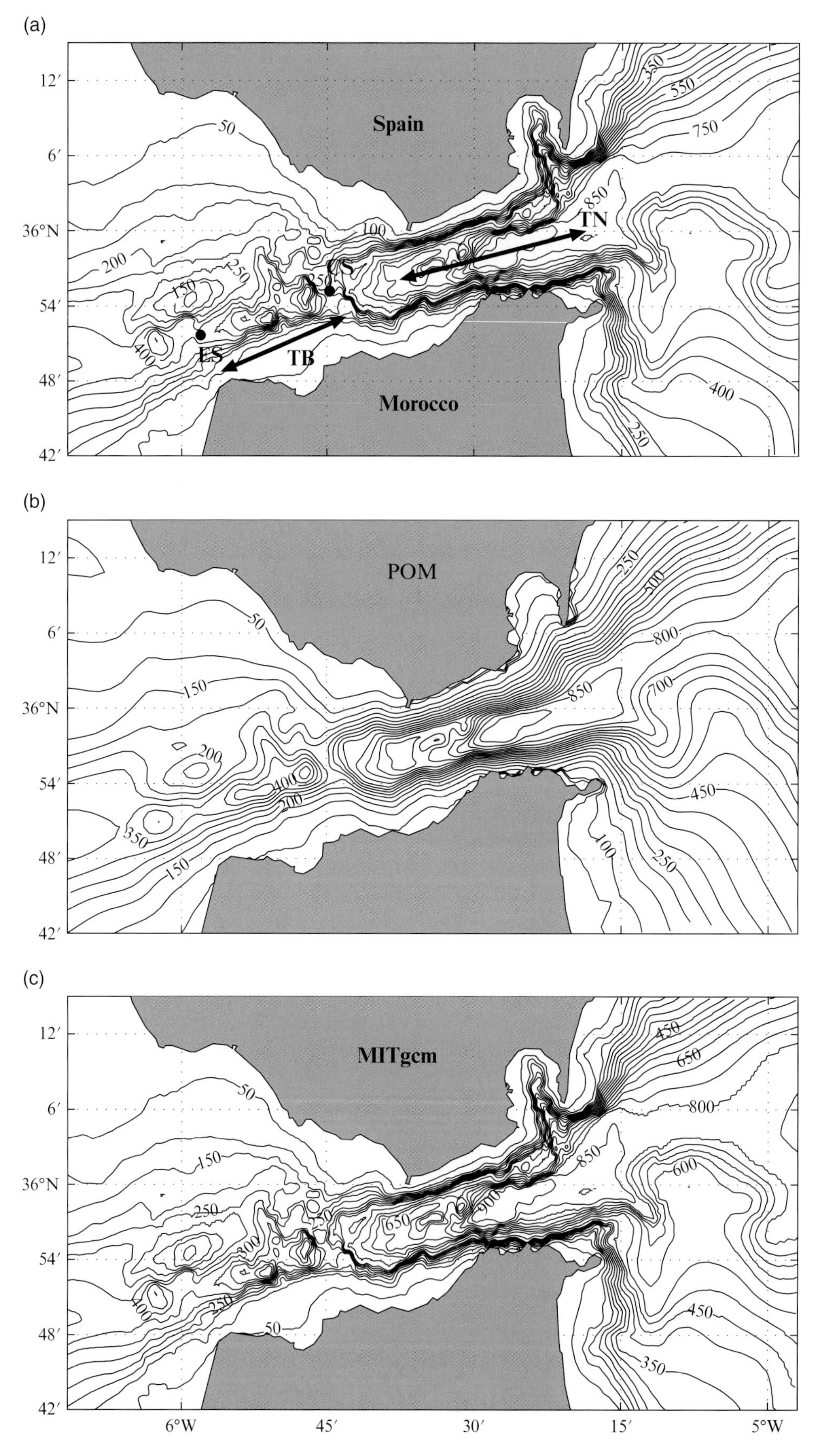

Figure 3.1 (a) Original bathymetric chart of the Strait of Gibraltar; (b) Bathymetry as represented in the σ-coordinate POM; (c) Bathymetry as represented in the z-level, partial cells MITgcm.

essentially based on the Princeton Ocean Model [POM; *Blumberg and Mellor*, 1987].

In the last 15 years, many models of different complexity have been implemented to study most of the aspects of the flow exchange through the SoG. However, the only three-dimensional model able to reproduce most of the features of the exchange flow was the one initially implemented by *Sannino et al.* [2007] and subsequently improved by *Sannino et al.*, [2009a]. The earliest version of the model was developed by *Sannino et al.* [2002] to study the mean flow exchange; subsequently the model was improved by *Sannino et al.* [2004] who introduced the tidal forcing. Recently, the model has been used to test the applicability of classical two-layer, one-dimensional hydraulic theory to the SoG [*Sannino et al.*, 2007, 2009a]. The model has been also used to estimate the Mediterranean water outflow at the western end of the SoG in a combined observational-modeling work [*Sanchez-Roman et al.*, 2009], and to assess kinematic properties of internal waves in the area [*Garrido et al.*, 2008]. The extensive use of POM in the study of different aspects of the flow exchange has led to an almost complete validation of the model. Results have been validated against most of the available *in situ* data.

According to the validation analysis so far performed, it appears that POM is able to capture, in a reasonable way, most of the main features of the exchange flow. However, there are some aspects of the strait hydrodynamics that could not be well reproduced by the model. This especially concerns the evolution of the internal tidal bore generated in the main sill area. After its generation, the bore progresses toward the Mediterranean, evolving into a series of short internal solitary waves of large amplitude [*Vlasenko et al.*, 2009]. These waves are strongly nonlinear and nonhydrostatic, thus their modeling requires fully nonhydrostatic codes such as, for example, MITgcm [*Marshall et al.*, 1997a] or SUNTANS [*Fringer et al.*, 2006].

Although the poor representation of the bore generation and propagation was an expected outcome for POM, there still remain some open questions regarding the effects produced by the hydrostatic assumption, the vertical and horizontal resolution adopted, and the parameterization used for mixing on the simulated hydraulic regime. Thus, the main goal of this study is the investigation of the effects produced by these factors on the simulated hydraulic behavior of the SoG by the nonhydrostatic assumption, the resolution adopted, and the parameterization used. To this purpose the exchange flow simulated by POM has been compared with the exchange flow simulated by a very high-resolution, fully nonhydrostatic model implemented for the strait region. The nonhydrostatic model is based on the z-coordinate MITgcm code (http://mitgcm.org). As pointed out by *Legg et al.* [2006], the nonhydrostatic version of the MITgcm, when implemented at very high resolution, is able to capture the largest-scale mixing processes responsible for entrainment. Thus, the model does not need specific parameterizations for the entrainment. The evaluation of the impact of these novel features on the water exchange and hydraulics regime of the SoG is one of the scopes of this paper. As will be demonstrated in the following, these particular features allow the simulation performed with MITgcm to be used as a benchmark against which the POM simulation and, more in general, any numerical model simulating the dynamics of the SoG at lower resolution can be compared. Thus, the overall objective of the present study is a systematic comparison of the main features simulated by POM, as for example the three-layer structure, the transports, and the hydraulic regime simulated, with those obtained by MITgcm.

The remainder of the paper is organized as follows: the two models are described in section 2 and validated in section 3; the comparison of the models in terms of the simulated internal wave field, three-layer properties, and hydraulics is shown in section 4; while conclusions will be discussed in section 5.

3.2. MODELS DESCRIPTION AND INITIALIZATION

The two models used in this work are the σ-coordinate Princeton Ocean Model [POM; *Blumberg and Mellor*, 1987] as implemented by *Sannino et al.* [2009a], and the height vertical coordinate Massachusetts Institute of Technology general circulation model [MITgcm; see *Marshall et al.*, 1997a, b, as implemented by *Garrido et al.*, 2012].

3.2.1. POM

The hydrostatic POM model as been implemented by *Sannino et al.* [2002] and *Sannino et al.* [2009a] to investigate different aspects of the circulation that take place in the SoG: mean and tidal exchange [*Sannino et al.*, 2002; *Sannino et al.*, 2004], time and spatial variability of the internal bore propagation [Garrido et al., 2008], estimation of the Mediterranean water outflow [*Sanchez-Roman et al.*, 2009], and hydraulic regimes [*Sannino et al.*, 2007; *Sannino et al.*, 2009a].

An extensive description of the model setting can be found in *Sannino et al.* [2004, 2009a]; some aspects relevant for the comparison with MITgcm are described in the following. The POM version used in *Sannino et al.* [2009a], is the one generally known as *pom*98 with the only exception for the advection scheme. As default POM uses a second-order centered (both spatially and temporally) scheme. It is well known that such a scheme is dispersive. Dispersion is more evident in presence of strong density gradients where it creates spurious temperature and salinity values (over- and undershooting problems). Thus, the presence of density gradients within the Strait of Gibraltar makes the centered scheme unable to simulate the water exchange. To overcome this problem, the second-order, sign-preserving Multidimensional Positive Definite Advection Transport Algorithm (MPDATA), as developed by *Smolarkiewicz* [1984] and implemented by *Sannino et al.* [2002], has been used. MPDATA is a flux corrected upstream scheme, that is, an upstream scheme characterized by a reduced implicit diffusion. The numerical diffusion is reduced through an iterative method based on antidiffusive velocities, which is applied to correct the excessive numerical diffusion of standard upstream scheme. The repeated procedure yields a positive definite advection algorithm with second-order accuracy. The number of iterations is optional; each additional iteration increases the solution accuracy and the computation time: the number of iterations chosen in *Sannino et al.* [2009a] was three.

The vertical mixing coefficients were obtained from the Mellor-Yamada turbulence scheme [*Mellor and Yamada*, 1982]. As demonstrated by *Ezer* [2005], the Mellor-Yamada scheme is able to explicitly capture the mixing processes responsible for entrainment, so there is no need for specific parameterizations of entrainment in POM. The horizontal momentum, heat, and salt small-scale mixing processes are parameterized via the Laplacian, along-sigma, velocity, and grid space dependent Smagorinsky diffusion scheme [*Smagorinsky*, 1963]. Normal velocities are set to zero along coastal boundaries. At the bottom, adiabatic boundary conditions are applied to temperature and salinity and a quadratic bottom friction, with a prescribed drag coefficient, is applied to the momentum flux. This is calculated by combining the velocity profile with the logarithmic law of the wall:

$$C_D = \max\left[2.5\times10^{-3}, k^2 \ln\left(\Delta z_b / z_0\right)\right] \quad (1)$$

where k is the Von Karman constant, z_0 is the roughness length set to 1 cm, and Δz_b is the distance from the bottom of the deepest velocity grid point.

The model grid extends longitudinally from the Gulf of Cadiz to the Alboran Sea. The grid has a nonuniform horizontal spacing (see Figure 6 of *Sannino et al.* [2009a]); the resolution is finer in the strait, where Δx (Δy) is 593 m (485 m) around Camarinal Sill (CS) (Figure 3.2a), while Δx (Δy) is 10 Km (20 Km) and 8 Km (15 Km) at the eastern and western ends, respectively. The vertical grid has 32 σ levels, logarithmically distributed at the surface and at the bottom, and uniformly distributed in the rest of the water column. In a water depth of 1000 m, the upper and lower six σ levels are concentrated in about 100 m, while the remaining levels are equally spaced (40 m). Model bathymetry has been obtained through a bilinear interpolation of data obtained merging the ETOPO2 bathymetry [*NOAA*, 2001] with the very high-resolution bathymetry chart of *Sanz et al.* [1992]. Moreover, to reduce the well-known pressure gradient error produced by σ coordinates in regions of steep topography [*Haney*, 1991], an additional smoothing was applied in order to reach values of $\delta H/H < 0.2$, where H is the model depth as suggested by *Mellor et al.* [1994]. The resulting model topography (Figure 3.1b) is different from the original one especially in the coastal regions where the continental slope has been significantly broadened. Two open boundaries are defined at the eastern and western ends of the computational domain. Here an Orlanski radiation condition [*Mellor et al.*, 1994] is used for the depth-dependent velocity, while a forced-Orlanski radiation condition [*Bills and Noye*, 1987] is used for the surface elevation and a zero gradient condition for the depth-integrated velocity. Boundary conditions for both temperature and salinity are specified using an upwind advection scheme that allows the advection of temperature and salinity into the model domain under inflow conditions.

The model starts from rest and is forced at the open boundaries through the specification of the surface tidal elevation that is characterized by the principal two semidiurnal and two diurnal harmonics: M_2, S_2, O_1, K_1. Amplitude and phase of these harmonics have been computed via the OTIS package [*Egbert and Erofeeva*, 2002]. Finally, the initial conditions for salinity and temperature have been taken from the climatologic Medar-MedAtlas Database [*MEDAR Group*, 2002] for the month of April.

The model was initially run without tidal forcing in order to achieve a steady two-way exchange system. Then the model was forced by tidal components in order to achieve a stable time-periodic solution. After this spin-up phase, the model was run for a further tropical month (27.321 days) that represents our reference experiment (Exp-POM). The term *time-averaged* that will be used in the remaining part of the paper refers to the average over this tropical month period.

3.2.2. MITgcm

MITgcm is the other model used in this work. MITgcm solves the fully nonlinear, nonhydrostatic Navier–Stokes equations under the Boussinesq approximation for an incompressible fluid with a spatial finite-volume discretization on a curvilinear computational grid. The model formulation includes implicit nonlinear free surface [*Campin et al.*, 2004], rescaled vertical height (z^*) coordinates [*Adcroft et al.*, 2004], and two-way nesting capabilities [*Sannino et al.*, 2009b]. An extensive online documentation is at http://mitgcm.org.

The model domain extends from 6.3°W to 4.78°W and is discretized by a nonuniform curvilinear orthogonal

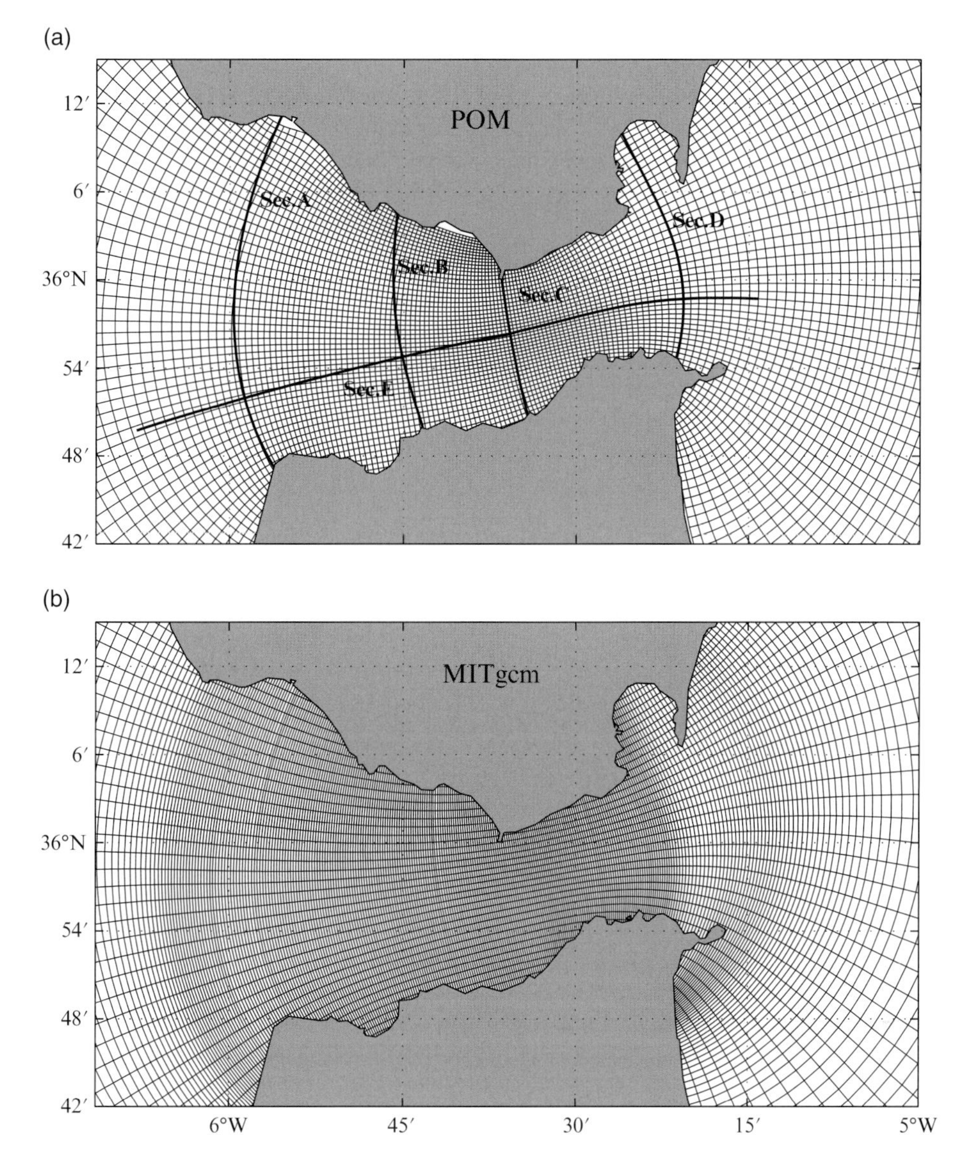

Figure 3.2 (a) Horizontal grid used in POM. Black lines indicate cross-sections referred in the text; (b) Horizontal grid used in the MITgcm (note that only 25% of the actual grid lines are shown).

grid of 1440 x 210 points (Figure 3.2b). Spatial resolution along the axis of the strait, Δx (across the strait axis, Δy), ranges between 46 and 63 m (175 and 220 m) in the area of CS. The mesh size is always less than 70 m (340 m) in the middle of the strait between Espartel Sill and CS, and less than 70 m (200 m) between CS and Tarifa Narrow. The vertical grid has 53 z-levels spaced 7.5 m in the upper 300 m, and their thickness gradually increases to a maximum of 105 m for the remaining 13 bottom levels.

The model topography (Figure 3.1c) has been obtained through a bilinear interpolation of the same initial data described in the POM section, however, in this case, no additional smoothing has been applied. The very high horizontal resolution adopted in MITgcm, together with the partial cell formulation, result in a very detailed description of the bathymetry. The only appreciable differences with respect to the original bathymetric data are confined on the eastern and western ends of the model in the region where the depth is greater than 800 m. No-slip conditions were imposed at the bottom and lateral solid boundaries. The selected tracer advection scheme is a third-order direct space-time flux limited scheme due to *Hundsdorfer et al.* [1995]. Following the numerical experiments conducted by *Vlasenko et al.* [2009] to investigate the 3-D evolution of LAIWs in the Strait of Gibraltar, the turbulent closure parametrization for vertical viscosity and diffusivity proposed by *Pacanowski and Philander* [1981] was used. Their Richardson-number-dependent expression reads:

$$\nu = \frac{\nu_0}{(1+\alpha \mathrm{Ri})^n} + \nu_b, \quad \kappa = \frac{\nu}{(1+\alpha \mathrm{Ri})} + \kappa_b, \tag{2}$$

where $\mathrm{Ri} = N^2(z)/(u_z^2 + v_z^2)$ is the Richardson number, $\nu_b = 1.5 \cdot 10^{-4}$ m² s⁻¹, $\kappa_b = 1 \cdot 10^{-7}$ m² s⁻¹ are background values, and $\nu_0 = 1.5 \cdot 10^{-2}$ m² s⁻¹, $\alpha = 5$ and $n = 1$ are adjustable parameters. Horizontal diffusivity coefficient is $\kappa_h = 1 \cdot 10^{-2}$ m² s⁻¹, whereas variable horizontal viscosity follows the parameterization of *Leith* [1968].

Initial and lateral boundary conditions used in MITgcm have been chosen in order to render it a one-way nested model of the POM simulation. MITgcm uses the same initial condition used in POM. Moreover, the two-way exchange through the strait is achieved by laterally forcing the model through the imposition of the mean baroclinic velocities and tracers extracted from the POM simulation. Tidal forcing was introduced by prescribing at the open boundaries the main diurnal (O_1, K_1) and semidiurnal (M_2, S_2) barotropic tidal currents (depth-averaged currents), always extracted from POM.

Wave reflections at the open boundaries are minimized by adding a Newtonian relaxation term to the tracer equations within the boundary area and implementing the flow relaxation scheme proposed by *Carter and Merrifield* [2007] for the velocity field. For consistency with POM, the same spin-up phase has been followed in the MITgcm simulation. As for Exp-POM, after this spin-up phase the MITgcm was run for a further tropical month (27.321 days) that represents our reference experiment (Exp-MIT). Here we stress that by construction, MITgcm represents a one-way nested model for the POM simulation.

3.3. MODELS VALIDATION

The simulation performed with POM has been validated against most of the available *in situ* data by *Sanchez-Roman et al.* [2009] and *Sannino et al.* [2009a]. *Sanchez-Roman et al.* [2009] compared the predicted and observed amplitude and phase of the diurnal and semidiurnal tidal components of the along-strait velocity field at different depths; the results were considered satisfactory with differences limited in most parts of the strait to less than 10 cm s⁻¹ in amplitude and 20° in phase. Moreover, comparing the predicted and observed amplitude and phase of the semidiurnal tidal components of the surface elevation, *Sannino et al.* [2009a] found that the maximum differences did not exceed 3.6 cm in amplitude (with a maximum error that did not exceed 18%) and 11° in phase.

Provided that the MITgcm model is a nested model of the POM simulation, barotropic tidal currents are similar in both models, except for small differences attributed to finer resolution and better representation of the bottom topography of the nested model. Table 3.1 shows results of harmonic analysis applied to barotropic velocity (depth-averaged) time series at ES (35°51.7′N, 5°58.6′W), CS (35°54.8′N, 5°44.7′W), TN (35°57.6′N, 5°33.0′), and GIB (35°59.7′N, 5°22.7′W). Differences between velocity amplitudes are 12.9 ± 8.2%, whereas tidal phases differ in 26.3 ± 10.5°. As expected, both models reproduce fairly similar barotropic tides, which in turn are in good agreement with *in situ* data. Despite the good agreement

Table 3.1 Tidal amplitudes and phases of barotropic currents simulated by the nesting (POM) and nested (MITgcm) models at ES (35°51.7′N, 5°58.6′W), CS (35°54.8′N, 5°44.7′W), TN (35°57.6′N, 5°33.0′), and GIB (35°59.7′N, 5°22.7′W).

Tidal Const.	ES	CS	TN	GIB
Amplitude POM				
M_2	56.74 ± 0.32	107.55 ± 1.08	60.96 ± 1.16	38.76 ± 0.51
S_2	20.88 ± 0.35	38.87 ± 1.28	22.20 ± 1.33	13.57 ± 0.60
K_1	12.08 ± 0.26	23.13 ± 0.55	12.12 ± 0.22	7.05 ± 0.24
O_1	14.20 ± 0.30	26.18 ± 0.50	13.26 ± 0.22	7.86 ± 0.23
Amplitude MITgcm				
M_2	56.97 ± 0.23	114.51 ± 0.38	54.56 ± 0.41	42.87 ± 0.39
S_2	20.54 ± 0.22	42.41 ± 0.41	18.83 ± 0.38	15.02 ± 0.30
K_1	9.33 ± 0.14	22.16 ± 0.34	10.15 ± 0.43	6.10 ± 0.23
O_1	11.11 ± 0.15	24.00 ± 0.35	10.34 ± 0.47	6.25 ± 0.27
Phase POM				
M_2	116.6 ± 0.3	117.1 ± 0.6	107.7 ± 1.1	114.9 ± 0.8
S_2	199.7 ± 0.9	198.0 ± 1.8	191.7 ± 2.9	195.0 ± 2.5
K_1	18.6 ± 1.4	6.5 ± 1.2	359.9 ± 1.1	357.9 ± 1.8
O_1	198.1 ± 1.2	184.3 ± 1.2	179.3 ± 0.1	174.9 ± 1.7
Phase MITgcm				
M_2	136.1 ± 0.2	136.8 ± 0.2	129.7 ± 0.4	127.9 ± 0.4
S_2	168.0 ± 0.7	168.7 ± 0.6	162.9 ± 1.3	161.4 ± 1.6
K_1	358.9 ± 0.9	354.9 ± 0.9	348.1 ± 2.5	336.6 ± 2.4
O_1	237.1 ± 0.9	227.3 ± 0.8	221.5 ± 2.5	210.7 ± 2.3

between MITgcm and POM barotropic tides, baroclinic tides present significant differences. As it will be shown in the next section, the fine resolution and nonhydrostatic formulation of MITgcm make the model capable of accurately resolving the generation and evolution of internal tides, including short internal lee waves generated near bottom obstacles, and propagating solitary waves of large amplitude, some of the most striking phenomena observed in the SoG [*Lacombe and Richez*, 1982].

3.4. RESULTS

In this section, the SoG circulation as simulated by the two models will be compared in terms of internal bore evolution, three-layer characteristics, and hydraulics.

3.4.1. Internal Bore Evolution

One of the most noticeable phenomena observed in the SoG is represented by the propagating internal tidal wave generated over CS. At its leading edge, the internal tide can be characterized as a tidal bore: a train of internal waves of about 100 meters amplitude and 1 km wavelength [*Garrido et al.*, 2008; *Lacombe and Richez*, 1982]. Figures 3.3 and 3.4 display the evolution of the salinity and velocity fields from the midstage of the flood tide to the midstage of the ebb tide of a tidal cycle as simulated by the POM and MITgcm model, respectively. The comparison of the two figures reveals important differences between the two models. In a late stage of the flood tide, POM exhibits two steep depressions of the isohalines at the lee side of ES and CS (Figures 3.3a, b). As discussed by *Sannino et al.* [2009a], these two internal features constitute two internal hydraulic jumps. The hydraulic jump at ES is quasi-permanent, whereas the one at the lee side of CS is intermittent since hydraulic control is lost in the majority of tidal cycles when barotropic tidal flow reverses. This occurs in panel 3c, which shows the release of the hydraulic jump toward the Mediterranean. Since nonhydrostatic effects are missing in POM, the moving hydraulic jump (or internal bore) cannot undergo its classical evolution into a series of short solitary waves as predicted by weakly nonlinear theories. The internal bore may only give rise to a shock wave since only the weak dispersion introduced by Earth's rotation, unable to balance the steepening effect of nonlinearity, is at work. Consequently, the baroclinic field beyond CS has to be considered unrealistic.

For the nonhydrostatic simulation (Figure 3.4), barotropic forcing produces, at first glance, the same overall picture with a double internal hydraulic jump at the lee side of ES and CS but with much better resolved fine baroclinic structures. Tangier Basin and the lee side of ES appear as places where short unsteady waves develop (Figures 3.4a–c). Further analysis of the baroclinic field reveals that in addition to the hydraulic jump located at the lee side of CS, another internal bore is generated upstream of the sill, just over the leading edge of CS crest. This bore is released with the relaxation of the barotropic flow evolving into a succession of internal solitary waves of depression (see *Garrido et al.*, 2012, for an exhaustive analysis). Figure 3.5 shows the baroclinic velocity field (along-strait component) reproduced by the two models across the internal wave train (see grey contour in Figures 3.3d and 3.4d). Baroclinic velocities are calculated as the difference between the total velocity and its barotropic component (depth-average value, which in this case is around 0.6 m/s). Throughout 8 kilometers, the total velocity field is dominated by the orbital velocities of a series of internal solitary waves (Figure 3.5a), which are as large as 1 m/s near the surface and around −0.6 m/s below the pycnocline. Although the magnitudes of velocity are similar, the velocity field is much coarser, and the countercurrent below 150-m depth is significantly underestimated. Differences in the vertical component of velocities are much more dramatic (Figures 3.5c, d).

As expected, vertical velocities associated with internal solitons are as large as 0.3 ms^{-1}, more than three times larger than those reproduced by the σ-coordinate model.

3.4.2. Three-layer definition and properties

As argued by *Bray et al.* [1995] and subsequently verified by *Sannino et al.* [2009a], the two-way exchange in the SoG is best represented by a three-layer system composed of an upper layer of Atlantic water, a lower layer of saltier and colder Mediterranean water, and an interface layer in between.

The same method applied in *Sannino et al.* [2009a] for classifying all salinity profiles as Atlantic layer (AL), interface layer (IL), and Mediterranean layer (ML) has been used for analyzing the MITgcm simulation. In particular, following *Bray et al.* [1995] the upper and lower limit of the halocline have been chosen as the upper and lower limits of the interface layer. Figure 3.6 shows the time-averaged thicknesses of the three layers together with the depth of the midpoint of the interface layer as obtained for MITgcm. By comparing such figure with the equivalent figure obtained for POM (see Figure 16 in *Sannino et al.* [2009a]), it appears that the main patterns are similar. However, the thickness of the AL is systematically larger in MITgcm than in POM, with values ranging from about 60 m west of CS to about 20 m along TN (Figure 3.7). Note that the difference along TN in terms of percentage of the MITgcm AL thickness is 100%. Such values are reached far from the coast; there POM exceeds MITgcm. The opposite behavior close to the coast is due to the different representation of the coastal bathymetry in the two models.

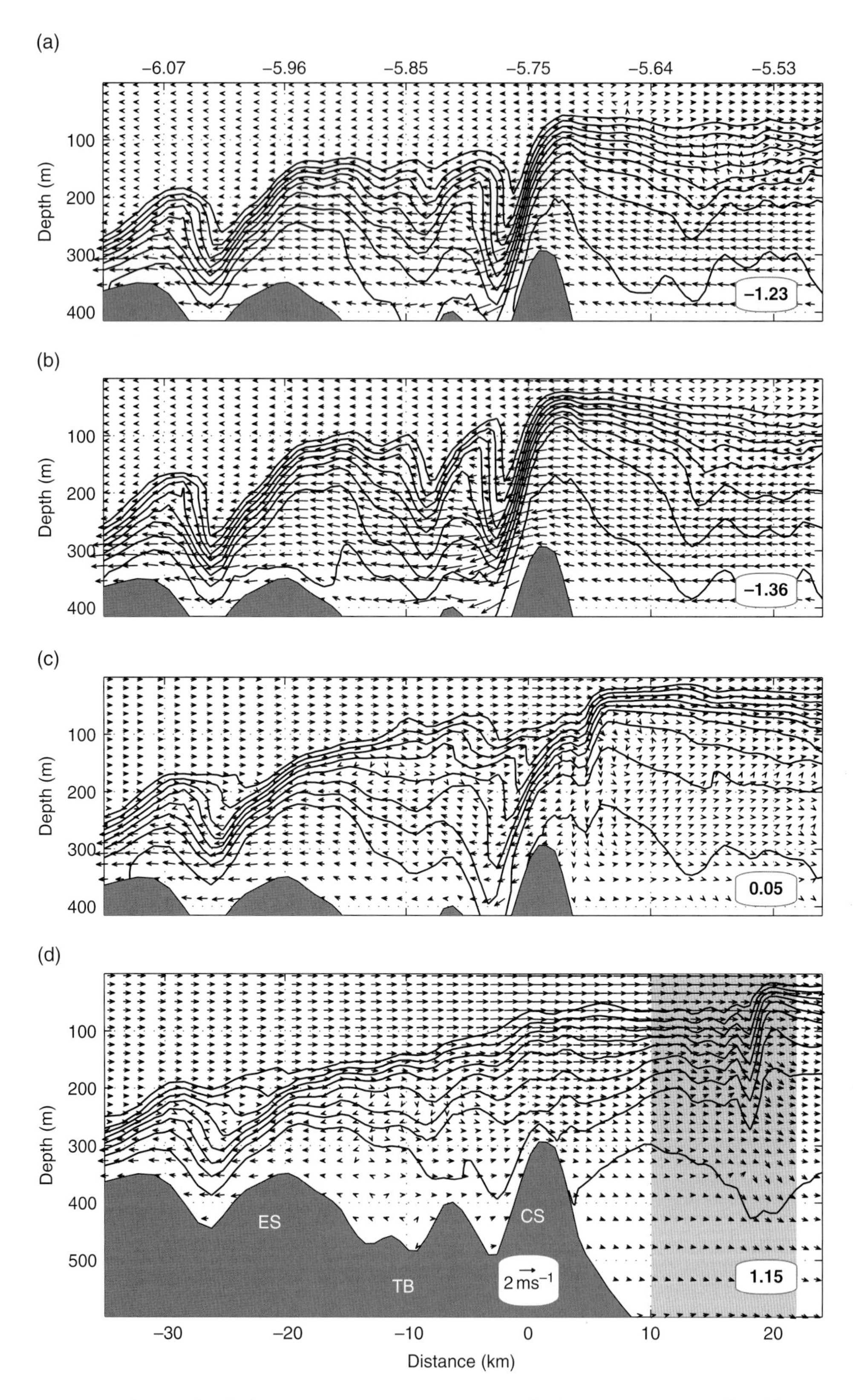

Figure 3.3 Time evolution of isohalines 36.40, 36.65, … , 38.65 and velocity currents simulated by POM during one tidal cycle of moderate tidal strength (Exp-POM). The barotropic velocity (in ms^{-1}) over Camarinal Sill is indicated at the lower right corner of the panels. Elapse times after (panel a) are 1:40 h (panel b), 4:40 h (panel c), and 7:00 h (panel d).

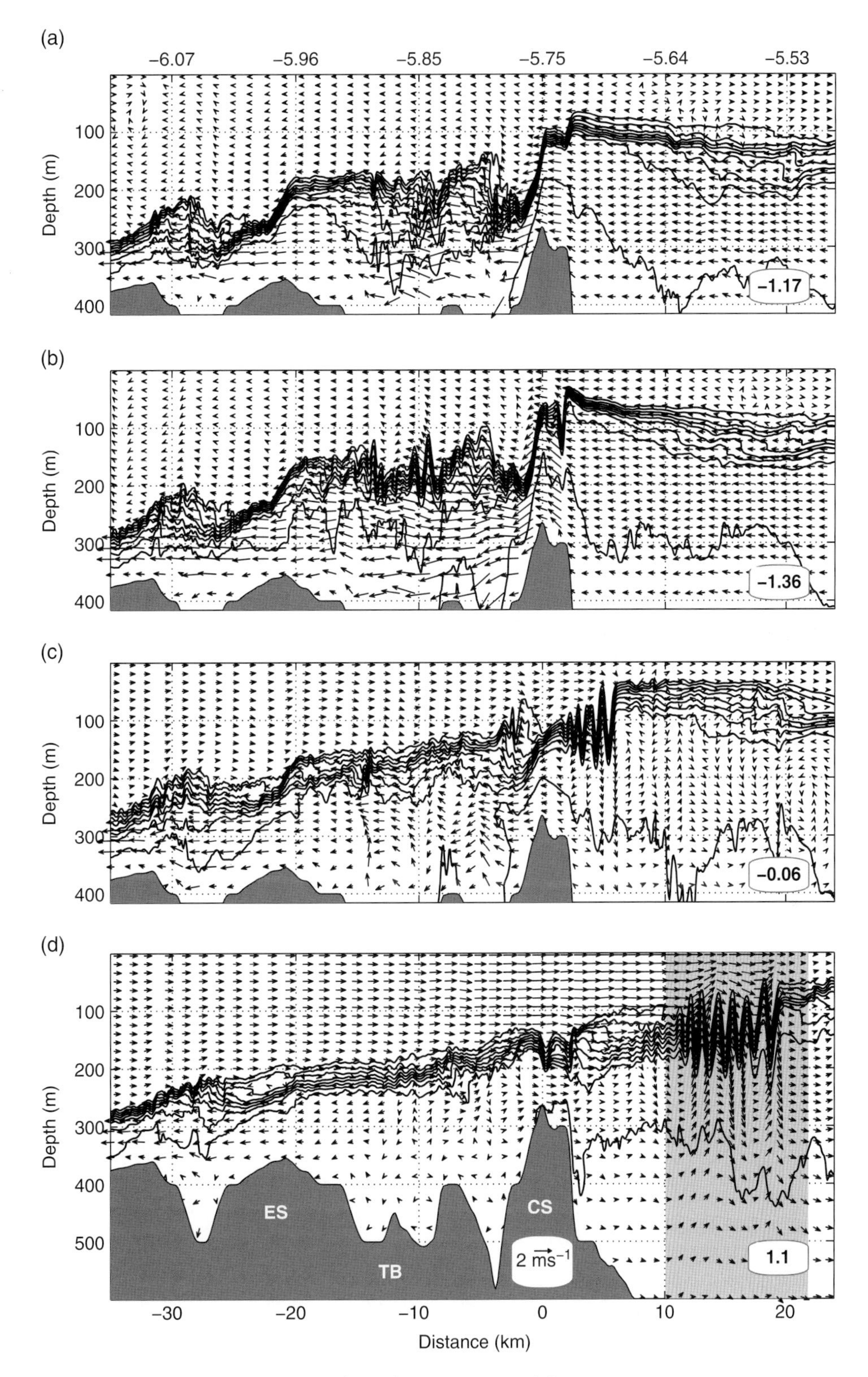

Figure 3.4 Same as Figure 3.3, using output from the MITgcm model (Exp-MIT).

As shown in Figure 3.8, POM bathymetry is systematically deeper than MITgcm. As for AL, the thickness of the ML is systematically higher in MITgcm than in POM (between 50 and 150 m), with some exceptions close to the northern coast, again due to the different representation of the bathymetry. The increased thickness for both the AL and ML in the MITgcm simulation implies a general reduction of the interface layer thickness with respect to

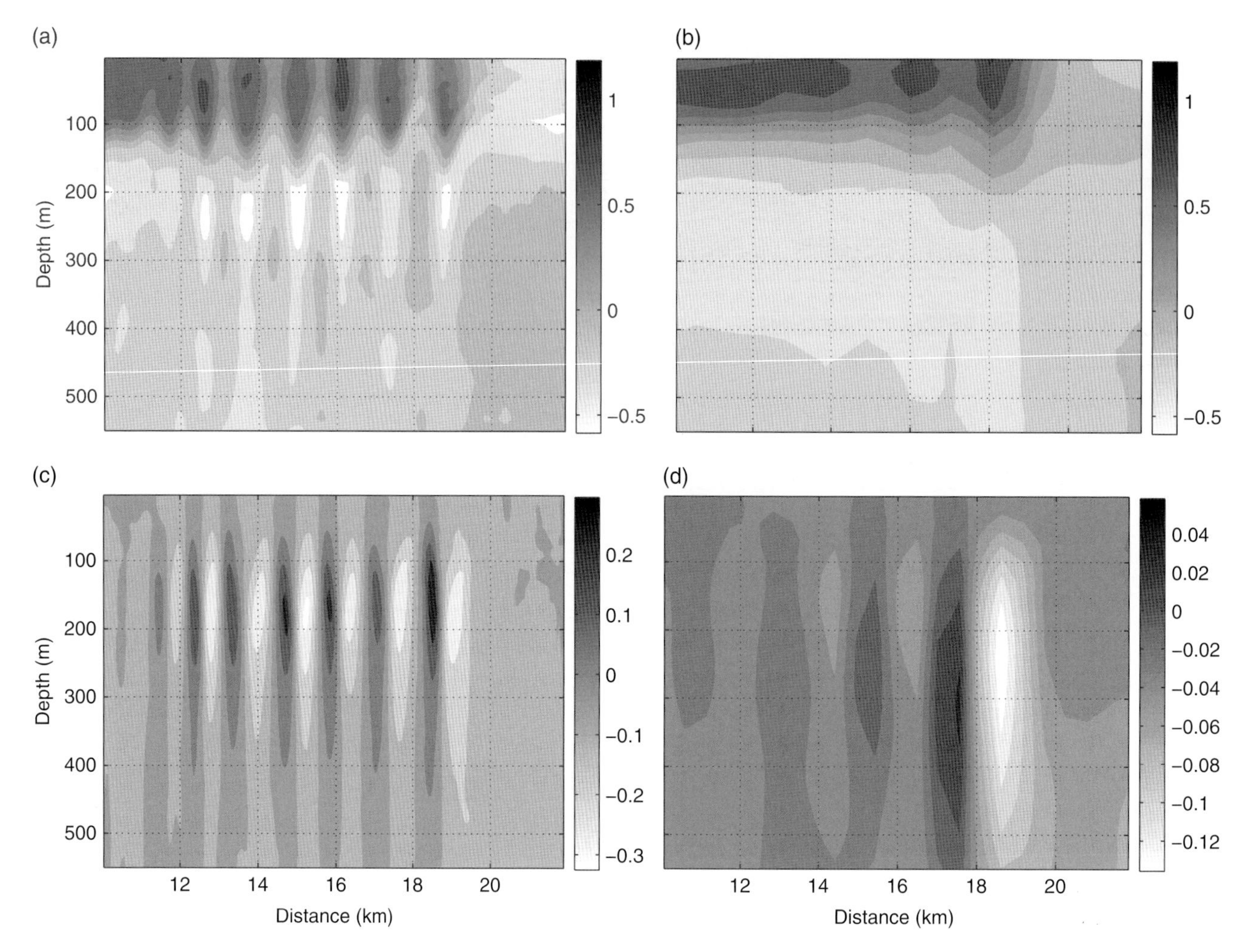

Figure 3.5 (a) Baroclinic horizontal velocity simulated by MITgcm during the arrival of an internal waves train at TN (see shaded rectangle in Figure 3.3d); (b) Same as (a) simulated by POM; (c) Vertical current simulated by MITgcm during the same instant of (a); (d) Same as (c) simulated by POM.

POM. Such reduction is more evident west of CS, along TB, and west of ES. The only exception occurs confined in the area near the northern coast of TN where the MITgcm predicts a thicker layer. Such difference is again attributed to the different representation of the bathymetry.

However, there is a good agreement between the mid-point interface depth simulated by POM and MITgcm (Figure 3.6). The differences hardly exceed 20 m, and are confined to points where the bathymetry disagrees.

Comparison of the two model results with those obtained by *Bray et al.* [1995] (Figure 6 in *Bray et al.* [1995]) shows good agreement with MITgcm whereas POM systematically overestimates the interface thickness. This difference can be attributed to the excess of spurious diapycnal mixing produced by POM that overshadows the naturally occurring mixing. The systematic spreading of the isohalines along the SoG (Figure 3.3) is additional evidence of such a strong diapycnal mixing affecting the POM simulation.

The different layer thicknesses have a direct effect on the water transport. For the MITgcm, the resulting transports for the three layers over the tropical month period are shown in Figure 3.9 for four different cross-strait sections located at ES, CS, Tarifa, and Gibraltar, respectively (sections A, B, C, and D in Figure 3.2). The Atlantic (*ALT*), interface (*ILT*), and Mediterranean layer transport (*MLT*) have been computed as in *Sannino et al.* [2009a], that is:

$$ALT(x,t) = \int_{y_{S1}(x)}^{y_{N1}(x)} \int_{dw_1(x,y,t)}^{up_1(x,y,t)} u(x,y,z,t)dzdy \quad (3)$$

$$ILT(x,t) = \int_{y_{S2}(x)}^{y_{N2}(x)} \int_{dw_2(x,y,t)}^{up_2(x,y,t)} u(x,y,z,t)dzdy \quad (4)$$

$$MLT(x,t) = \int_{y_{S3}(x)}^{y_{N3}(x)} \int_{dw_3(x,y,t)}^{up_3(x,y,t)} u(x,y,z,t)dzdy, \quad (5)$$

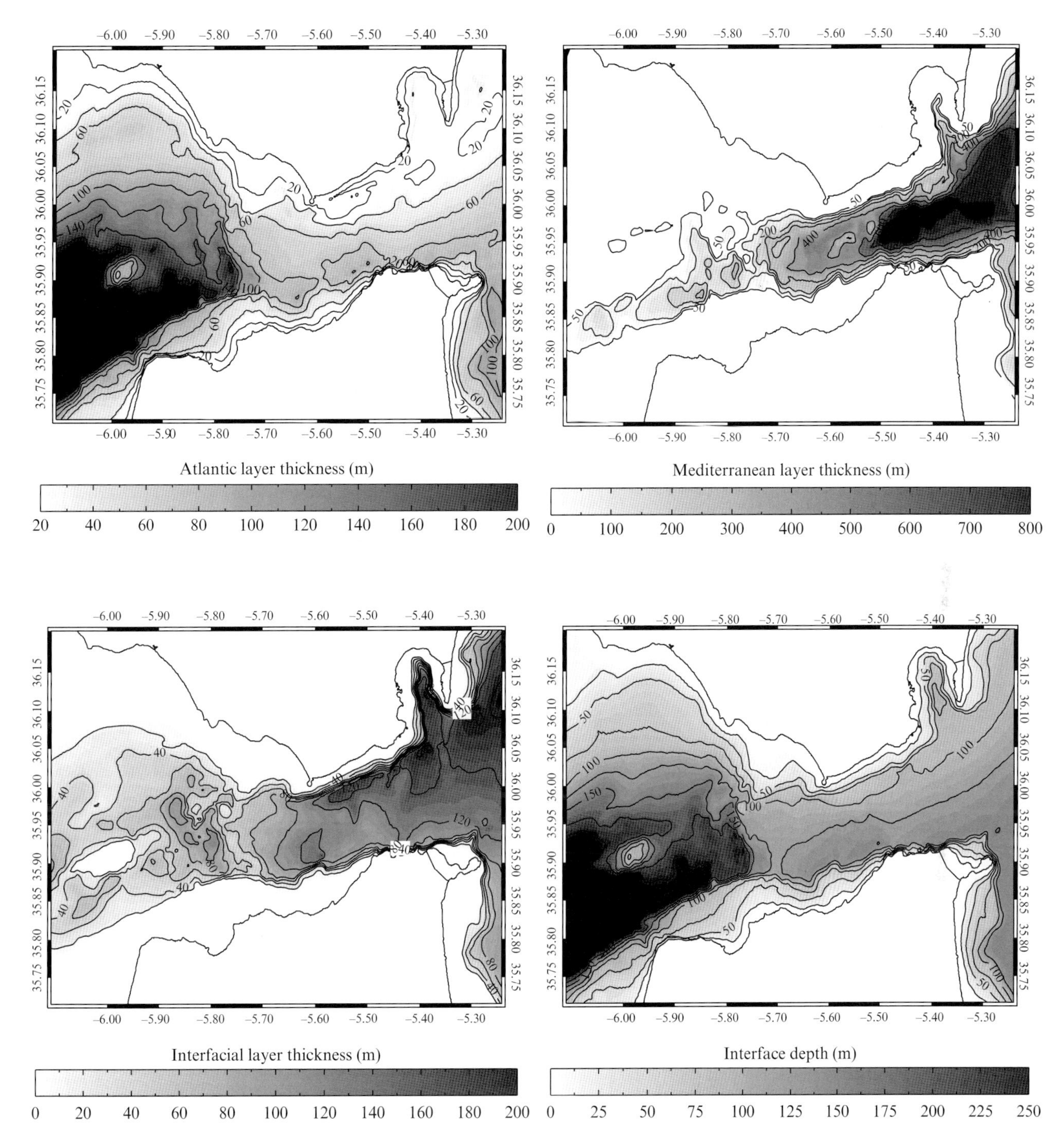

Figure 3.6 Time-averaged Atlantic, interface, and Mediterranean layer thickness, and depth of the midpoint of the interface layer as simulated by the MITgcm numerical model (Exp-MIT).

where up_n and dw_n are the instantaneous depths of the upper and lower bounds of the nth layer, while y_{Sn} and y_{Nn} represent the southern and northern limit of the cross section x and nth layer.

Figure 3.9 shows that the Atlantic layer carries water eastward with a small fraction of the transport periodically directed in the opposite direction. This small fraction reduces progressively from west to east, becoming null after crossing CS. An opposite behavior is exhibited by *MLT* where the principal direction is westward, and the eastward fraction reduces gradually from section D to section A where it is reduced to zero. *ILT* is comparable with *MLT* (*ALT*) transport at the western (eastern) side of the strait, but never exceeds the Atlantic or

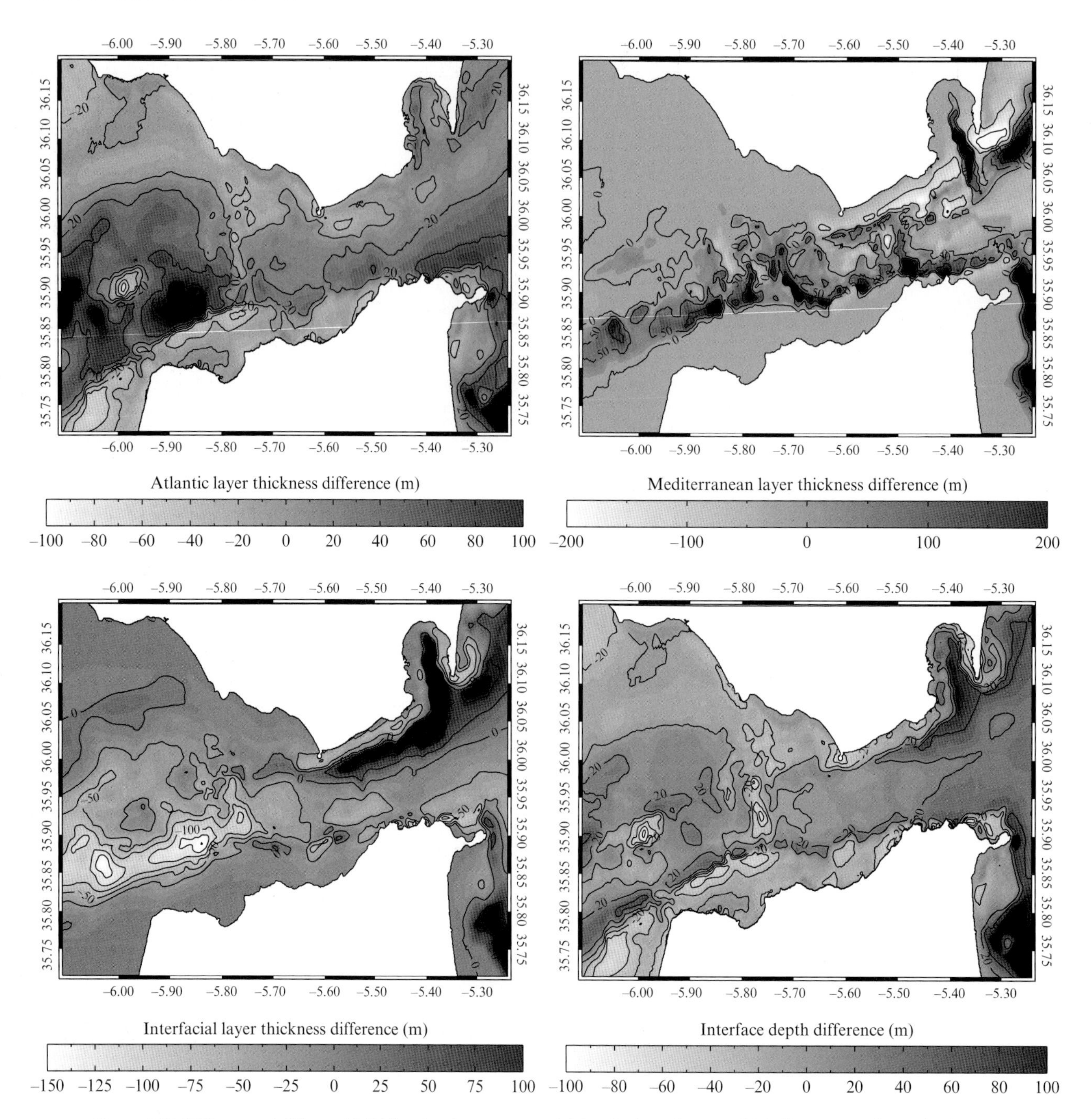

Figure 3.7 Difference MITgcm-POM for the time-averaged Atlantic, interface, and Mediterranean layer thickness, and depth of the midpoint of the interface layer.

Mediterranean contribution. This is an important difference with respect to the results obtained by *Sannino et al.* [2009a] where the interface layer transport was about two times larger than the Mediterranean over Espartel Sill, and three times larger than the Atlantic at the eastern limit of the SoG (see Figure 18 of *Sannino et al.* [2009a]).

3.4.3. Hydraulics

A long-standing question about the SoG concerns whether, and to what extent, the narrows and the two major sills hydraulically control the exchange flow. *Armi and Farmer* [1988] provide a detailed 2-D, two-layer system analysis of control locations based on observations

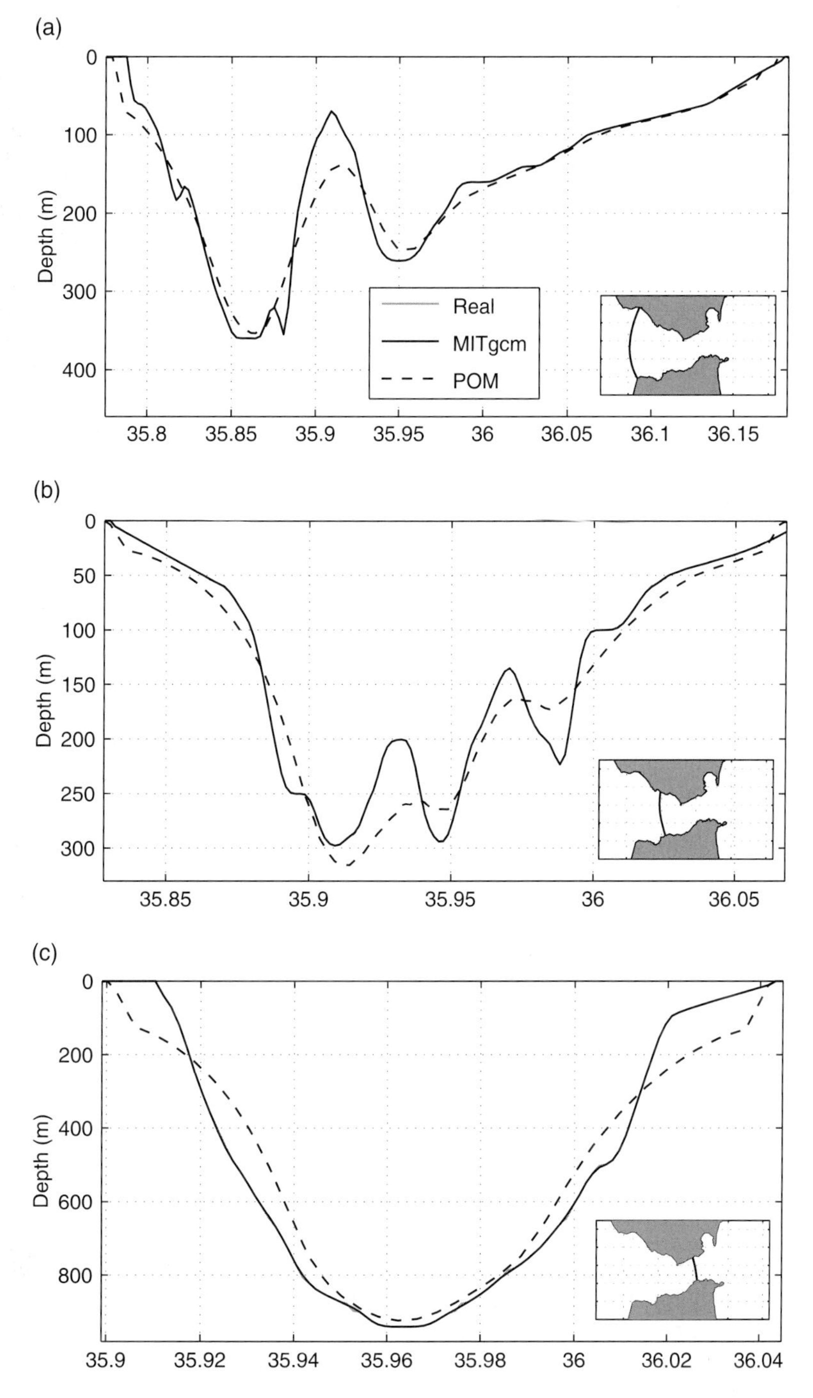

Figure 3.8 Original bottom topography (gray line) representation by partial cells (black solid line), and representation by σ-levels (dashed line) at (a) Espartel cross section, (b) Camarinal cross section, and (c) Tarifa Narrows cross section. Note that the actual partial cell representation fits perfectly the original bottom topography.

taken along or near the strait center line. They observed two permanent controls: one located at ES and the second within TN. The location of the control along TN is modified by the eastward propagating bore released at CS. They observed also a periodic control at CS due to the tidal action. The existence of two permanent controls would imply maximal exchange between the Atlantic and Mediterranean and overmixing within the Mediterranean. *Sannino et al.* [2009a] attempted to verify these conclusions by undertaking careful assessment of the hydraulic conditions at various sections of the SoG based on a three-layer representation of the flow with transversally varying velocity within each layer.

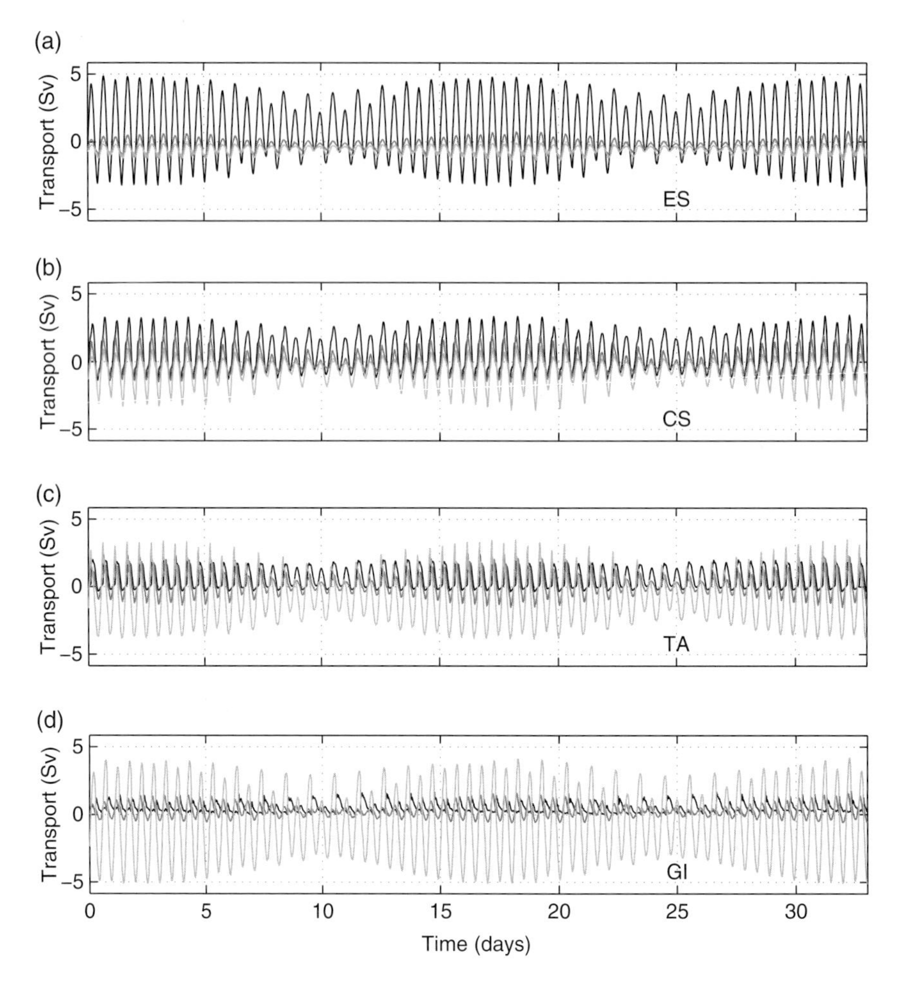

Figure 3.9 (a) Time evolution of Atlantic layer transport (blue line), interfacial-mixed layer (green line), and Mediterranean layer transport (red line) at Espartel section; (b) Same as (a) for Camarinal Sill section; (c) Same as (a) for Tarifa section; (d) Same as (a) for Gibraltar section. For color detail, please see color plate section.

We now examine the extent to which their conclusions are supported by MITgcm. The three-layer formulation was undertaken because of uncertainties about where, in two-layer formulation, the interface is to be located. (Episodic controls are located at ES and CS.) The existence of two permanent controls would imply maximal exchange between the Atlantic and Mediterranean and overmixing within the Mediterranean. *Sannino et al.* [2009a] attempted to verify these conclusions by undertaking careful assessment of the hydraulic conditions at various sections of the SoG based on a three-layer representation of the flow with transversally varying velocity within each layer.

We now further examine the extent to which their conclusions are supported by MITgcm. As discussed in *Sannino et al.* [2009a] and *Pickart et al.* [2010], the linear long waves of the three-layer system are associated with two vertical modes akin to the first and second baroclinic modes of a continuously stratified system. There are two waves for each mode and, in the absence of background flow, the two waves propagate in opposite directions. In the presence of weak background velocity, this situation continues to hold, and we say that the flow is subcritical with respect to each mode. If the velocity in one or more layers increases, the flow may become critical, meaning that the phase speed of one of the waves of a particular mode is brought to zero. The condition for critical flow is (see Figure 3.10):

$$\tilde{F}_1^2 + \left(\frac{1-r}{r} + \frac{w_3}{w_2}\right)\tilde{F}_2^2 + \tilde{F}_3^2 - \frac{w_3}{w_2}\tilde{F}_1^2\tilde{F}_2^2 - \tilde{F}_1^2\tilde{F}_3^2 - \frac{1-r}{r}\tilde{F}_2^2\tilde{F}_3^2 = 1, \tag{6}$$

where

$$\tilde{F}_1^2 = \left(\frac{1}{w_2}\int_{y_1L}^{y_1R}\frac{g'_{21}H_1}{u_1^2}dy_1\right)^{-1}, \quad \tilde{F}_2^2 = \left(\frac{1}{w_2}\int_{y_2L}^{y_2R}\frac{g'_{32}H_2}{u_2^2}dy_2\right)^{-1},$$
$$\tilde{F}_3^2 = \left(\frac{1}{w_3}\int_{y_3L}^{y_3R}\frac{g'_{32}H_3}{u_3^2}dy_3\right)^{-1}, \tag{7}$$

$g'_{21}=g(\rho_2-\rho_1)/\bar{\rho}$, $g'_{32}=g(\rho_3-\rho_2)/\bar{\rho}$, $r=\dfrac{\rho_2-\rho_1}{\rho_3-\rho_1}$ and w_n is the width of the interface overlying layer n. Note that $\tilde{F}_1^2$, $\tilde{F}_2^2$, and $\tilde{F}_3^2$ are generalized versions of layer Froude numbers and cannot be thought of as ratios of intrinsic to advection speeds.

When the speed of one or more layers is increased, the two waves belonging to a particular mode may move in the same directions, in which case we say that the flow is supercritical with respect to that mode. The connection with traditional hydraulics maybe clouded by the fact that the phase speeds can be complex, and *Sannino et al.* [2009a] therefore use the terms *provisionally subcritical* and *provisionally supercritical* to describe the appropriate regimes. This caveat is to be understood in what follows, so we will drop the modifier *provisional*.

The object of the hydraulic analysis is to map out regions of subcritical and supercritical flow along with the location control sections. There the flow is critical, equation 6 is satisfied, and a transition between regimes exists. To distinguish between states that are critical with respect to the first, as opposed to second, internal mode, one can evaluate the sign of

$$\alpha=\frac{w_2}{w_3\tilde{F}_2^2}\left[\frac{r}{1-r}\left(\tilde{F}_1^2-1\right)+\tilde{F}_2^2\right],$$

which gives the ratio of the lower to upper interface displacements due to the wave. If α is negative, the flow is classified as controlled for the second baroclinic mode. On the contrary, a positive α indicates criticality for the first mode. The remaining case occurs when both negative and positive α are possible. Such a situation is classified as two-modes controlled (see *Garrido et al.* [2012] for a rigorous derivation).

Figure 3.11 shows the frequency of occurrence, over the tropical month period, of supercritical flow with respect to only one mode (Figure 3.11 intermediate panel), and both modes (Figure 3.11 upper panel) along

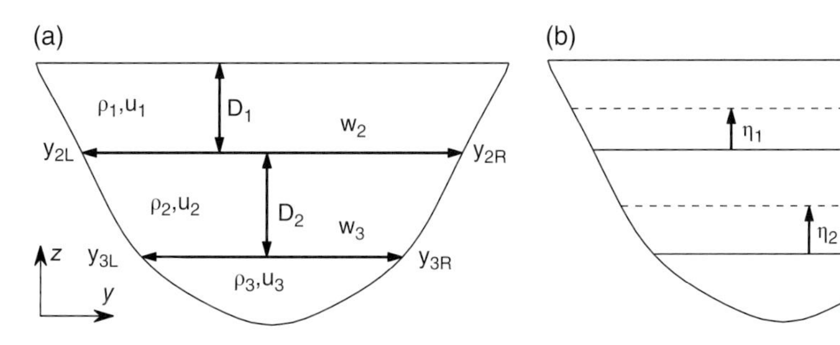

Figure 3.10 Definition sketch for a three-layer flow.

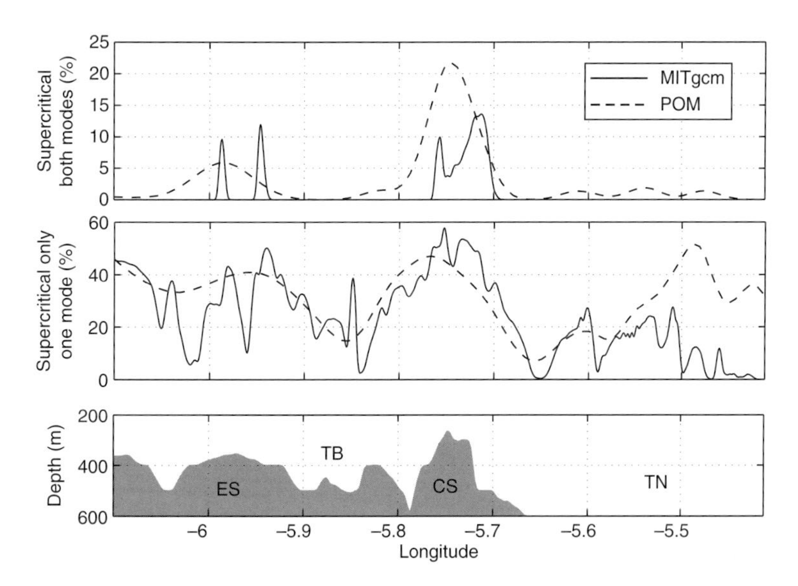

Figure 3.11 Frequency of occurrence, over the tropical month period, of supercritical flow with respect to bath modes (upper panel) and one mode (intermediate panel) along the strait as obtained by POM (dashed line) and MITgcm (solid line). Lower panel shows the bottom topography along the central axis of the strait.

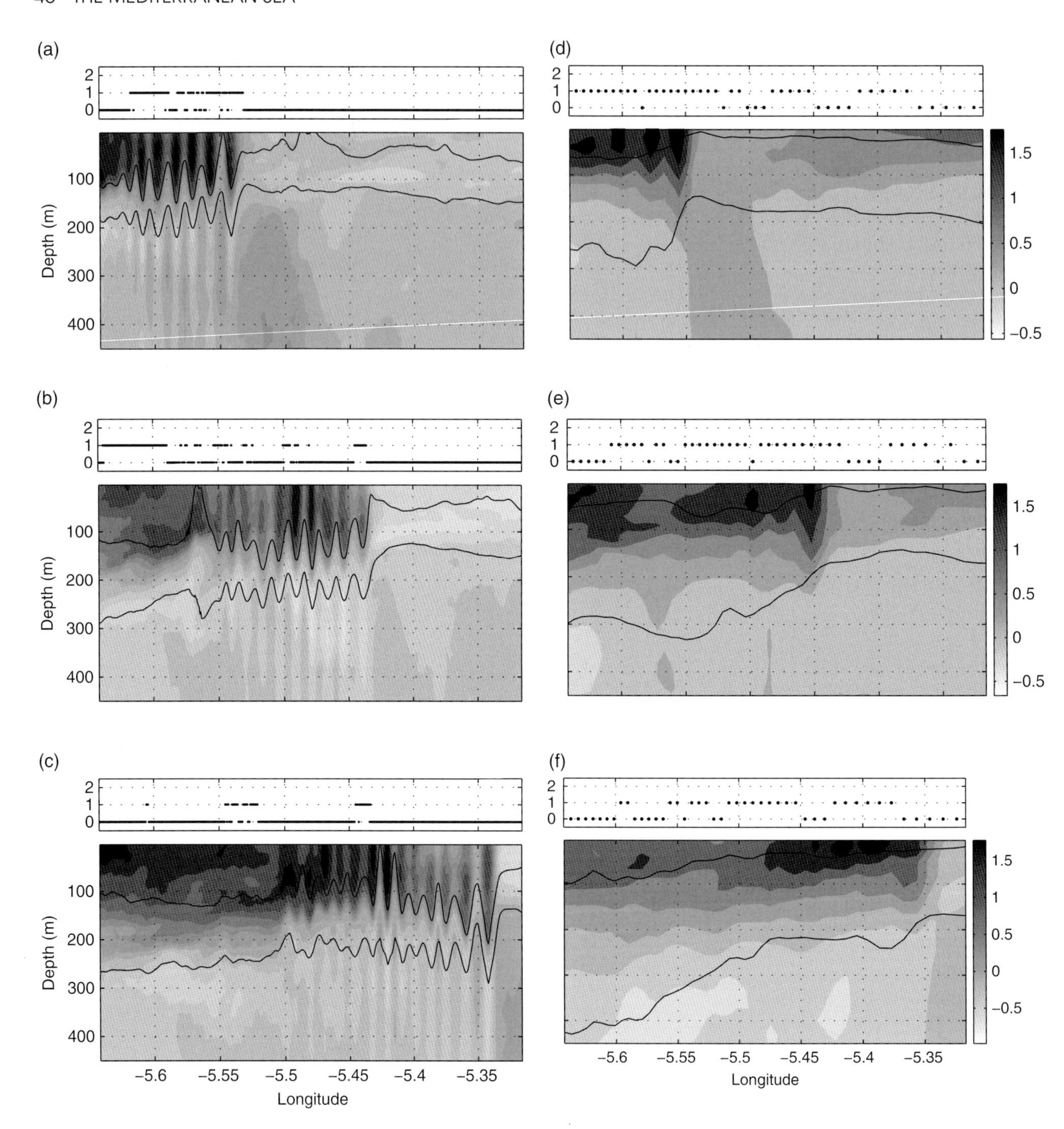

Figure 3.12 Evolution of the horizontal velocity field along longitudinal Section E (See Figure 3.2) during the arrival of an internal wave train to TN. Thick solid lines indicate the layer interfaces in the three-layer system. The sequence corresponds to the 13th day of simulation. Elapse time between upper and lower frames is 1.33 hours. Panels on the top of each frame indicate the flow criticality; zero: subcritical flow; one: only one internal mode controlled; two: both internal modes controlled. Left panels are for MITgcm, while right panels are for POM.

the strait as obtained by POM and MITgcm. Note that in both models the flow is much more prone to be supercritical with respect to only one mode. Moreover, for both models we observe that over CS and ES the flow criticality displays its maximum frequency.

That being said, some differences between the two models are still evident. In general, MITgcm displays a marked along-strait variability related to the finer description of the bathymetry. Moreover, when both modes are supercritical, MITgcm predicts lower values all along the

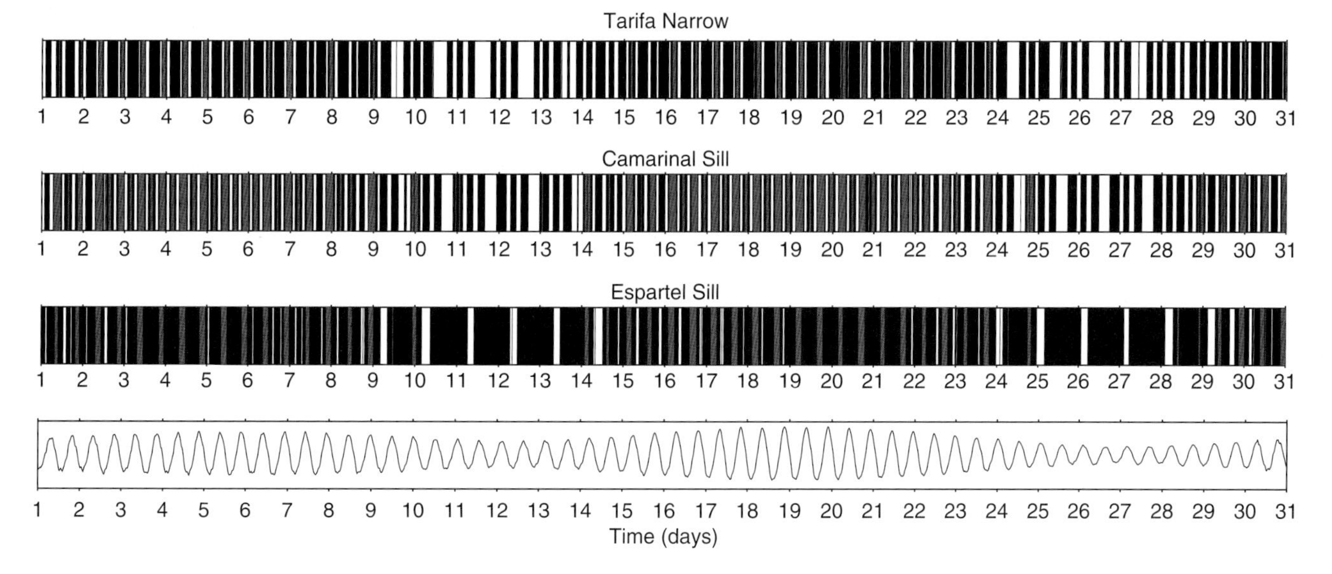

Figure 3.13 Bars indicating the presence of provisional supercritical flow as simulated by MITgcm with respect to one mode (black), and with respect to both modes (gray) in the three main regions of the strait: (from top) Tarifa Narrow, Camarinal Sill, and Espartel Sill; (bottom panel) time reference referred to the tidal elevation at Tarifa.

strait compared to POM, except for ES where MITgcm exceeds POM. When the flow is supercritical with respect to just one mode, the major differences are confined along TN. In particular, POM predicts higher frequencies compared to MITgcm. Due to the importance that a control in TN can play in determining the final hydraulic regime reached in the SoG, it is interesting to analyze in detail such differences.

Figure 3.12 shows the evolution of a bore through TN as simulated by MITgcm and POM. The bore propagation is presented as a temporal sequence of three along-strait velocity fields. To highlight the shape and position of the bore, two black lines representing the upper and lower bound of the interface thickness have been superimposed. The related hydraulic control is also indicated in each figure. As shown before, the upper layer is systematically thinner in POM, while the interface layer is thinner in MITgcm. Such differences, together with differences in the velocity fields, lead to the determination of different hydraulic behavior. POM predicts a quasi-permanent supercritical flow. East of the bore the velocity of the AL in POM is systematically higher than in MITgcm. In contrast to POM, MITgcm achieves the control only when the eastward bore is present in TN.

For a complete understanding of the hydraulic regimes in terms of maximal and submaximal exchange, the simultaneous presence of supercritical flow regions through the strait have to be explored. As in *Sannino et al.* [2009a], the three most likely regions of CS, ES, and TN will be analyzed. In Figure 3.13 a bar-plot similar to Figure 19 of *Sannino et al.* [2009a] shows the evolution of the hydraulic control in these three regions according to MITgcm. Black bars indicate the presence of supercritical flow with respect to only one mode, while gray bars are used when the flow is supercritical with respect to both modes. The frequency of appearance of supercritical flow, with respect to one and both modes, over the entire tropical month period is about 46%, 73%, and 92% at TN, CS, and ES, respectively.

Moreover, while the flow is supercritical with respect to both modes for only 4% in TN, the percentage increases up to 30% at CS. A slightly lower percentage is found at ES. Thus, while at CS the flow is supercritical with respect to one and both modes with approximately the same percentage, the flow is principally controlled with respect to only one mode both at TN and ES. Similar percentages for both modes controlled were found also by *Sannino et al.* [2009a]. On the contrary, comparing the values obtained when only one mode is controlled, it appears that the percentage for TN has undergone a substantial reduction with respect to *Sannino et al.* [2009a] (from 70% to 42%); a more limited reduction occurs at ES (from 74% to 62%); and a similar value is obtained for CS (from 41% to 48%).

In conclusion, the hydraulic control section in TN, deemed necessary for maximal exchange conditions in the *Armi and Farmer* [1988] model, occurs with a significantly lower frequency in the MITgcm. If one requires that both modes be supercritical, which would correspond to supercritical flow in their two-layer model, the percentage of time over which control occurs in TN (grey bars in Figure 3.13) is quite low. It would therefore appear that maximal control is largely expunged within the MITgcm.

Table 3.2 Main characteristics and parameterizations used for the seven experiments performed.

	Model	Model grid	Non-hydrostatic	Vertical diffusivity	Horizontal diffusivity	MPDATA iterations
Exp-POM	POM	$362\times53\times32$	NO	*Mellor and Yamada* [1982]	TPRNI =1	3
EXP1	POM	$362\times53\times32$	NO	Null	TPRNI =1	3
EXP2	POM	$362\times53\times32$	NO	Null	TPRNI =1	5
EXP3	POM	$362\times53\times32$	NO	Null	TPRNI =0	5
Exp-MIT	MITgcm	$1440\times210\times53$	YES	*Pacanowski and Philander* [1981]	Laplacian: $1\cdot10^{-2}$ $m^2\ s^{-1}$	Not present in MITgcm
EXP4	MITgcm	$1440\times210\times53$	NO	*Pacanowski and Philander* [1981]	Laplacian: $1\cdot10^{-2}$ $m^2\ s^{-1}$	Not present in MITgcm
EXP5	MITgcm	$362\times53\times46$	NO	*Pacanowski and Philander* [1981]	Laplacian: $1\cdot10^{-2}$ $m^2\ s^{-1}$	Not present in MITgcm

3.4.4. Sensitivity Experiments

A set of five additional sensitivity experiments was carried out. Three experiments were conducted to investigate the origin of diapycnal mixing present in the POM simulation. The fourth experiment was performed to evaluate the impact of the nonhydrostaticity on the simulated hydraulic regime. Finally, another experiment was carried out to examine the influence of the horizontal and vertical resolution on the simulated exchange flow. In Table 3.2 are reported the main model characteristics and parameterizations used for the entire set of experiments.

3.4.4.1. Advection and diapycnal mixing The reference experiment (Exp-POM) is first repeated with the parameter controlling vertical diffusivity (KH in *pom*98) set to zero (EXP1). In the second experiment (EXP2), KH remains zero and the number of corrective iterations of the MPDATA advection scheme is increased from three to five. The third experiment (EXP3) is the same as EXP2 except that the horizontal diffusivity is also set to zero (TPRNI = 0 in *pom*98). A qualitative comparison shows that in EXP1 and EXP2 the spurious diapycnal mixing is only slightly reduced below that of Exp-POM (see Figures 3.14 and 3.15, respectively), whereas a more drastic improvement is obtained in EXP3 (Figure 3.16). These results suggest that the numerically induced mixing is so large in Exp-POM that no additional horizontal diffusivity is needed. Recently, *Marchesiello et al.* [2009] demonstrated that spurious diapycnal mixing can arise from the advection scheme adopted. In particular, they demonstrated that implicit diffusion in diffusive advection schemes, as for example MPDATA, is large enough to produce excessive diapycnal mixing even in high-resolution σ models. We suspect that this is exactly the case.

3.4.4.2. Hydrostatic vs. nonhydrostatic regime To investigate the impact of the nonhydrostaticity on the simulated hydraulic regime, an additional experiment has been performed with MITgcm in which the reference experiment (Exp-MIT) is rerun in hydrostatic mode (EXP4). Figure 3.17 shows the resulting salinity and velocity fields for the midstage of the ebb tide of a tidal cycle of moderate strength obtained for EXP4. Comparison of Figure 3.17 and its nonhydrostatic counterpart (Figure 3.4d), shows that the internal bore is reduced to a simple shock wave when the pressure is rendered hydrostatic. However, such a limitation has no impact on the final simulated hydraulic regime. This can be confirmed through the comparison of the evolution of the hydraulic control along TN as simulated by Exp-MIT (Figure 3.12a, b, c) and EXP4 (Figure 3.18a, b, c), respectively. It is interesting to note that the hydraulic control along TN moves from west to east in a similar way in both experiments. In particular for Exp-MIT, the control in TN is linked to the internal bore, while for EXP4 it is linked to the shock wave.

3.4.4.3. Resolution A further experiment (EXP5) was performed to examine the influence of the horizontal and vertical resolution on the simulated exchange flow (Figure 3.19). In this experiment, Exp-MIT was repeated in hydrostatic mode and with the same horizontal grid as for Exp-POM. In the vertical, the model used 46 unevenly distributed vertical z-levels with decreasing resolution from the ocean surface to the bottom. The first 20 vertical levels are concentrated within the first 300 m of the water column. Model topography is obtained through a bilinear interpolation of the same initial data used for Exp-MIT. Qualitative comparison of simulation with its lower resolution counterpart (Figure 3.19d) shows that higher horizontal resolution has an impact principally on the steepness of the shock wave, which however does not alter the final simulated hydraulic regime. The latter remains very similar to that of EXP4 and consequently to Exp-MIT.

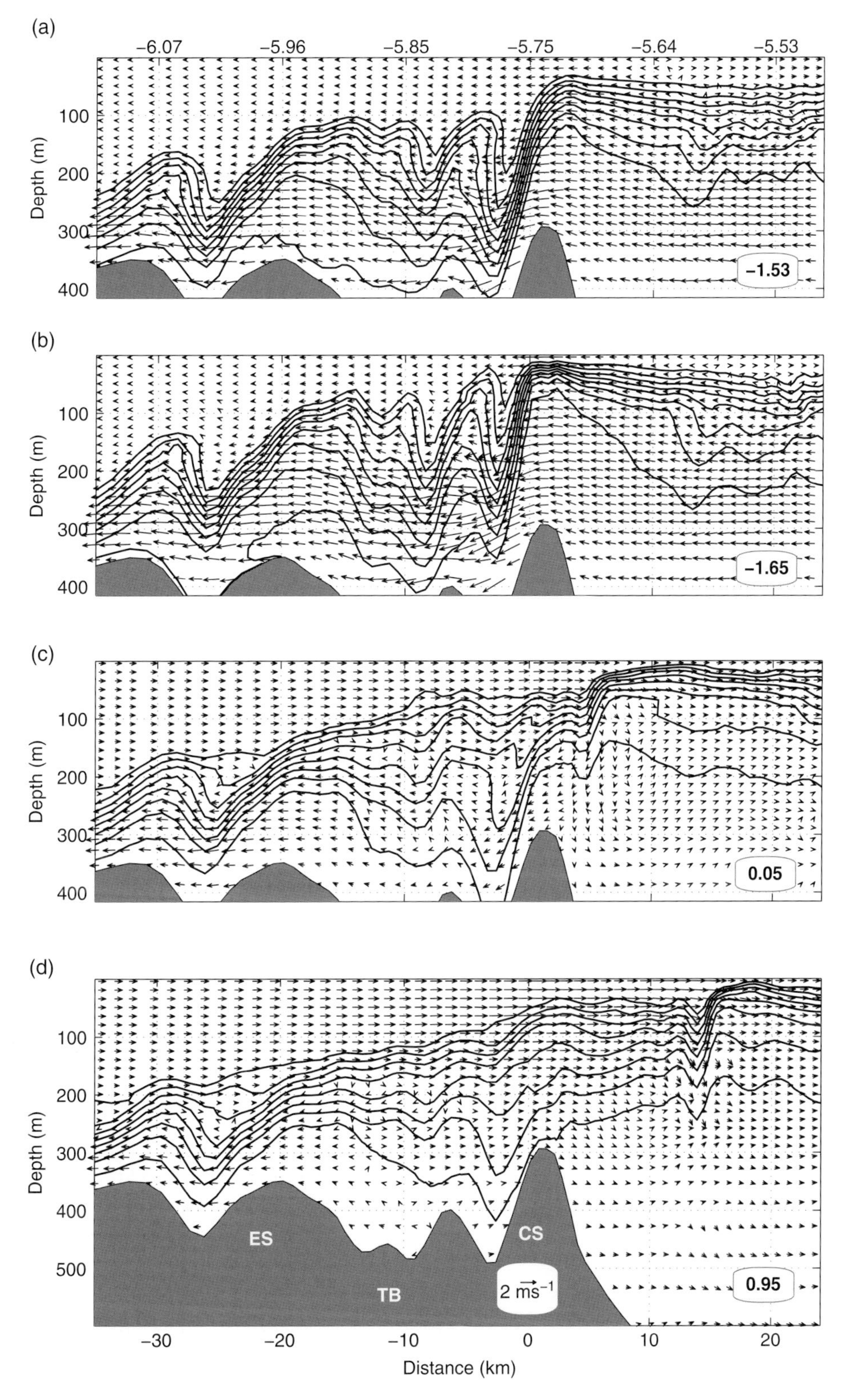

Figure 3.14 Time-evolution of isohalines 36.40, 36.65, … , 38.65 and velocity currents simulated by POM with vertical diffusivity set to zero (EXP1) during one tidal cycle of moderate tidal strength. The barotropic velocity (in ms^{-1}) over Camarinal Sill is indicated at the lower right corner of the panels. Elapse times after (panel a) are 1:40 h (panel b), 4:40 h (panel c), and 7:00 h (panel d).

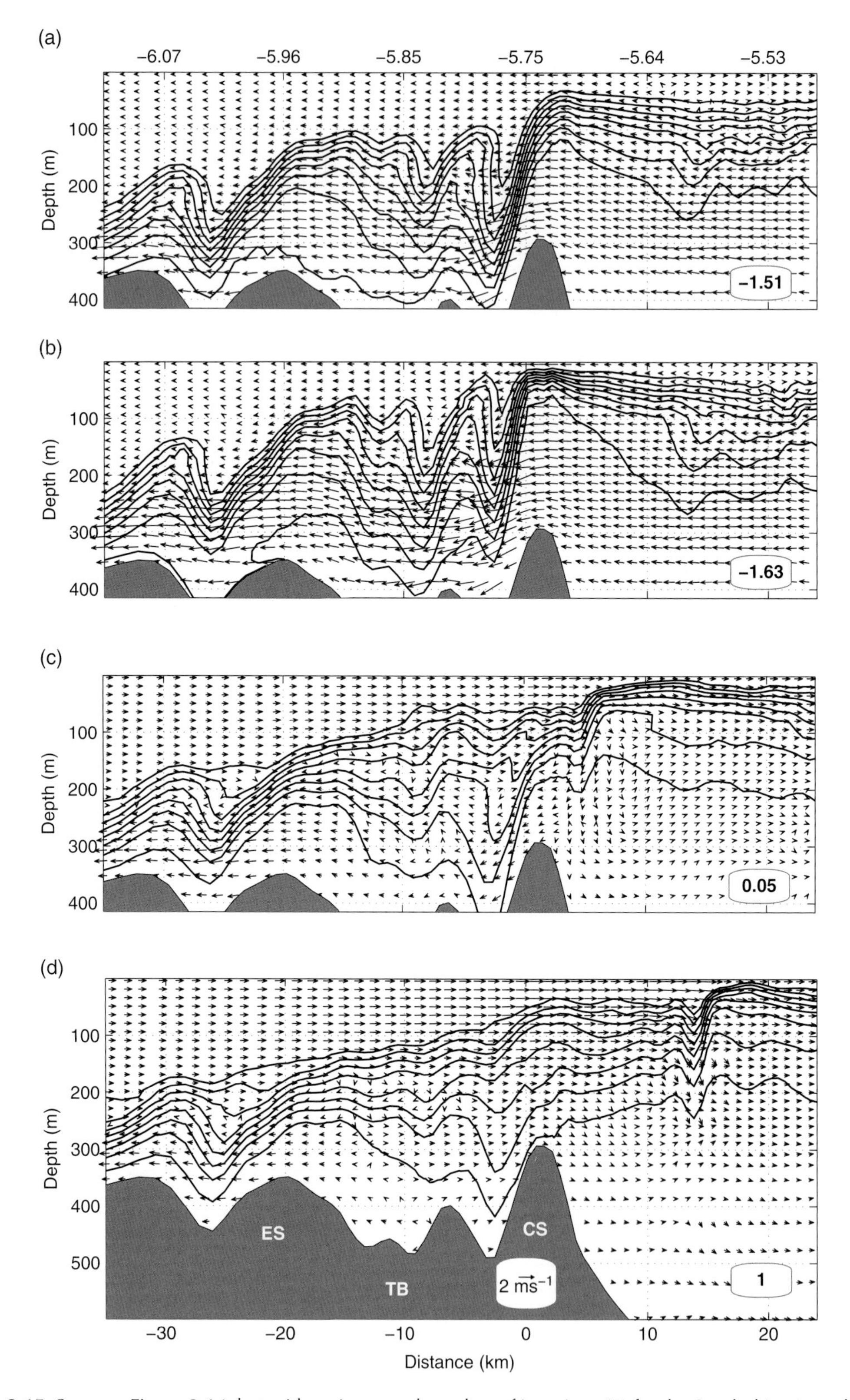

Figure 3.15 Same as Figure 3.14, but with an increased number of iterations (5) for the Smolarkiewicz advection scheme (EXP2).

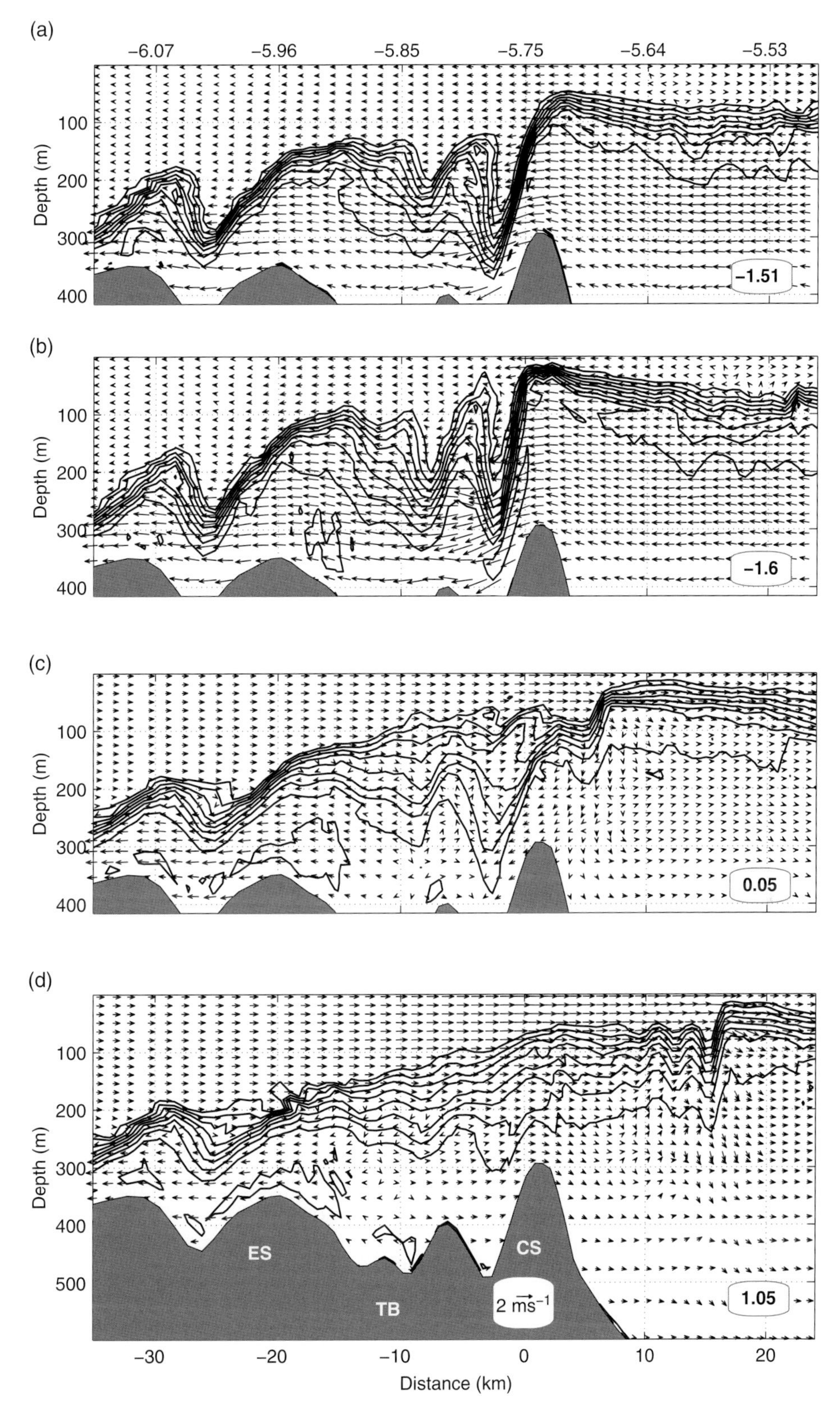

Figure 3.16 Same as Figure 3.15, but with the horizontal diffusivity set to zero (EXP3).

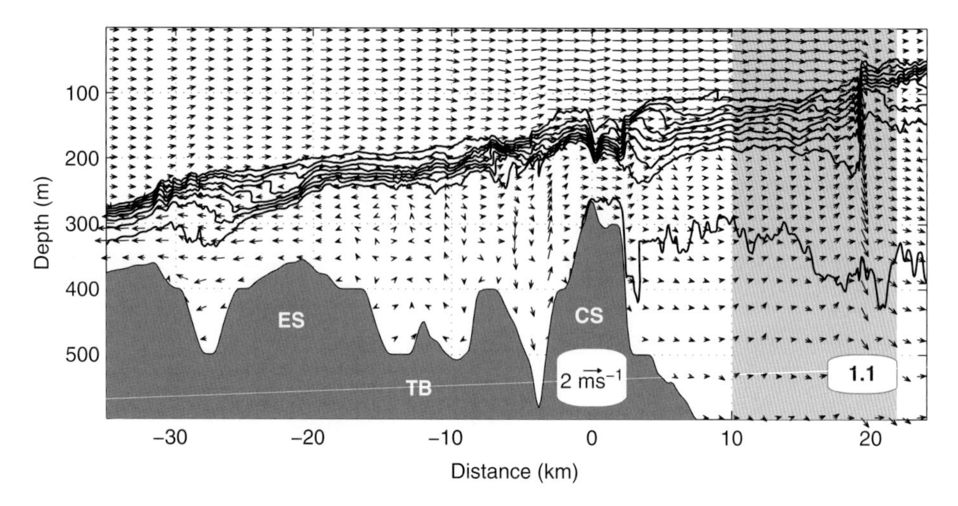

Figure 3.17 Same as Figure 3.4, but with the nonhydrostatic feature off.

3.5. DISCUSSION AND CONCLUSION

The flow exchange through the SoG simulated by a high-resolution σ-coordinate hydrostatic model has been compared with that simulated by a very high-resolution, z-coordinate, nonhydrostatic model. The hydrostatic model used is POM as implemented by *Sannino et al.* [2009a]. Differently from the original version of POM (generally known as *pom*98), the model implemented for the SoG used the MPDATA advection scheme as implemented by *Sannino et al.* [2002]. MPDATA has been subsequently included in the official version of POM (known as *pom2k*).

We stress that the implementation of MPDATA was absolutely necessary because of the presence of strong density contrasts associated with intense tidal currents present in the SoG that made the dispersive second-order–centered advection scheme (the only advection scheme present in *pom*98), unable to reproduce the water exchange. The POM implementation has been positively validated by different authors against most of the *in situ* data. However, due to the scarcity in space of long and high frequency temporal measurements, an overall analysis of the model results was still necessary.

A common method to assess the validity of a numerical simulation consists in comparing the model results against a sort of numerical "true" solution, that is, a solution obtained by a very high-resolution numerical model characterized by more robust physical parameterizations. Thus, in this paper we have compared the hydrostatic model results with those obtained by a very high-resolution z-coordinate nonhydrostatic model. The nonhydrostatic model used is the MITgcm implemented for the SoG at a horizontal resolution of about 50 m on the middle of the strait. MITgcm was initialized and laterally forced as a one-way nested model of the POM simulation.

The two models differ primarily in their horizontal resolution. POM has a maximum resolution in the strait of about 300 m, while MITgcm reaches in the same region of the strait about 50 m. In other words, MITgcm adopts a resolution six times finer than POM. A less-marked difference is in the vertical discretization. POM has 32 σ-levels, and MITgcm 53 z-levels. In the middle of the strait, in a water depth of 500 m, POM has a resolution of about 8 m in the upper and lower 50 m, and a resolution of 20 m for the remaining 400 m. MITgcm uses 53 z-levels spaced 7.5 m in the upper 300 m and with spacing that gradually increases to a maximum of 105 m for the remaining 13 bottom levels.

As expected, nonhydrostatic effects have an impact on the simulated internal wave field. In the hydrostatic POM model, the internal bore does not undergo its classical evolution into a series of short easterly propagating solitary waves as predicted by weakly nonlinear theories. The MITgcm nonhydrostatic simulation does simulate this process in a way that is qualitatively faithful. Differences also have been found in the way the two models reproduce the three-layer hydraulic regime and transports. MITgcm displays a more marked along-strait variability of supercritical flow regions due to the finer description of the bathymetry (Figure 3.11). However, looking at the frequency of occurrence of supercritical flow with respect to both modes, not only in single sections but in larger areas around the three main hydraulic regions, it can be seen that the MITgcm predicts a similar supercritical flow along TN (gray bars in Figure 3.13).

On the contrary, looking at the frequency of occurrence of supercritical flow with respect to one mode, it can be seen that the maximum difference between the two model simulations is confined to TN. Here MITgcm predicts a frequency 28% less than POM (from 70% to 42%), while a minor reduction affected ES (from 74% to 62%), and

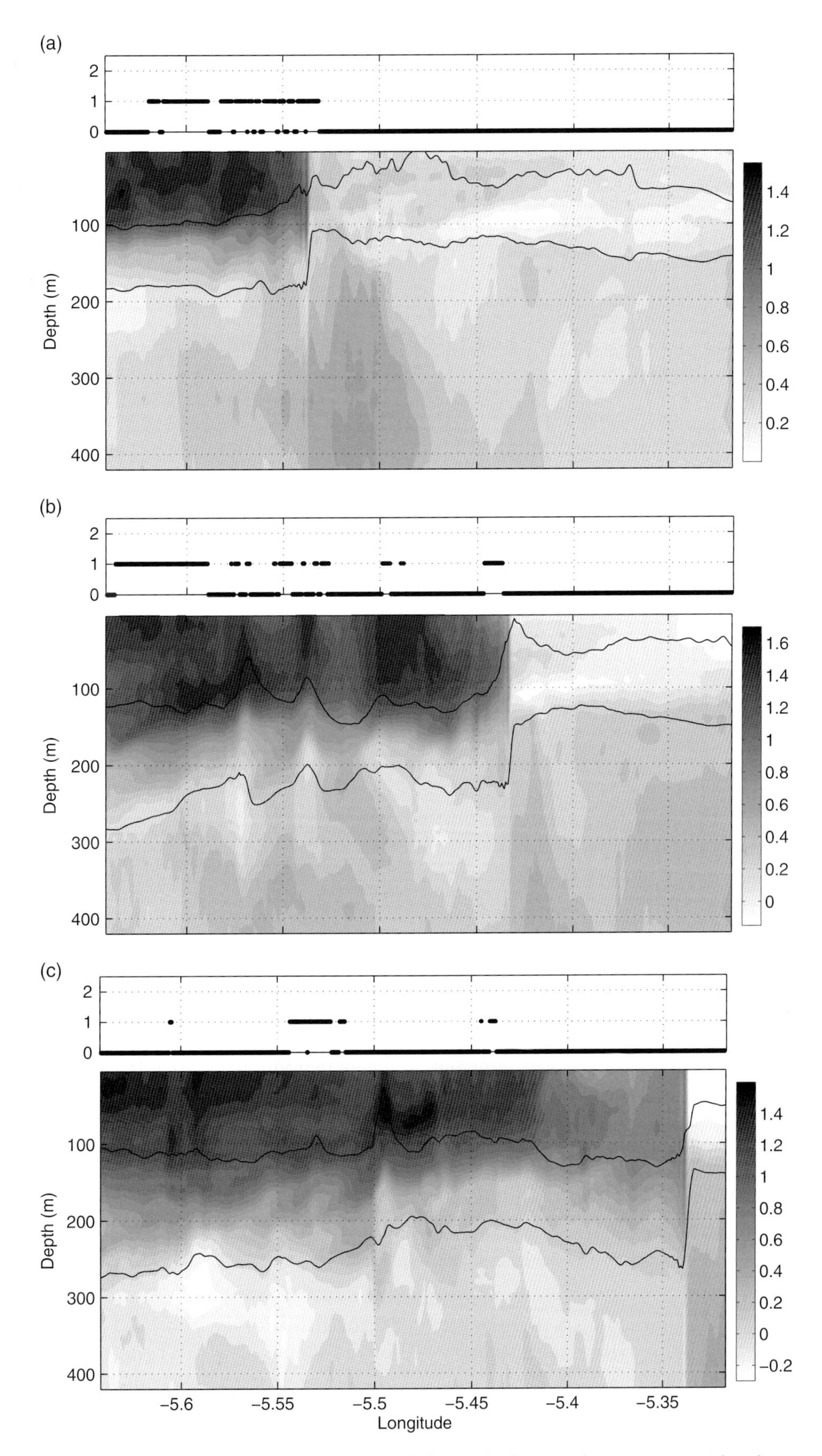

Figure 3.18 Same as left panels in Figure 3.12, but with the nonhydrostatic feature not considered.

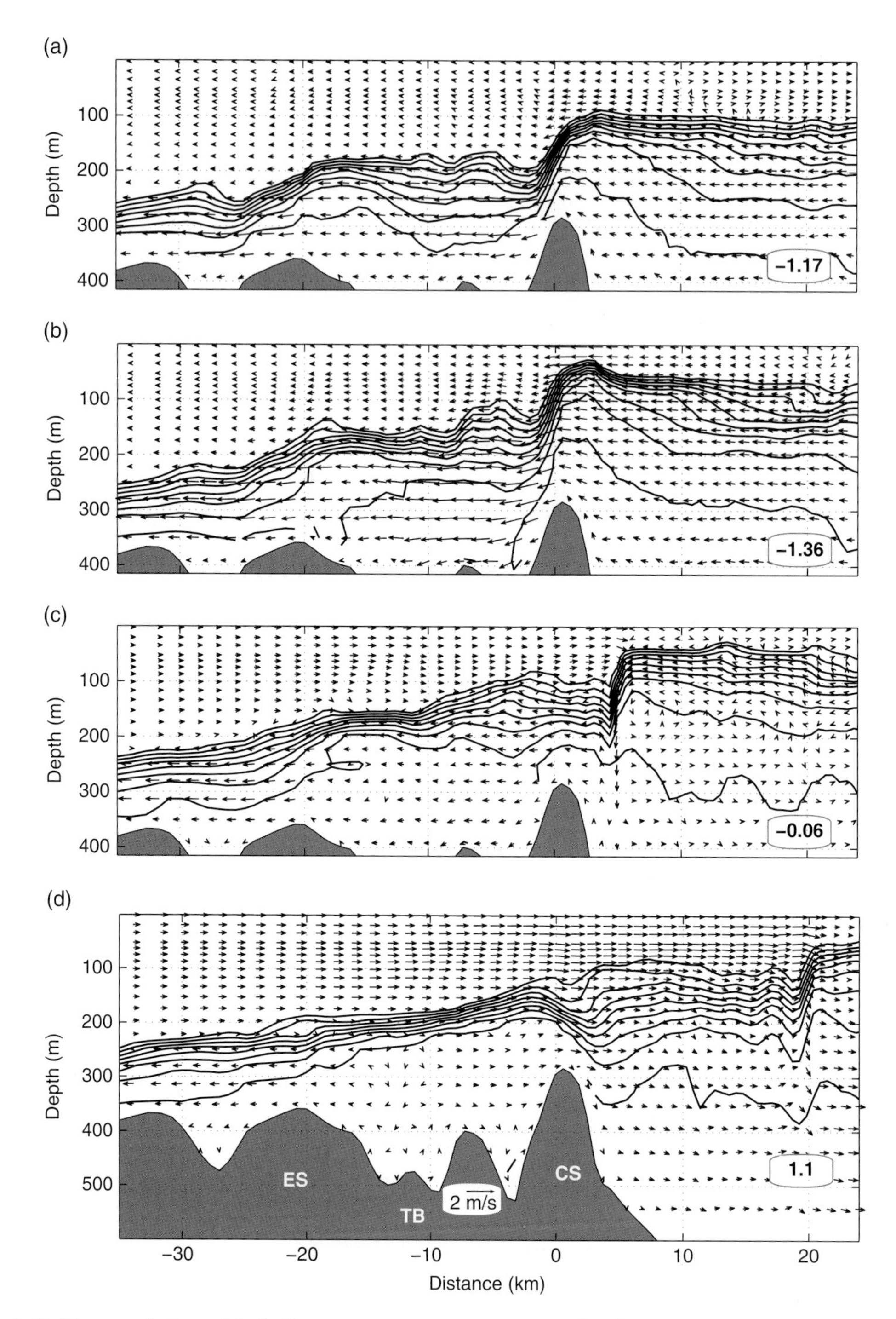

Figure 3.19 Time-evolution of isohalines 36.40, 36.65,…,38.65 and velocity currents simulated by MITgcm using the model grid initially used for POM (EXP5) during one tidal cycle of moderate tidal strength. The barotropic velocity (in ms^{-1}) over Camarinal Sill is indicated at the lower right corner of the panels. Elapse times after panel a are 1:40 h (panel b), 4:40 h (panel c), and 7:00 h (panel d).

almost similar values were obtained for CS (from 41% to 48%). The larger difference is due to the systematic underestimation of the upper layer along TN produced by POM (Figure 3.6). This is particularly clear comparing the evolution of a propagating bore across a section in TN together with the associated hydraulic regime of the flow as simulated by MITgcm and POM, respectively (Figure 3.12). It is shown that, in contrast to POM, MITgcm achieves the control only when the eastward bore is present in TN. The different hydraulic behavior is the result of a different representation of the three-layer thickness (Figure 3.7). While the three-layer thickness simulated by MITgcm is in good agreement with that derived from observed salinity profiles (Figure 6 in *Bray et al.* [1995]), POM overestimates the interface layer thickness and consequently underestimates both the upper and lower layers' thickness. The overestimation is attributed to the excess of spurious diapycnal mixing produced by POM that overshadows the naturally occurring mixing. The large spreading of the isohalines along the SoG present in the POM simulation (Figure 3.3), in comparison with those obtained by MITgcm (Figure 3.4), clearly shows that the simulated internal wave propagation induces a greater level of spurious diapycnal diffusion in POM.

To investigate the origin of such spurious diapycnal mixing, three additional experiments were performed to examine the sensitivity of the POM results to the choice of the coefficients governing the tracers diffusivity parameterization (EXP1-3). From these experiments, the numerically induced mixing was found to be due to the adopted advection scheme (MPDATA).

An additional experiment was carried out to explore the impact of the nonhydrostaticity on the simulated hydraulic regime. MITgcm was rerun in hydrostatic mode (EXP4). Results indicated that the hydraulic regime was not affected by the hydrostatic limitation.

A further experiment was performed to examine the influence of the horizontal and vertical resolution on the simulated exchange flow (EXP5). MITgcm was implemented on the same horizontal grid as for Exp-POM, and with a z-level distribution that ensured a vertical resolution similar to Exp-POM. Results indicated that the hydraulic regime remained very similar to that of EXP4 and, consequently, to Exp-MIT.

To conclude, the results of the experiments can be used to derive some general rules to correctly model the exchange flow through the SoG. First, the horizontal resolution similar to the one adopted in Exp-POM is enough for simulating the hydraulic regime of the SoG. Second, the nonhydrostatic formulation is not strictly necessary for simulating the hydraulic regime. Finally, the application of diffusive advection schemes, such as MPDATA in high-resolution σ-coordinate models used to simulate a tidally forced large internal wave, can potentially be a source of spurious diapycnal mixing. As a general recommendation in these cases, we suggest maintaining horizontal diffusion as small as possible in order to alleviate, at least in part, the spurious mixing.

Acknowledgments. We are grateful to the CRESCO supercomputing facilities located at ENEA (http://www.cresco.enea.it). Pratt is supported by National Science Foundation Grant OCE-0927017. The authors would also like to thank the anonymous referees whose comments helped improve this manuscript.

REFERENCES

Adcroft A., and J. M. Campin (2004), Rescaled height coordinates for accurate representation of free-surface flows in ocean circulation models, *Ocean Modelling*, *7*, 34, 269–284.

Armi, L., and D. M. Farmer (1988), The flow of Atlantic water through the Strait of Gibraltar, *Progress in Oceanography 21*, 1–105.

Bills, P., and J. Noye (1987), *Numerical Modelling: Applications to Marine Systems*, Elsevier Science Publishers.

Blumberg, A. F., and G. L. Mellor (1987), *A description of a three-dimensional coastal ocean circulation model*, n. s. heaps Edition, American Geophysical Union, Washington D.C.

Bray, N., J. Ochoa, and T. H. Kinder (1995), The role of the interface in exchange through the Strait of Gibraltar, *J. Geophys. Res.*, *100*, 10755–10776.

Bryden, H., and H. Stommel (1984), Limiting processes that determine basic features of the circulation in the Mediterranean Sea, *Oceanologica Acta*, *7* (3), 289–296.

Campin, J. M., A. Adcroft, C. Hill, and J. Marshall (2004), Conservation of properties in a free-surface model, *Ocean Modelling*, *6*, 34, 221–244.

Candela, J., C. Winant, and A. Ruiz (1990), Tides in the Strait of Gibraltar, *J. Geophys. Res.*, *95*, 7313–7335.

Carter, G. S., and M. A. Merrifield (2007), Open boundary conditions for regional tidal simulations, *Ocean Modelling*, *18* (3–4), 194–209.

Egbert, G. D., and L. Erofeeva (2002), Otis-osu tidal inversion software, www.coas.oregonstate.edu/research/po/research/tide/otis.html.

Ezer, T. (2005), Entrainment, diapycnal mixing and transport in three-dimensional bottom gravity current simulations using the Mellor-Yamada turbulence scheme, *Ocean Modelling*, *9*, 151–168.

Farmer, D. M., and L. Armi (1988), The flow of Mediterranean water through the Strait of Gibraltar, *Prog. Oceanogr.*, *21*, 1–105.

Fringer, O., M. Gerritsen, and R. Street (2006), An unstructured-grid, finite-volume, nonhydrostatic, parallel coastal ocean simulator, *Ocean Modelling*, *14*, 139–278.

Garrido, J. C. S., G. Sannino, L. Liberti, and L. Pratt (2012), Numerical modeling of three-dimensional stratified tidal flow over Camarinal Sill, Strait of Gibraltar, jgr 116 (C12026), doi:10.1029/2011JC007093.

Garrido, J. C. S., J. G. Lafuente, F. C. Aldeanueva, A. Baquerizo, and G. Sannino (2008), Time-spatial variability observed in velocity of propagation of the internal bore in the Strait of Gibraltar, jgr 113 (C7).

Griffies, S., R. Pacanowski, and R. Hallberg (2000), Spurious diapycnal mixing associated with advection in a z-coordinate ocean model, *Mon. Wea. Rev.*, *128* (3), 538564.

Haney, R. L. (1991), On the pressure gradient force over steep topography in sigma coordinate ocean models, *J. Phys. Oceanogr.*, *21*, 610–619.

Hundsdorfer, W., B. Koren, M. Vanloon, and J. Verwer (1995), A positive finite-difference advection scheme, *J. Computational Physics*, *117* (1), 35–46.

Lacombe, H., and C. Richez (1982), *Hydrodynamics of Semi-Enclosed Seas*, Elsevier, Amsterdam.

Legg, S., R. Hallberg, and J. Girton (2006), Comparison of entrainment in overflows simulated by z-coordinate, isopycnal and non-hydrostatic models, *Ocean Modeling*, *11* (1–2), 69–97.

Leith, C. (1968), Parameterization of vertical mixing in numerical-models of tropical oceans, *Physics of Fluids*, *10*, 1409–1416.

Marchesiello, P., L. Debreu, and X. Couvelard (2009), Spurious diapycnal mixing in terrain-following coordinate models: The problem and a solution, *Ocean Modelling*, *26* (3–4), 156–169.

Marshall, J., A. Adcroft, C. Hill, L. Perelman, and C. Heisey (1997a), A finite-volume, incompressible Navier Stokes model for, studies of the ocean on parallel computers, *J. Geophys. Res.*, *102* (C3), 5753–5766.

Marshall, J., C. Hill, L. Perelman, and A. Adcroft (1997b), Hydrostatic, quasi-hydrostatic, and nonhydrostatic ocean modeling, *J. Geophys. Res.*, *102* (C3), 5733–5752.

MEDAR Group (2002), Medatlas/2002 database, Mediterranean and Black Sea database of temperature salinity and bio-chemical parameters, *Climatological Atlas*, IFREMER Edition.

Mellor, G., and T. Yamada (1982), Development of a turbulence closure model for geophysical fluid problems, *Reviews of Geophysics and Space Physics*, *20* (4), 851–875.

Mellor, G., T. Ezer, and L. Oey (1994), The pressure gradient conundrum of sigma coordinate ocean models, *J. of Atmospheric and Oceanic Technology*, *11* (4 part 2), 1126–1134.

NOAA (2001), (ETOPO2) 2-minute gridded global relief data.

Orlanski, I. (1976), A simple boundary condition for unbounded hyperbolic flows, *J. Computational Physics*, *21*, 251–269.

Pacanowski, R., and S. Philander (1981), Parameterization of vertical mixing in numerical-models of tropical oceans, *J. Phys. Oceanogr.*, *11* (11), 1443–1451.

Pickart, R. S., L. J. Pratt, D. J. Torres, T. E. Whitledge, A.Y. Proshutinsky, K. Aagaard, T. A. Agnew, G. W. K. Moore, and H. J. Dail (2010), Evolution and dynamics of the flow through Herald Canyon in the western Chukchi Sea, *Deep-Sea Research II*, *57*, 5–26.

Pratt, L. (2008), Critical conditions and composite Froude numbers for layered flow with transverse variations in velocity, *J. Fluid Mech.*, *602*, 241–266.

Sanchez-Roman, A., G. Sannino, J. Garcia-Lafuente, A. Carillo, and F. Criado-Aldeanueva (2009), Transport estimates at the western section of the Strait of Gibraltar: A combined experimental and numerical modeling study, *J. Geophys. Res.*, 114.

Sannino, G., A. Bargagli, and V. Artale (2002), Numerical modeling of the mean exchange through the Strait of Gibraltar, *J. Geophys. Res.*, *107* (8), 9 1–24.

Sannino, G., A. Bargagli, and V. Artale (2004), Numerical modeling of the semidiurnal tidal exchange through the Strait of Gibraltar, *J. Geophys. Res.*, *109*, C05011, doi:10.1029/2003JC002057.

Sannino, G., A. Carillo, and V. Artale (2007), Three-layer view of transports and hydraulics in the Strait of Gibraltar: A three-dimensional model study, *J. Geophys. Res.*, *112*, C03010, doi:10.1029/2006JC003717.

Sannino, G., L. Pratt, and A. Carillo (2009a), Hydraulic criticality of the exchange flow through the Strait of Gibraltar, *J. Phys. Oceanogr.*, *39* (11), 2779–2799.

Sannino, G., M. Herrmann, A. Carillo, V. Rupolo, V. Ruggiero, V. Artale, and P. Heimbach (2009b), An eddy-permitting model of the Mediterranean Sea with a two-way grid refinement at the Strait of Gibraltar, *Ocean Modelling*, *30*, 1 56–72.

Sanz, J., J. Acosta, M. Esteras, P. Herranz, C. Palomo, and N. Sandoval (1992), Prospeccin geofsica del estrecho de Gibraltar (resultados del programa hrcules 1980–1983), *Publ. Espec. Inst. Esp. Oceanogr.*, *7*, 48.

Smagorinsky, J. (1963), General circulation experiments with primitive equations, i: The basic experiment, *Mon. Wea. Rev.*, *91*, 99–164.

Smolarkiewicz, P. (1984), A fully multidimensional positive definite advection transport algorithm with small implicit diffusion, *J. Comput. Phys.*, *54*, 325–362.

Vlasenko, V., J. C. Sanchez Garrido, N. Stashchuk, J. Garcia Lafuente, and M. Losada (2009), Three-dimensional evolution of large-amplitude internal waves in the Strait of Gibraltar, *J. Phys. Oceanogr.*, *39*, (9), 2230–2246.

Wesson, J., and M. Gregg (1994), Mixing at Camarinal Sill in the Strait of Gibraltar, *J. Geophys. Res.*, *99* (C5), 9847–9878.

4

Mixing in the Deep Waters of the Western Mediterranean

Harry Bryden[1], Katrin Schroeder[2], Mireno Borghini[3], Anna Vetrano[3], and Stefania Sparnocchia[4]

4.1. INTRODUCTION

In 2005 and 2006, there were major deep water formation events in the northwestern Mediterranean [*Schroeder et al.*, 2006; *Font et al.*, 2007; *Schroeder et al.*, 2008; *Smith et al.*, 2008; *Grignon et al.*, 2010; *Schroeder et al.*, 2010]. Deep convection occurred over a large area of the northwestern Mediterranean in marked contrast to the localized deep water formation in the Gulf of Lion that has been traditionally observed [*MEDOC*, 1970]. A large volume of new dense water was formed, which spread out from its formation region and filled the western Mediterranean below about 2200 m, so a 600–800-m thick layer of new deep water was observed throughout the western basin by 2006. Significantly, the new deep water is warmer, saltier, and denser than the old deep water so it can be readily identified in deep CTD profiles. From March 2006 through March 2010, there has been no evidence for deep water formation in the western Mediterranean: winters have been mild; Argo profiles and glider missions have not exhibited wintertime mixed layers as deep as 1000 m. Thus, for the period since 2006, we have a natural tracer in the form of warmer, saltier deep water with which to examine the evolution of the new deep water and its mixing with the older deep water above that is colder and fresher. Here we use repeat surveys across the southwestern Mediterranean (Figure 4.1) to describe the temporal evolution of the new deep water and its mixing with the old deep waters.

In terms of the overall mixing environment, the Mediterranean Sea has relatively small tides and relatively weak winds compared with the greater ocean environment, so vertical diffusion due to mechanical mixing is likely to be generally smaller than in the open ocean. Thermohaline mixing processes, however, are likely to be of greater importance in the Mediterranean than in the open ocean since the 1 Sv ($10^6\,m^3/s$) of relatively fresh, warmer Atlantic water inflow through the Strait of Gibraltar must ultimately be converted into a 1 Sv outflow of salty, colder Mediterranean water. Within the western Mediterranean, the vertical distribution of salinity and temperature exhibits maximum salinity and temperature at about 400-m depth associated with the core of Levantine intermediate water (LIW) that is formed in the eastern Mediterranean and flows into the western Mediterranean through the Sicily Channel. Because western Mediterranean deep water (WMDW) is naturally fresher and colder than the LIW, salinity and temperature both decrease downward below the LIW core toward the deep water. In this halocline-thermocline between the core of LIW and the deep water, warmer saltier waters overlie colder fresher waters and in such a region salt finger mixing processes can be effective mixing agents transporting salt, heat, and density downward. Indeed, within this

[1] *National Oceanography Centre Southampton, University of Southampton, Empress Dock, Southampton United Kingdom; Istituto di Scienze Marine—ISMAR, Consiglio Nazionale delle Ricerche (CNR), Venice, Italy*

[2] *Istituto di Scienze Marine—ISMAR, Consiglio Nazionale delle Ricerche (CNR), Venice, Italy*

[3] *Istituto di Scienze Marine—ISMAR, Consiglio Nazionale delle Ricerche (CNR), Lerici (SP), Italy*

[4] *Istituto di Scienze Marine—ISMAR, Consiglio Nazionale delle Ricerche (CNR), Trieste, Italy*

The Mediterranean Sea: Temporal Variability and Spatial Patterns, Geophysical Monograph 202. First Edition.
Edited by Gian Luca Eusebi Borzelli, Miroslav Gačić, Piero Lionello, and Paola Malanotte-Rizzoli.

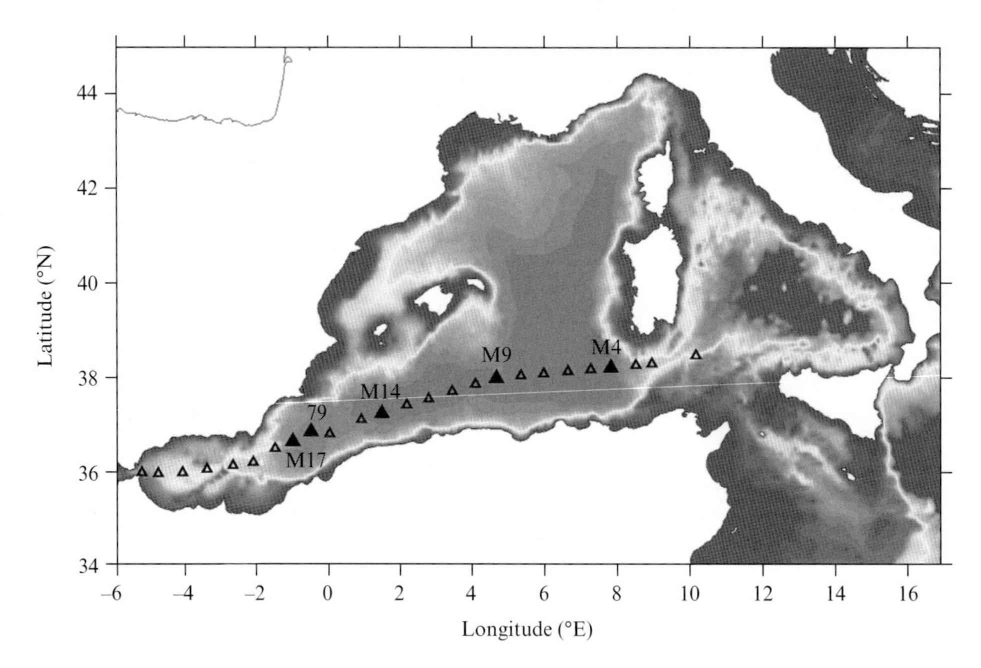

Figure 4.1 Map of CTD stations that were occupied by Urania during 2004, 2005, 2006, 2008, and 2010. These stations are part of regular repeat surveys of the western Mediterranean Sea since 2004 with support from Consiglio Nazionale delle Ricerche (CNR). Here we focus on stations M4–M17. Station 79 was used by *Borghini et al.* (2012) to estimate the overall changes in temperature and salinity of the Mediterranean waters as they are about to exit the Mediterranean. For color detail, please see color plate section.

halocline-thermocline it is common to observe staircase structures in the western Mediterranean (*Johannessen and Lee* [1974]; *Krahmann* [1997]; and note the steps above 1600 dbar in all profiles in Figure 4.2) and these staircase structures are widely considered to be signatures of salt finger mixing [*Stern and Turner*, 1969].

Historically, the deep waters have been observed to have nearly uniform properties below about 1600 m (note the 2004 profiles in Figure 4.2). The injection of new warmer, saltier, denser deep waters formed in 2005 and in 2006 has led to a deep transition zone between the colder, fresher old deep water above the warmer saltier new deep water below. This transition zone has the potential for double diffusive convective mixing processes that transport heat and salt upward but density downward. These double diffusive mixing processes are thought to be not as effective as salt fingers in effecting vertical mixing [*Schmitt*, 1994]. In summary for the deep western Mediterranean, we expect vertical diffusion due to mechanical mixing to be small; downward mixing of heat salt and density to be substantial in the halocline-thermocline 400–1500-m depth where salt finger processes operate; and upward mixing of heat and salt to be relatively small in the transition zone between older deep waters above and new deep waters below where double diffusive convective processes may operate.

4.2. EVOLUTION OF THE DEEP WATERS

Repeated surveys of the western Mediterranean have been made since 2004 on board R/V *Urania* of the Italian Consiglio Nazionale delle Ricerche (CNR). We concentrate this analysis on the section of stations M4–M17 (Figure 4.1) because they have been consistently occupied. The evolution of the deep water properties with focus on the new deep water is exhibited by the potential temperature and salinity profiles at M9 (Figure 4.2). The old deep water prior to the 2005 and 2006 formation events is evident in the 2004 profiles with nearly constant potential temperature of 12.825°C and salinity of 38.45 below 2000 dbar. In 2005, a small layer of warmer, saltier bottom water appears at M9 below 2750 dbar with potential temperature of 12.86°C and salinity of 38.475. By 2006, this new deep water is observed below 2250 dbar, with a sharp interface separating it from the old deep water. The sharp interface suggests that new deep waters are still arriving at station M9. The old deep water that in 2006 is limited to the depth range 1800–2250 dbar intriguingly appears to be getting warmer and saltier in its characteristics from 2004 to 2005 to 2006. By 2008, the interface between old and new deep waters is more gradual suggesting mixing is occurring. By 2010, the new deep water extends up to about 1900 dbar and its interface with the old deep water again suggests mixing. The same basic pattern is evident at all stations M4–M17.

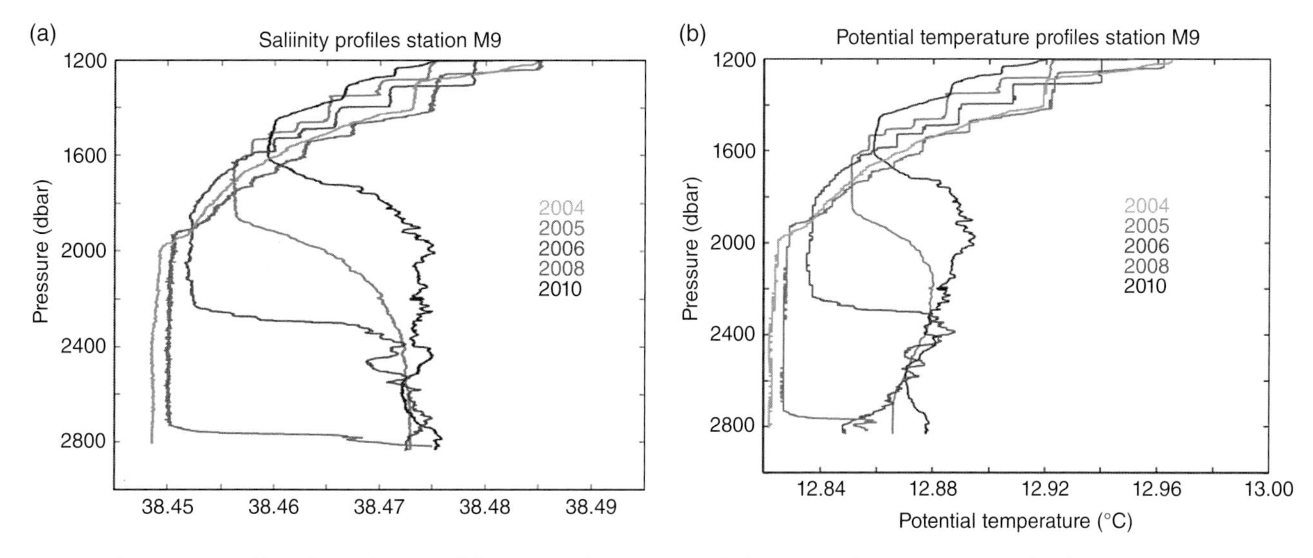

Figure 4.2 Profiles of (a) salinity and (b) potential temperature below 1200 dbar at station M9 for the 2004, 2005, 2006, 2008, and 2010 surveys. For color detail, please see color plate section.

Above the deep waters, there is a halocline-thermocline that represents the transition between LIW and the deep water. The signature of LIW throughout the western Mediterranean is a core of high-salinity (38.55), high-temperature water (13.25°C) at about 400 dbar. Thus, from 400 dbar down to the colder, fresher western Mediterranean deep water, there is a halocline-thermocline in which steplike features in salinity and temperature are commonly observed (see Figures 4.2a and 4.2b) and associated with salt finger processes. In the laboratory, salt fingers transfer heat, salt and density downward [*Turner*, 1973]. Large steplike features were first observed in the Tyrrhenian Sea in the halocline-thermocline beneath the core of LIW [*Johannessen and Lee*, 1974], and they have been commonly observed throughout the western Mediterranean [e.g., *Krahmann*, 1997].

The transition region between the halocline-thermocline and the deep waters changes as well from 2004 to 2010 (Figure 4.2). Initially in 2004, the halocline-thermocline penetrated down to 2000 dbar but it slowly withdrew to 1900 in 2005, 1800 in 2006, 1600 in 2008, and 1450 dbar in 2010. At the bottom of the halocline-thermocline, there is a well-mixed layer that we label as old deep water in 2004, 2005, and 2006; but in 2008 and 2010, this well-mixed layer appears to be the result of processes that accumulate heat and salt in a well-mixed layer at the base of the halocline-thermocline.

4.3. MIXING ESTIMATES

The region between the bottom of the halocline-thermocline and the new deep water appears to be growing warmer and saltier. Initially from 2004 to 2006, new deep waters arrived from the northern deep water formation sites to the southern stations creating a nearly homogeneous layer of new deep water below 2300 dbar. But after 2006, the transition region between the halocline-thermocline and the deep waters became warmer and saltier (Figure 4.3a and 4.3b). This transition region is the only part of the water column below 500 dbar where the properties diverge between 2006, 2008, and 2010. As this transition region becomes warmer and saltier, it takes on the characteristics of the new deep water so the layer of new deep water grows thicker with time. By the 2010 survey, the new deep water characteristics extended from the bottom up to 1900 dbar.

We use the amount of temperature and salinity increase in this transition zone to quantify the amount of mixing between the new deep water and the older waters above it. We are not sure whether the arrival of new deep water had completely stopped by the time of the 2006 survey: profiles at stations 5–10 exhibit sharp interfaces at the top of the new deep water indicating that the new deep water is still arriving; and the comparison of profiles at stations 15 and 17 for 2006 and 2008 indicate that the arrival of new deep waters does continue after the 2006 survey. For these reasons, we focus on the temperature and salinity increases from 2008 to 2010 to quantify the amount of mixing.

For each station, we plotted the 2008 and 2010 potential temperature and salinity profiles (Figure 4.3a and 4.3b for Station M9) and identified the upper pressure surface where the properties are the same in both 2008 and 2010 and the lower pressure surface where the properties are the same in both years. The upper surface is effectively the bottom of the halocline-thermocline in 2010, while the lower surface is the top of the new deep water in 2008. We then plotted the 2010–2008 differences

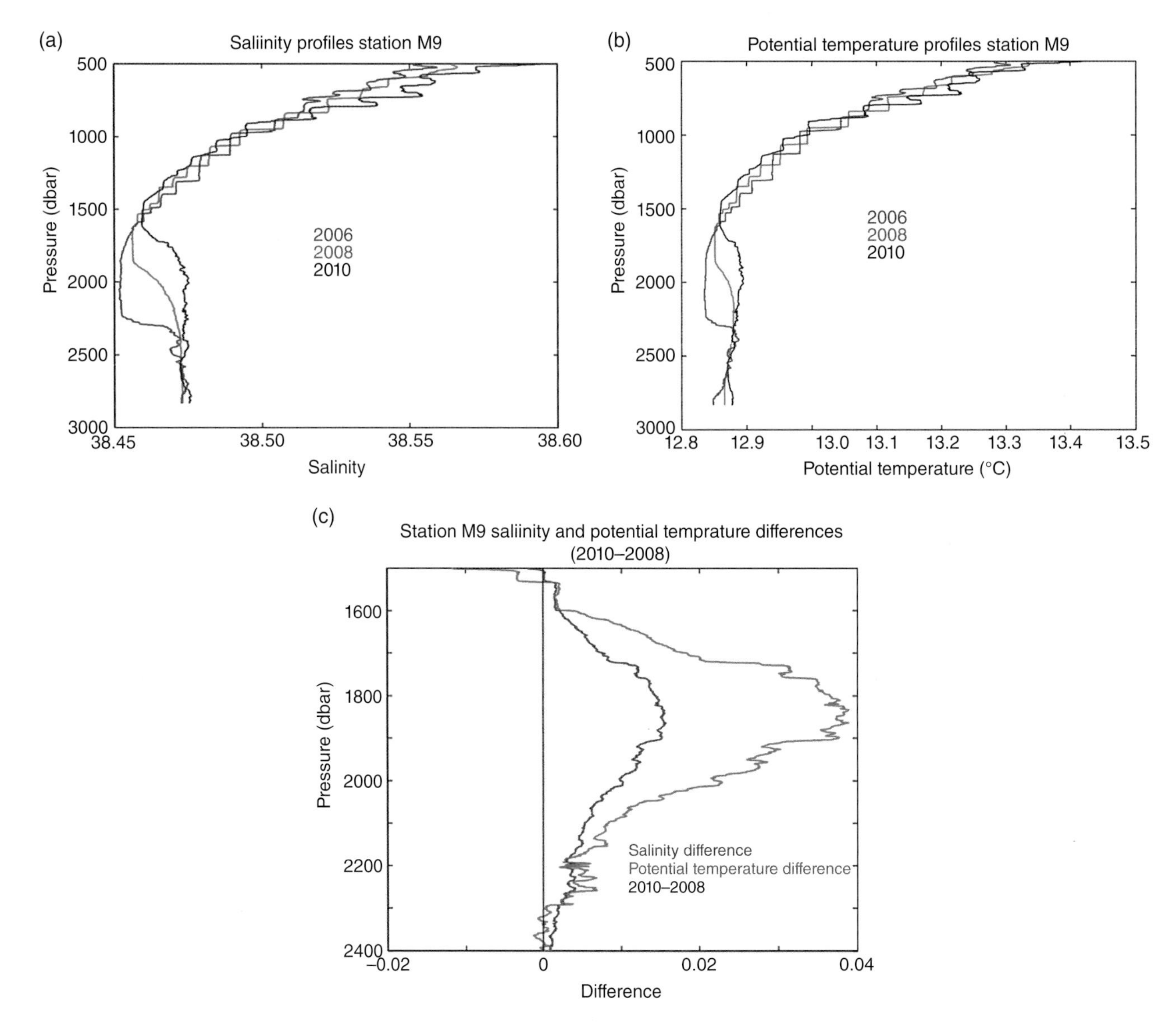

Figure 4.3 Profiles of (a) salinity and (b) potential temperature below 500 dbar at station M9 for 2006, 2008, and 2010 surveys. The halocline-thermocline from 500 dbar to 1500 dbar exhibits steplike features that are usually associated with salt finger processes; the deep waters below 2300 dbar exhibit similar characteristics of new deep water formed in 2005–2006. The only part of the water column where changes are immediately evident is the transition zone between the bottom of the halocline-thermocline and the top of the new deep water. Here the transition waters become progressively warmer and saltier from 2006 to 2008 to 2010. (c) Profiles of the 2010 minus 2008 differences in salinity and potential temperature in the transition zone between the bottom of the halocline-thermocline and the top of the new deep water at station M9 are used to quantify the amount of salt and heat flux convergences into this transition zone, tabulated in Table 4.1. For color detail, please see color plate section.

in potential temperature and salinity (Figure 4.3c), noted the maximum differences, and also calculated the depth-integrated differences (called Sbulk and ThBulk) between the upper and lower surfaces (Table 4.1). For all stations, the 2010–2008 differences are positive, that is, warmer and saltier in 2010, indicating that heat and salt have been added to the transition region between the bottom of the halocline-thermocline and the top of the new deep waters.

There are inherent uncertainties in salinity of order 0.002 and temperature of 0.001°C between surveys that lead to uncertainties in bulk differences of order 1 psu m^1 and 0.5°Cm. The maximum differences are clearly above these uncertainties and nearly all of the bulk differences exceed their uncertainty thresholds. Thus, there is clear warming and salinification from 2008 to 2010 in the depth interval between the bottom of the halocline-thermocline and the top of new deep water.

Table 4.1 Differences in Salinity and Potential Temperature 2010 Minus 2008 for Stations across the Southern Western Mediterranean Sea.

	Top	Bottom	ΔSmax	Ratio	ΔThmax	Sbulk	Ratio	ThBulk
Station	(dbar)	(dbar)		ΔTh/ΔS	°C	psu m	Th/S	°Cm
4	1570	2070	0.012	2.3	0.026	3.50	2.1	7.46
5	1670	2125	0.010	2.2	0.022	2.46	2.1	5.09
6	1600	2130	0.011	2.3	0.027	2.87	2.2	6.23
7	1540	2100	0.010	2.4	0.023	2.39	2.0	4.80
8	1785	1920	0.006	1.9	0.012	0.55	1.7	0.
9	1520	2300	0.014	2.5	0.039	5.65	2.3	12.78
10	1500	2250	0.014	2.3	0.032	5.56	2.1	11.57
11	1600	2300	0.009	2.8	0.025	3.04	2.6	7.92
12	1550	2260	0.006	2.5	0.014	2.07	2.4	4.95
13	1500	2230	0.004	2.7	0.010	0.97	2.9	2.79
14	1450	2200	0.012	2.6	0.030	4.18	2.5	10.30
15	1750	2350	0.006	2.5	0.014	1.04	2.0	2.07
16	1500	2100	0.011	2.7	0.029	4.21	2.7	11.45
17	1450	2050	0.010	2.5	0.024	3.10	2.5	7.73
Average				2.4		2.97	2.3	6.86

Note: The pressure intervals over which differences are estimated are given by top and bottom pressures. The maximum differences over this interval are significant above the uncertainties in salinity of 0.002 and in temperature of 0.001°C. The bulk differences are vertical integrals of the differences over the pressure interval for each station. For these bulk values, we find it useful to use the psu symbol for salinity so the units of the vertical integral are clear. The ratios are intended to provide an indication of whether these changes represent increases or decreases in density: for ratios less than 3.1 $=\beta/\alpha$ (= haline contraction coefficient divided by thermal expansion coefficient) these salinity and temperature increases represent increases in density.

This southern region of the western Mediterranean is considered to be relatively "quiet," without strong tides or wind mixing, compared with the ocean. Whilst we would not want to estimate mixing from a single station, the set of 14 stations from Sicily to the eastern entrance to the Alboran Sea exhibiting similar changes gives confidence that the mixing process is not locally determined, for example by nearby bottom topographic features. Thus, an average of these 14 estimates provides a realistic estimate of mixing in the deep waters. From the average depth-integrated changes over 14 stations, there is a convergent salt transport of 3.0 psu m and convergent temperature transport of 6.9 °C m into the zone between the bottom of the halocline-thermocline and top of the new deep water. Dividing these transports by the 21 months between the 2008 and 2010 stations yields a salinity flux convergence of $F_S = 5.35 \times 10^{-8}$ psu m s^{-1} and a heat flux convergence of $F_T = 12.4 \times 10^{-8}$°C m s^{-1}. Multiplying the transports by gravity and haline contraction ($\beta = 7.35 \times 10^{-4}$ psu^{-1}) and thermal expansion ($\alpha = 2.36 \times 10^{-4}$°C^{-1}) coefficients, evaluated at $S = 38.475$, $\theta = 12.88$°C and $p = 2000$ dbar, yield buoyancy flux convergences of $g\beta F_S = 3.97 \times 10^{-10}$ W kg^{-1} (due to salinity convergence) and $g\alpha F_T = 2.94 \times 10^{-10}$ W kg^{-1} (due to heat convergence) for a net density flux convergence of 1.0×10^{-10} W kg^{-1}. The buoyancy flux ratio is then $\alpha F_T / \beta F_S = 0.74$.

It is worth noting that salinity is formally a ratio and has no units. When vertically integrating salinity to obtain bulk values, we find it useful to add the symbol psu to represent salinity in the integrated values so the units are clear and further calculations are understandable.

4.4. MIXING PROCESSES

We think the mixing between the bottom of the halocline-thermocline and the new deep water is a result of salt finger processes in the halocline-thermocline that are transferring heat and salt downward. Qualitatively, there is evidence for salt fingers in the steplike features within the halocline-thermocline (Figures 4.3a and 4.3b) and it appears to be the bottom of the halocline-thermocline that is eroding (Figures 4.2a and 4.2b). In the halocline-thermocline, salt finger processes act to transfer heat, salt, and density downward. Below the bottom of the halocline-thermocline, in the transition zone down to the warmer, saltier deep waters, salt finger processes cannot operate. Furthermore, diffusive mixing within this transition zone or double diffusive instabilities between the overlying colder, fresher old deep waters and the warmer, saltier new deep waters would transfer heat and salt upward. Thus, in the zone between the bottom of the halocline-thermocline and the top of the new deep water, there is a convergence between the downward salt finger fluxes of heat and salt from the halocline thermocline and the upward fluxes of heat and salt due to diffusive processes.

Double diffusive convection between colder fresher waters above warmer saltier waters in the transition region and diffusion across the deep interface would each erode the high salinity and temperature signature of the new deep waters, and we do not observe such erosion; instead we observe a growth in the thickness of the deep water. Also, double diffusive convection processes are thought to be less effective than salt finger processes [*Turner,* 1973]. Finally, with a traditional diffusion coefficient of $1 \times 10^{-4} m^2 s^{-1}$ [e.g., *Munk,* 1966], the upward diffusive fluxes across the deep interface are an order of magnitude smaller than the flux convergences we estimate here. Thus, we think diffusion and diffusive convection processes contribute less to the heat and salt flux convergences than do the downward heat and salt fluxes associated with salt finger processes in the halocline-thermocline.

If we relate the bulk salt and heat transports F_S and F_T derived above to the stratification of the halocline-thermocline where the vertical salinity gradient is about 8.3×10^{-5} ‰ m^{-1} and the vertical potential temperature gradient is about 3.9×10^{-4} °C m^{-1},

$$F_S = k_S dS/dz \quad \text{and} \quad F_T = k_T d\theta/dz$$

we estimate a salinity mixing coefficient $k_S = 6.5 \times 10^{-4} m^2 s^{-1}$ and a temperature mixing coefficient, $k_T = 3.2 \times 10^{-4} m^2 s^{-1}$. Both are larger than canonical values of $1 \times 10^{-4} m^2 s^{-1}$ [e.g., *Munk,* 1966] for the deep ocean, but perhaps of reasonable size for an active salt finger region. *Stern and Turner* [1969] estimated $k_S = 5 \times 10^{-4} m^2 s^{-1}$ for laboratory experiments when they set up a background stratification and *Lambert and Sturges* [1977] estimated $k_S = 5.7 \times 10^{-4} m^2 s^{-1}$ from observations in a salt finger region of the North Atlantic.

4.5. DISCUSSION

In our view, there is a relatively constant downward flux of temperature and salinity through the halocline-thermocline. At the bottom of the halocline-thermocline, the vertical salinity and temperature gradients effectively vanish in the deep water. Thus, the downward heat and salt transports converge into the top of the deep water making it warmer and saltier. Under normal circumstances where the deep water is relatively homogeneous, these salt and heat fluxes would increase the salinity and temperature of a 1000-m thick deep water layer by 0.0017 psu and 0.0039°C over a year in the absence of new deep water injection. Such changes would be difficult to observe on a year-to-year basis given instrumental uncertainties.

The situation since 2006 when the new deep water is distinct from the old deep water so that the layer below the halocline-thermocline and above the new deep water has a thickness of only 400 m has focused the downward flux convergences into a more limited depth region. Hence, they become more evident as seen here in the evolution of the deep water properties from 2006 to 2008 to 2010 (Figures 4.3a and 4.3b).

The downward density flux associated with salt finger mixing and calculated here from the sum of the heat and salt fluxes multiplied by thermal expansion and haline contraction coefficients is quite small, only $0.00056 kg m^{-3} m$ over 21 months. Over a year, the density of a 1000-m thick deep water layer would increase by only $0.00032 kg m^{-3}$, equivalent to only $0.016 kg m^{-3}$ over 50 years.

Whilst it may appear that we are relying on local changes to estimate downward fluxes at each station, we argue that the downward fluxes of heat and salt through the halocline-thermocline are occurring all over the western Mediterranean transforming the water mass properties in the zone between the bottom of the halocline-thermocline and the top of the new deep water. The waters in this zone are surely flowing around the basin so one might technically insist that the local changes at each station are due to lateral advection of a new water mass. But the relative uniformity in the size and structure of the salinity and temperature increases in this zone across the section of 14 stations (Table 4.1) is strong evidence that local advective effects are not the determining mechanism for the observed changes and that downward heat and salt fluxes through the bottom of the halocline-thermocline associated with salt finger processes represent the ultimate cause of the changes.

Efforts to quantitatively estimate the size of salt finger fluxes of heat and salt have concentrated on analyzing the thickness of individual steps and the gradients at the interfaces and measuring the level of turbulent dissipation near the sharp interfaces [*Schmitt,* 1994; *Schmitt et al.*, 2005; *Hebert,* 1988] did examine changes in layer temperature and salinity below the center of a Mediterranean eddy (Meddy) to define the heat and salt fluxes associated with salt fingers. In the open ocean, his was the only study we could find related to our approach to quantify flux convergences/divergences from observed temporal changes in layer temperature and salinity. *Hebert* [1988] argued that lateral intrusions into the core of the Meddy, mechanical mixing processes, as well as the heaving of density surfaces were relatively minor issues with his estimates of changes in layer temperature and salinity. The principal uncertainty was the definition of the bottom boundary for the flux divergence calculations. Overall, *Hebert* [1988] found that the changes in layer temperature and salinity were an order of magnitude smaller than the fluxes based on laboratory formulae. The advantage of our Mediterranean situation is that the upper and lower boundaries for the flux convergence zone are well defined by the bottom of the halocline-thermocline and the top of the new deep water.

In terms of quantitative estimates of salt finger fluxes in the Mediterranean, *Zodiatis and Gasparini* [1996] estimated heat and salt fluxes across salt fingering interfaces in the Tyrrhenian Sea where the steplike features were first observed by *Johannessen and Lee* [1974] using theoretical formulae developed by *Kunze* [1987]. They found buoyancy flux convergences due to salinity of about $1.5 \times 10^{-10}\,W\,kg^{-1}$ (range 0.38 to $3.99 \times 10^{-10}\,W\,kg^{-1}$) and due to temperature of $1.0 \times 10^{-10}\,W\,kg^{-1}$ (range 0.28 to $2.81 \times 10^{-10}\,W\,kg^{-1}$), a factor of 3 lower than we estimate here for the southern western Mediterranean. They also noted a progressive increase in the temperature and salinity of the 600–1500-m depth interval of about 0.009°C per year and 0.002 per year over 19 years, values that are close to what we estimate here for the changes in deep water temperature and salinity.

If these observed heat and salt flux convergences do indeed represent downward salt finger fluxes, such convergences may offer an accurate method for determining the fluxes of heat and salt in salt finger regimes. Previous estimates of salt finger fluxes in the open ocean are usually made uncertain by the amount of mechanical mixing present in the ocean. Here, in the quiet Mediterranean with weak circulation, low winds, and small tides, the amount of mechanical mixing becomes less of an issue. We encourage attempts to use microstructure profilers to measure the amount of mechanical and thermodynamic mixing in the Mediterranean Sea to assess whether these observed convergences are indeed due to salt finger processes. If the convergences are the result of salt finger processes, relatively simple observations of convergences from repeat surveys in different salt finger regimes could offer a way to quantify the size of salt finger fluxes under different environmental conditions and thereby lead to an accurate parameterization for the effects of salt fingers in circulation models.

On decadal time timescales, *Borghini et al.* [private communication] have shown that the western Mediterranean has become steadily warmer and saltier over the past 50 years. They indicate that the average salinity below 200 m has increased by 0.07 and the average potential temperature has increased by 0.19°C from 1961 to 2010. Some of this warming and salinification is due to the arrival of warmer and saltier LIW. But the deep water has also increased in temperature and salinity by similar amounts. The heat and salinity fluxes that we associate here with salt finger mixing down through the halocline-thermocline operate at a similar rate: the estimated yearly changes of 0.0017 and 0.0039°C for a 1000-m thick deep water layer multiplied by 50 years would yield increases in deep water salinity of 0.08 and 0.20°C, not dissimilar to the observed changes noted by *Borghini et al.* Downward fluxes of heat and salt through the halocline-thermocline associated with salt finger mixing may be the cause of observed increases in deep water temperature and salinity over the past 50 years.

REFERENCES

Borghini, M., H. L. Bryden, K. Schroeder, S. Sparnocchia, and A. Vetrano (2012), The Mediterranean is getting warmer and saltier, In preparation.

Font, F., P. Puig, J. Salat, A. Palanques, and M. Emelianov (2007), Sequence of changes in NW Mediterranean deep water due to exceptional winter of 2005, *Sci. Mar.*, *71*, 339–346.

Grignon, L., D. A. Smeed, H. L. Bryden, and K. Schroeder (2010), Importance of the variability of hydrographic preconditioning for deep convection in the Gulf of Lion, NW Mediterranean, *Ocean Sci.*, *6*, 573–586.

Hebert, D. (1988), Estimates of salt-finger fluxes, *Deep-Sea Res.*, *35*, 1887–1901.

Johannessen 0. M., and 0. S. Lee (1974), A deep stepped thermohaline structure in the Mediterranean, *Deep-Sea Res.*, *21*, 629–639.

Krahmann, G. (1997), Horizontal variability of thermohaline staircases in the western Mediterranean. Double-Diffusive Processes, 1996 Summer Study Program in Geophysical Fluid Dynamics, S. Meacham and D. Tucholke, eds., Woods Hole Oceanogr. Inst. Tech. Rep. WHOI-97-10, 331–347.

Kunze, E. (1987), Limits on growing, finite-length salt fingers: A Richardson number constraint, *J. Mar. Res.*, *45*, 533–556.

Lambert, R. B., and W. Sturges (1977), A thermohaline staircase and vertical mixing in the thermocline, *Deep-Sea Res.*, *24*, 211–222.

MEDOC Group (1970), Observation of formation of deep water in the Mediterranean Sea, 1969, *Nature*, *227*, 1037–1040.

Munk, W. H. (1966), Abyssal recipes, *Deep-Sea Res.*, *13*, 707–730.

Schmitt, R. W. (1994), Double diffusion in oceanography, *Ann. Rev. Fluid Mech.*, *26*, 255–285.

Schmitt, R. W., J. R. Ledwell, E. T. Montgomery, K. L. Polzin, and J. M. Toole (2005), Enhanced diapycnal mixing by salt fingers in the thermocline of the tropical Atlantic, *Science*, *308*, 685–688.

Schroeder, K., A. Ribotti, M. Borghini, R. Sorgente, A. Perilli, and G. P. Gasparini (2008), An extensive western Mediterranean deep water renewal between 2004 and 2006, *Geophys. Res. Lett.*, *35*, L18605, doi:10.1029/2008GL035146.

Schroeder, K., G. P. Gasparini, M. Tangherlini, and M. Astraldi (2006), Deep and intermediate water in the western Mediterranean under the influence of the Eastern

Mediterranean Transient, *Geophys. Res. Lett.*, *33*, L21607, doi:10.1029/2006GL027121.
Schroeder K., S. A. Josey, M. Herrmann, L. Grignon, G. P. Gasparini, and H. L. Bryden (2010), Abrupt warming and salting of the Western Mediterranean Deep Water after 2005: Atmospheric forcings and lateral advection, *J. Geophys. Res.*, *115*, C08029 doi:10.1029/2009JC005749.
Smith, R. O., H. L. Bryden, and K. Stansfield (2008), Observations of new western Mediterranean deep water formation using ARGO floats, 2004 –2006, *Ocean Sci.*, *4*, 133–149.
Stern, M. E., and J. S. Turner (1969), Salt fingers and convecting layers, *Deep-Sea Res.*, *16*, 497–511.
Turner, J. S. (1973), *Buoyancy Effects in Fluids*, Cambridge University Press.
Zodiatis, G., and G. P. Gasparini (1996), Thermohaline staircase formations in the Tyrrhenian Sea, *Deep-Sea Res. I*, *43*, 655–678.

5

The 2009 Surface and Intermediate Circulation of the Tyrrhenian Sea as Assessed by an Operational Model

E. Napolitano[1], R. Iacono[1], and S. Marullo[2]

5.1. INTRODUCTION

The Tyrrhenian Sea (TYS hereafter), the deepest basin of the western Mediterranean Sea, has a complex circulation, driven both by the local atmospheric forcing and the exchanges occurring at its three openings: the Sardinia and Corsica channels and the Sicily Strait (see Figure 5.1). The large-scale flow patterns through these openings are now relatively well established (see, e.g., the review by *Millot* [1999] and the more recent work by *Vetrano et al.* [2004]), but a detailed understanding of the basin dynamics is still lacking.

During the 1990s, the classical picture of a simple wind-driven, large-scale circulation [*Krivosheya*, 1983] was supported by numerical studies using simplified models of the TYS dynamics [*Artale et al.*, 1994; *Pierini and Simioli*, 1998], and by the first, coarse resolution GCM simulations of the Mediterranean Sea circulation [*Roussenov et al.*, 1995; *Zavatarelli and Mellor*, 1995; *Korres et al.*, 2000]. The TYS surface circulation predicted by these models basically consisted in a basinwide cyclonic cell during winter and spring, including some quasi-permanent gyres in the western part of the basin. These numerical works also showed a reversal of the circulation in summer in the eastern TYS that appeared consistent with the change of sign of the large-scale wind stress curl occurring in summer in that region.

The 1990s also marked the beginning of systematic satellite observations over the Mediterranean Sea (for the TYS, see the early work by *Marullo et al.* [1994]), revealing the presence of a rich mesoscale activity poorly resolved by the cited GCM's (note that the Rossby radius of deformation in the Mediterranean Sea is of only 10–15 Km, that is, 3–4 times smaller than what is typical in the world ocean). In the last decade, a new generation of higher resolution GCMs has been developed, but these models have not been used yet to investigate the TYS dynamics in detail. Some recent studies [*Rio et al.*, 2007; *Budillon et al.*, 2009; *Vetrano et al.*, 2010; *Rinaldi et al.* 2010], however, have provided further information on the TYS, revealing a complex circulation in the southeastern part of the basin, and the presence of some persistent vortices. In particular, in the reconstruction of the spring 2004 circulation by *Vetrano et al.* [2010], based both on observations and on numerical results, the central and eastern TYS are filled with large cyclonic and anticyclonic structures, which are also found in typical spring altimeter data, and could therefore be robust features around which the spring flow organizes. In a very recent work by *Iacono et al.* [2012], this picture has been confirmed and extended, through a careful analysis of 18 years of altimeter data, which has allowed the authors to characterize the average surface circulation in the whole basin, and to study its seasonal variability.

Less is known about the intermediate circulation, the reference scheme still being that proposed by *Krivosheya* [1983], within the western TYS, the same cyclonic centers characterizing the surface layer, and a wide cyclonic cell in the southern part of the basin, which gets smaller in summer. Recent works [*Vetrano et al.*, 2010; *Menna and Poulain*, 2010] support the existence of a basinwide cyclonic cell, but show that the intermediate circulation is also characterized by the presence of many eddies, some of which are strongly correlated with structures observed in the surface layer.

[1] *ENEA—C. R. Casaccia, Roma, Italy*
[2] *ENEA—C. R. Frascati, Frascati, Italy*

The Mediterranean Sea: Temporal Variability and Spatial Patterns, Geophysical Monograph 202. First Edition.
Edited by Gian Luca Eusebi Borzelli, Miroslav Gačić, Piero Lionello, and Paola Malanotte-Rizzoli.

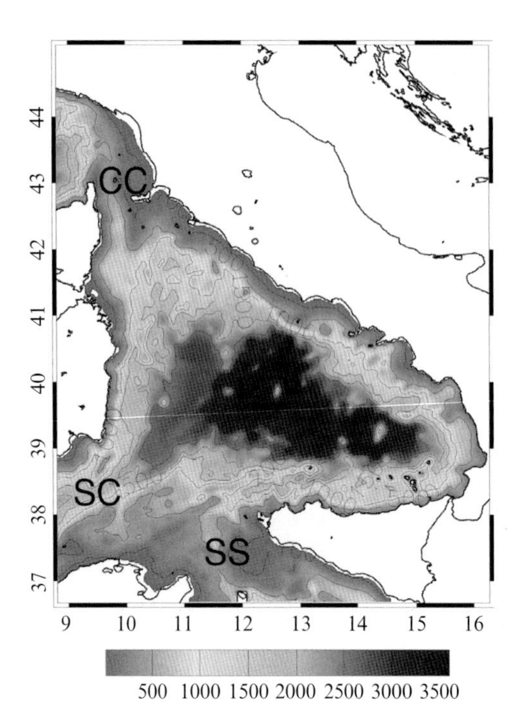

Figure 5.1 The Tyrrhenian Sea, with its three openings: the Corsica Channel (CC), the Sardinia Channel (SC), and the Sicily Strait (SS). Bathymetry is indicated by colors. The domain represented exactly coincides with the computational domain of the operational model TYREM, whose outputs are discussed in the text. For color detail, please see color plate section.

The purpose of the present work is to gain new insights on the surface and intermediate dynamics of the TYS, through the analysis of the results produced in the first year of life (2009) of an operational, high-resolution model of the TYS circulation. The regional ocean model has been developed in the context of PRIMI (PRogetto pilota Inquinamento Marino da Idrocarburi), a national project funded by the Italian Space Agency (ASI), aimed at the realization of a prototypal system for monitoring and forecasting the evolution of oil spills in the Italian seas. During 2009, the model has produced both weekly forecasts and hindcasts. The latter make use of reanalyses, both of the circulation of the "father" oceanic model, and of the surface forcings, and consequently provide our best assessment of the TYS three-dimensional circulation.

The paper is organized as follows. The basic features of the operational model of the TYS are described in section 2; the weekly hindcasts produced by the model are used in section 3 to characterize the seasonal variability of the surface and intermediate circulation in 2009, and in section 4 to discuss the main aspects of water masses distributions; finally, a summary of the main results is given in section 5.

5.2. MODEL DESCRIPTION

5.2.1. The Numerical Model

The operational model of the TYS circulation (TYrrhenian REgional Model, TYREM hereafter) is based on the Princeton Ocean Model [POM, *Blumberg and Mellor*, 1987; *Mellor*, 2004] a three-dimensional, free surface, primitive equation, hydrostatic model with sigma coordinates in the vertical direction. POM is a well-documented model that has been used in a number of oceanographic studies in the past 20 years, as well as in the development of several forecasting systems [e.g., *Oddo et al.* 2006; *Korres and Lascaratos*, 2003].

The model domain (see Figure 5.1) extends into the Sicily Strait and in the Ligurian Sea, and is covered by a regular longitude-latitude grid of 360 x 376 points. This yields an average horizontal resolution of about 2 km (1/48°), which is sufficient to resolve both mesoscale and submesoscale eddies. The vertical grid consists of 40 sigma levels that are smoothly distributed along the water column, with appropriate thinning designed to better resolve the surface and intermediate layers. The model has three open boundaries: a zonal boundary located at 36.68°N, across the Sicily Strait, and two meridional boundaries along the 8.81°E line, across the Sardinia Channel and from Corsica up to the Ligurian coast.

The bathymetry is interpolated from the DBDV data set (http://www.navo.navy.mil/) that has a 1' resolution. To reduce the pressure gradient errors inherent to the use of sigma coordinates over steep topography [*Haney*, 1991], a Laplacian smoothing has been applied, so to have a maximum $\Delta H/H$ smaller than 0.1 (about 0.08), as suggested in *Mellor et al.* [1994]. Thanks to the high horizontal resolution, the smoothed model bathymetry, shown in Figure 5.1, is still quite realistic, even though some isolated topographic features, such as the Marsili and the Vavilov Seamounts, are lower than in reality. The residual pressure gradient error has been estimated integrating the model with closed boundaries, no surface forcing, and no initial horizontal density gradient. After 8 months, the maximum intensity of erroneous currents induced by the sigma coordinates was of about 1 cm/s.

The POM model contains an embedded second-order turbulence closure scheme [*Mellor and Yamada*, 1982] that provides vertical exchange coefficients. In TYREM, we have included a modification of this closure scheme proposed by *Kantha and Clayson* [1994]. Horizontal diffusion is parameterized following *Samgorinsky* [1963].

5.2.2. Boundary Conditions and Hindcast Procedure

TYREM is nested one way with MFS, an operational forecasting model of the whole Mediterranean Sea circulation [*Tonani et al.*, 2008], which makes use of *z*

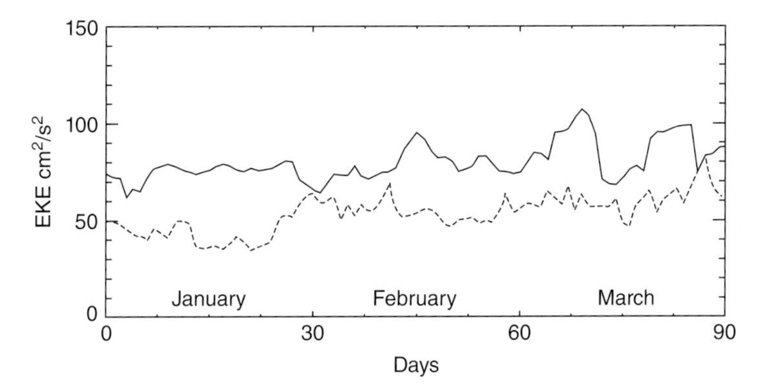

Figure 5.2 Winter 2009 time series of the daily EKE (average over the basin) computed from TYREM (solid line) and MFS (dashed line).

levels in the vertical, and has a lower horizontal resolution of 1/16° x 1/16° (the TYREM horizontal resolution of 1/48° x 1/48° yields a grid spacing ratio of 3:1). One-way nesting means that TYREM periodically takes from MFS initial conditions and boundary conditions at the three openings, interpolates them on the finer grid, and uses them to evolve its own dynamics, which is not used, however, to provide feedbacks to the "father" model.

The nesting procedure is designed in such a way to ensure that the volume transports across the open boundaries of TYREM match those of MFS; to do so, we use the same treatment for the barotropic and baroclinic velocities described in *Zavatarelli and Pinardi* [2003]. The baroclinic velocities at the open boundaries of TYREM are prescribed, and are obtained from those of MFS through bilinear interpolation. On the other hand, the boundary condition on the barotropic velocity at the open boundaries is that by *Flather* [1976], modified as in *Marchesiello et al.* [2001]. On the open boundaries, temperature and salinity from the MFS model are advected into the model domain when the flow is into the model domain.

The heat and momentum fluxes needed at the sea-air interface are computed through standard bulk formulae (see Appendix at the end of this chapter for details). The atmospheric data (air temperature, relative humidity, cloud cover, and wind components) needed to compute these fluxes are taken from the European Center for Medium Range Forecasting (ECMWF); they have a horizontal resolution of 1/4° and a frequency of 6 hrs. The water flux resulting from the equilibrium between evaporation and precipitation has been parameterized as a salt flux, while the evaporation flux has been estimated from the computed latent heat flux. The precipitation is present in the father model, and therefore in the initial conditions, but is set to zero in TYREM, since it would have very small effect on the short timescale of the forecast (see, e.g., *Lermusiaux and Robinson* [2001]).

TYREM produces both weekly forecasts and hindcasts. The hindcasts, which are used for this work, are performed as follows. Each Tuesday, the MFS system provides analysis fields (velocity, temperature, salinity, and elevation) for the previous 15 days through a run with data assimilation. These fields, together with forcing data from the EMCWF analyses, are used to provide initial and boundary conditions for a TYREM hindcast run over the same period. Since in the first week of simulation there is an adjustment phase during which spurious external gravity waves produced by the initialization decay, only the second week of simulation is used for the analysis to follow.

Although an analysis of the skill of TYREM is out of the scope of the present paper, it is interesting to give an idea of the effects of the higher resolution. For the winter months (January–March), we have computed the daily basin average of the eddy kinetic energy (EKE) from the velocity field at 10 m of depth, both for TYREM and MFS (the EKE is computed with respect to the winter mean flow). Figure 5.2 shows the result. The TYREM EKE values are consistently higher than those computed from MFS, and close to the average EKE values estimated in *Rinaldi et al.* [2010] from drifter data (73 cm^2/s^2), for the period 2001–2004. This indicates that TYREM resolves relevant scales of motion not resolved by the father model.

5.3. THE SEASONAL VARIABILITY OF THE CIRCULATION DURING 2009

5.3.1. The Surface Circulation

Before discussing the model results, it is useful to look at the structure of the wind stress over the basin. The 2009 wind stress is quite typical: in February (Figure 5.3a) its main feature is the "Mistral jet" blowing northwesterly over most of the basin, while in August (Figure 5.3b) the wind blows zonally in the area of the Bonifacio Strait (the narrow passage between Sardinia and Corsica) and

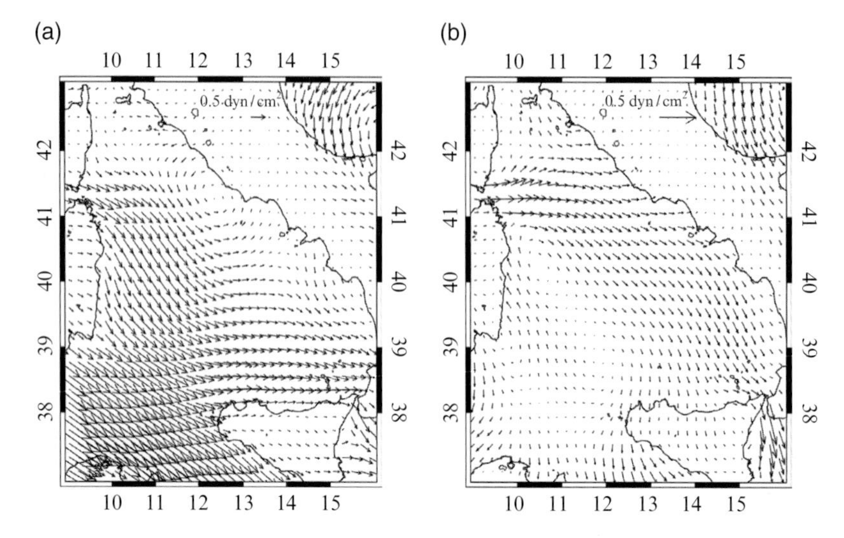

Figure 5.3 2009 wind stress from ECMWF: (a) February, (b) August.

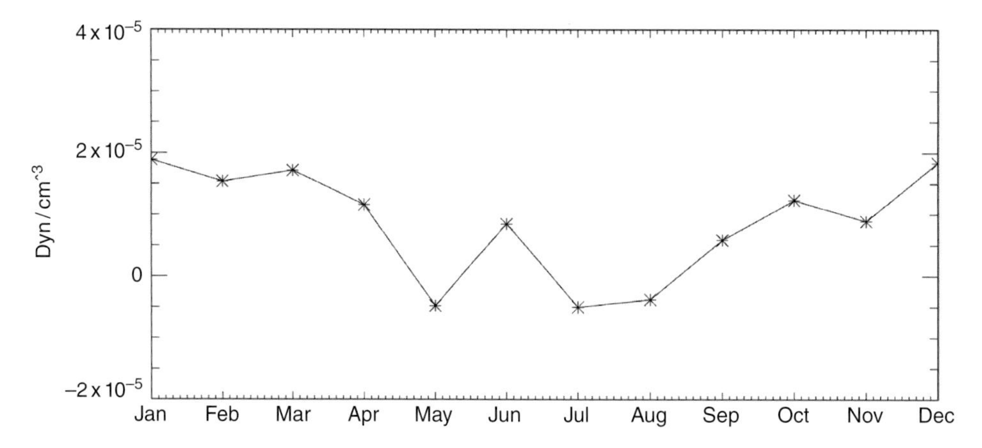

Figure 5.4 Time variation of the basin average of the wind-stress curl.

forms a wide anticyclonic cell in the southern part of the basin. Figure 5.4 shows the basin average of the wind stress curl, which is positive from autumn to midspring and weakly negative in the summer months. The positive June value is atypical; analysis of the last two decades of ERA interim data shows that, on average, the June wind-stress curl is almost vanishing.

Let us now examine some monthly mean surface circulation maps derived from the TYREM hindcasts, representative of the different seasons. The February map (Figure 5.5a) illustrates the winter situation, with the surface circulation dominated by a global cyclonic cell formed by Atlantic water (AW hereafter) and sustained by the positive wind-stress curl. The Algerian Current, after crossing the Sardinia Channel, experiences a well-known bifurcation [*Herbaut et al.*, 1996] around 38°N, 12°E with one branch entering the Sicily Strait, while the other progresses farther east into the TYS. The latter flows over the Sicilian continental slope, meanders around the Aeolian islands, and then veers northwestward, following the Italian coast up to 42°N. Here the current moves away from the coast, rounding the Bonifacio cyclonic gyre, which is quite small, and then experiences another bifurcation at about 10°E, with a major branch going north and exiting the basin from the Corsica Channel, while the other moves southward, bordering the eastern coast of Sardinia. Several mesoscale vortices are present in the central part of the basin; in particular, a wide cyclonic region (1) around 39°N and between 13°E and 14E, and a central dipole 2–3. The western region is occupied by the expected quasi-permanent structures, that is. the Bonifacio cyclone-anticyclone couple (4–5) and the wide cyclonic region between Sardinia and Sicily (6), which encompasses several submesoscale structures.

It may be noted that some meanders are apparently developing all along the path of the cyclonic AW stream. Before discussing the other maps of Figure 5.5, it is interesting to follow the evolution of this pattern from February to March. Figure 5.6, focusing on the eastern part of the basin, shows that this evolution is quite

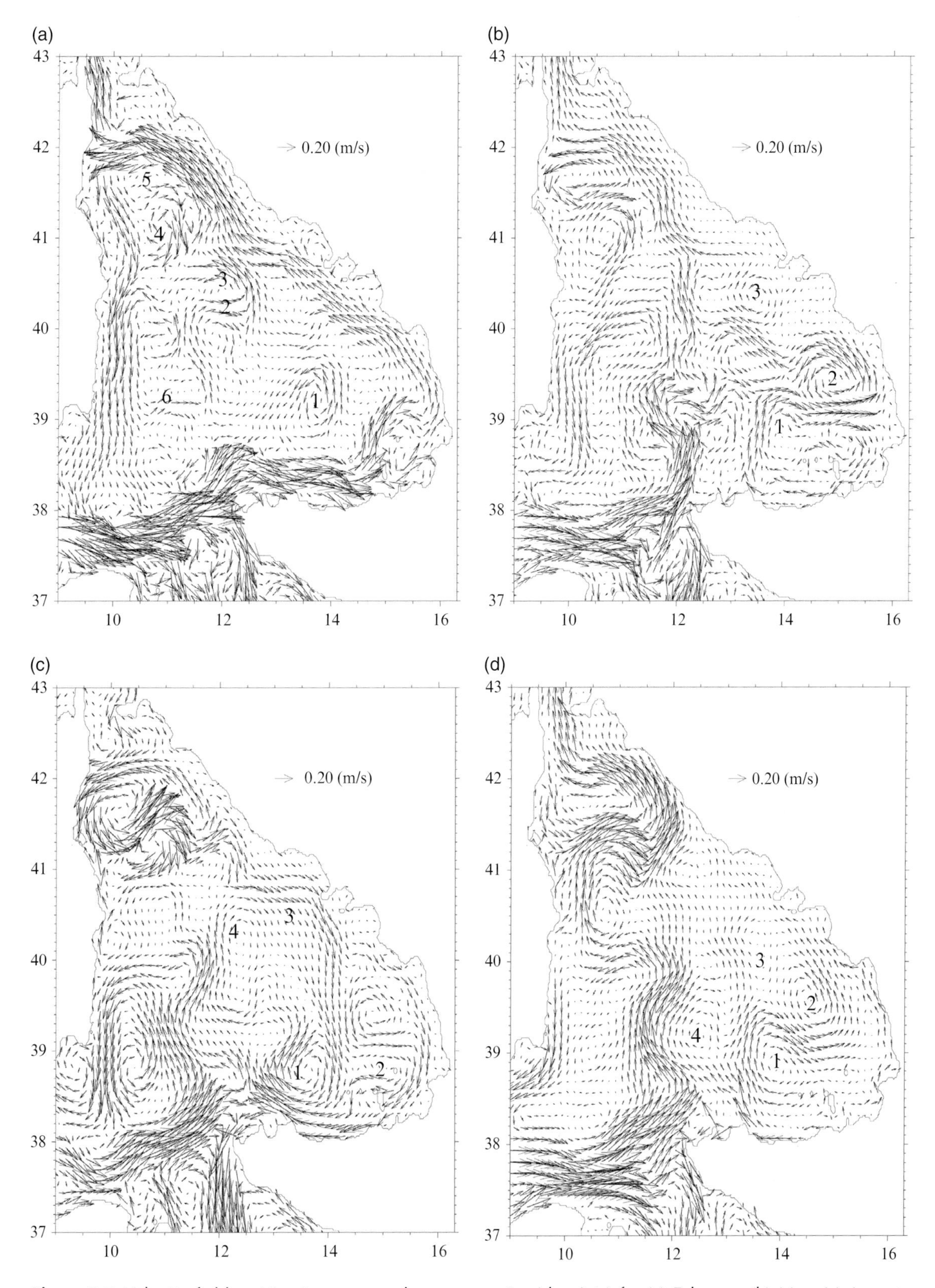

Figure 5.5 Velocity field at 10m (arrows are drawn every 6 grid points) for (a) February, (b) May, (c) August, (d) October.

dramatic, and clearly reveals an instability of the AW stream, which produces wide anticyclones along the coast and somewhat smaller cyclones on the offshore side of the stream. These vortices persist and, in some cases, strengthen in April (not shown). The picture here is in very good agreement with the description of the average winter-spring circulation made in *Iacono et al.* [2012], where evidence of a similar instability has been obtained from the analysis of altimeter data. In fact, it is shown in *Iacono et al.* [2012] that this pattern is a robust feature of the winter-spring TYS circulation.

To give a qualitative validation of the March circulation patters produced by TYREM, in the last panel of Figure 5.6 we show the Sea Level Anomaly (SLA) for the week of March 4 to March 10, from the AVISO dataset (see details below). The SLA (indicated by contours) and the circulation map of March 10 show very good agreement. All the main structures are found in both maps, approximately in the same locations: the two anticyclonic cores above Sicily, centered at about 12.5°E and 14°E, with a cyclone in between more offshore; the anticyclonic core off the Calabrian coast, with a cyclonic companion offshore; the anticyclonic area off Naples, with two cores; the anticyclone-cyclone couple around 41°N, 12°E; and finally the anticyclone along the coast further north. We note however, that the good agreement with the SLA may be partly due to the fact that SLA tracks are assimilated by MFS, and therefore influence the initial conditions of the TYREM simulations. For this reason, further independent experimental evidence of the structures and patterns we have discussed would be useful. We have therefore looked at the Sea Surface Temperature (SST) fields for the same period. The last panel of Figure 5.6 shows the SST field for the week 4–10 March, indicated by colors (the SST map has been obtained combining MODIS data from the Terra and Aqua satellites, from all night passes during the week in consideration). Despite the weak horizontal gradients, which sometimes make it difficult to identify the dynamic signatures, some of the structures we have discussed can be seen quite clearly along the Italian coast.

Let us now go back to Figure 5.5, and examine the May surface circulation (Figure 5.5b). The first thing to notice is that the winter basinwide cyclonic cell is now lacking. A smaller cell is present in the western part of the basin, delimited, at about 12°E, by a northward AW stream originating near the Sicily western tip, in the same area in which the winter bifurcation occurs. The stream crosses the whole basin in the meridional direction, reaching the Bonifacio cyclone, which is wider than in winter, and then exiting through the Corsica Channel. To our knowledge, the existence of this more direct AW path in spring has not been reported before in the literature. In the eastern part of the basin the circulation is dominated by large eddies, which result from the evolution of some of the vortices created during late winter: a wide anticyclonic area to the north of Sicily, a cyclonic area just above, and again an anticyclonic area off Naples, more displaced offshore than in winter.

The August surface flow field (Figure 5.5c) shows that a further reorganization of the flow has taken place, guided by the change of the wind-stress curl over the eastern TYS. Three distinct regions are now apparent: a large anticyclonic region that occupies the central and eastern parts of the basin, and encompasses at least three anticyclones (1,2,3); the Bonifacio couple, with the cyclone extending zonally toward the Italian coast; and finally the southwestern cyclonic region. It is worth noting that this three-lobed structure of the large-scale summer circulation is similar to that obtained by *Artale et al.* [1994] with a simple wind-driven, quasi-geostrophic model, even though our numerical results reveal many finer-scale details of the circulation. A last point to stress is that in this typical summer situation there is virtually no surface outflow at Corsica (as observed from mooring data by *Gasparini et al.* [2008]), while the AW entering from the Sardinia Channel mostly feeds the cyclonic area between Sardinia and Sicily.

In the October map (Figure 5.5d), there is no longer evidence of the anticyclonic cell in the eastern TYS, and no AW inflow in the south. The AW stream entering from the Sardinia Channel recirculates in the wide southwestern cyclonic region, while to the north the Bonifacio cyclone-anticyclone system has reached its maximum extension and strength. The eastern part of the basin is occupied by four main structures (1,2,3,4), three of which (1,2,3) are still those present in spring and summer. No well-defined surface stream is present inside the basin.

It is interesting to compare the surface circulation predicted by the model with altimeter data available for the same period. The data are from the AVISO dataset; they are provided by Ssalto/Duacs and distributed by AVISO (http://www.aviso.oceanobs.com/) with support from CNES, the French space agency. The AVISO dataset, starting at the end of 1992, now covers almost two decades, and, as shown in *Iacono et al.* [2012], contains precious information about the TYS surface circulation and its variability. Here we look, in particular, at the weekly Updated Delayed maps of Absolute Dynamic Topography (ADT), provided on a regular 1/8° x 1/8° grid. The ADT is obtained by adding to the SLA the Mean Dynamic Topography (MDT), a reconstruction of the average circulation based not only on the SLA, but also on all the other data (e.g., drifters, hydrological data) available over a reference period [see *Rio et al.*, 2007].

The ADT maps, for the same four months of Figure 5.5, are shown in Figure 5.7, with a geostrophic reconstruction of the circulation superposed, also provided by

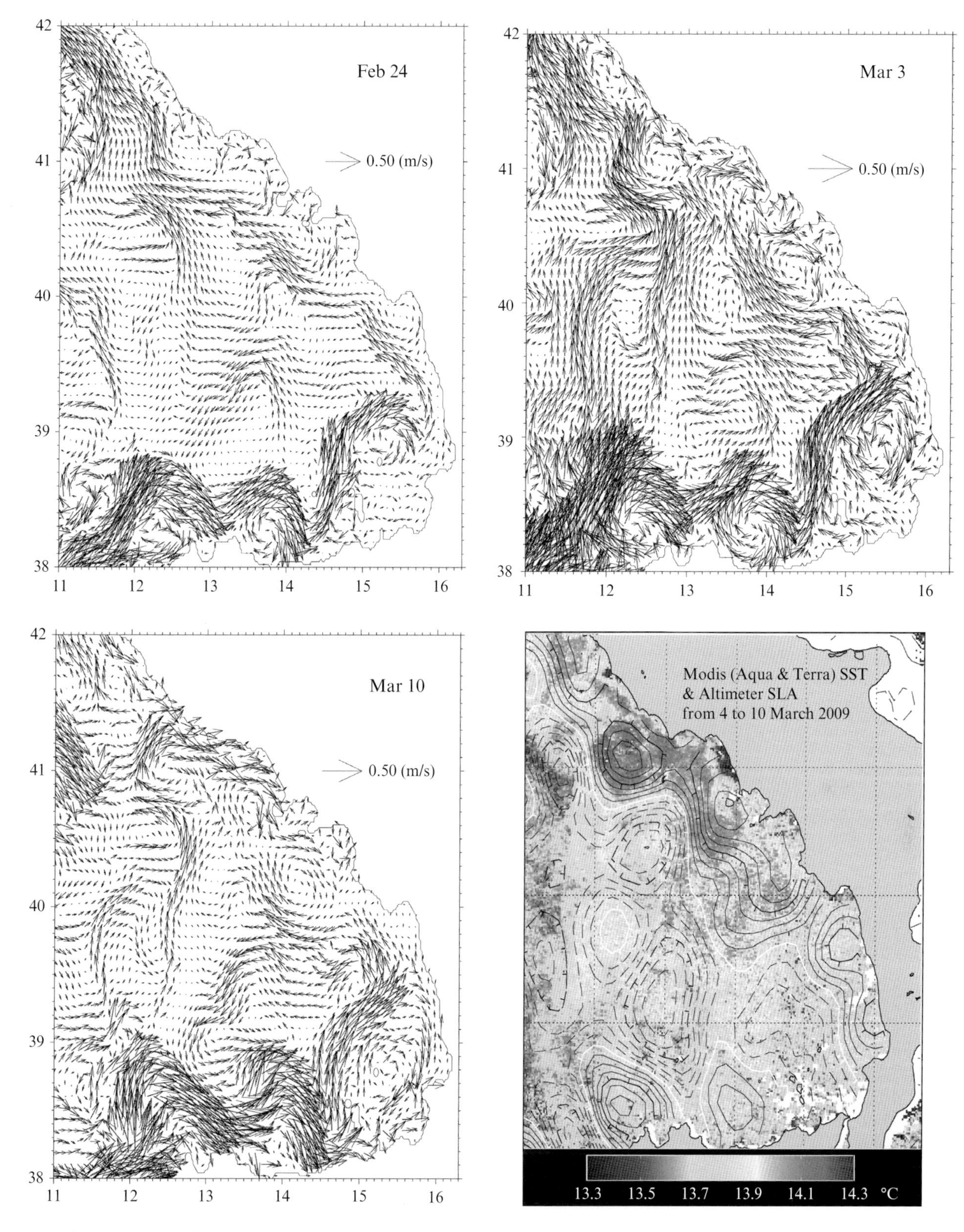

Figure 5.6 Weekly averaged TYREM velocity fields at 10 m (zooms in the eastern TYS). The fourth panel shows the SLA for the week 4–10 March (contours), with the SST of the same week superposed (colors). For color detail, please see color plate section.

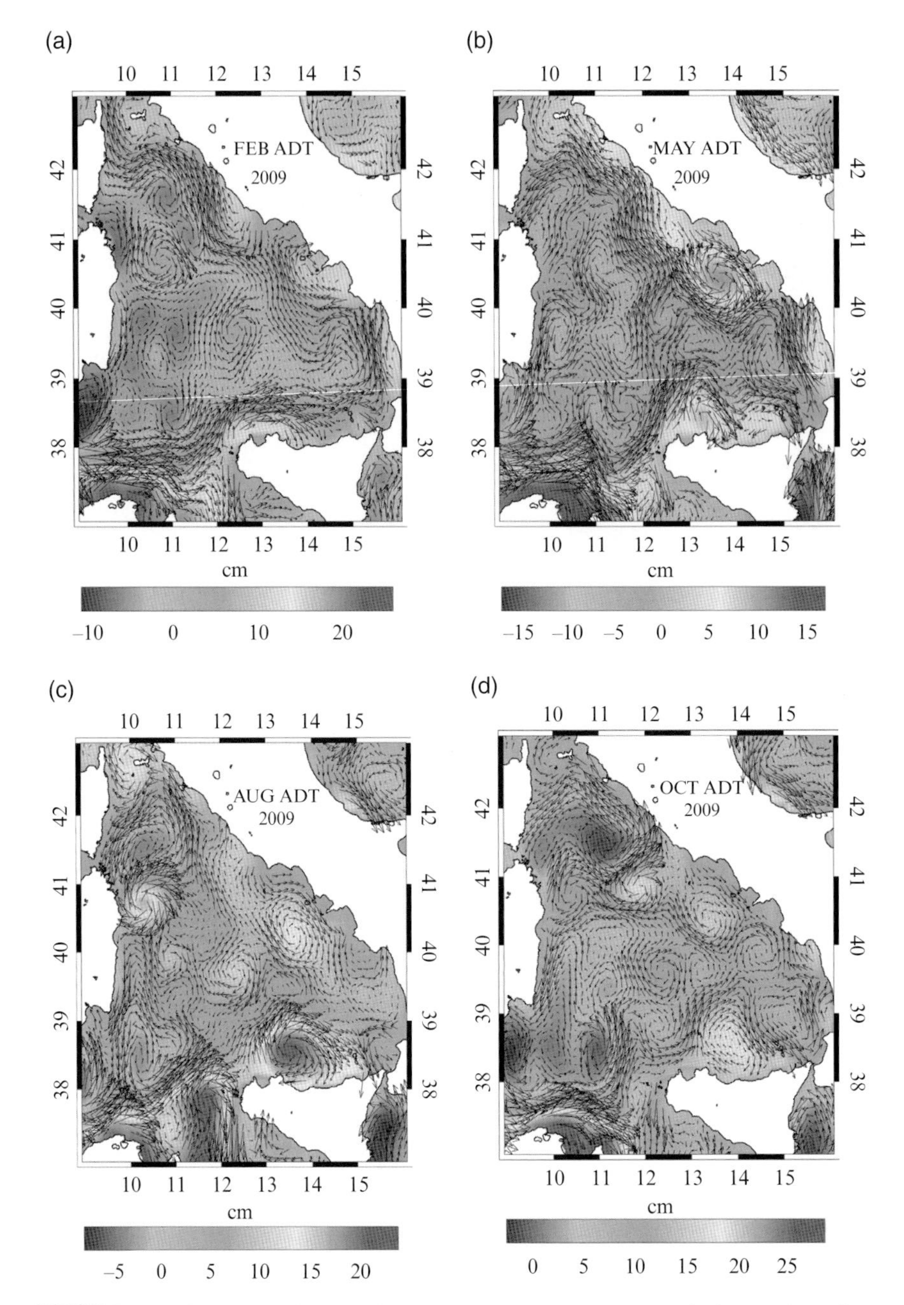

Figure 5.7 Monthly mean of the ADT for (a) February, (b) May, (c) August, and (d) October. The corresponding average geostrophic circulations are also shown. For color detail, please see color plate section.

AVISO. Comparison with Figure 5.5 shows that the TYREM surface circulation is in good agreement with the AVISO reconstruction in all seasons, even though there are some differences. In February (Figure 5.7a), for example, the geostrophic flow along the Italian coast is more meandering than the TYREM surface velocity field, and the positions of some eddies are slightly different. In May (Figure 5.7b), the main features of the ADT are quite close to those of Figure 5.5b; in particular, the direct northward branch of AW at about 12°E, and the eddies 1,2,3. In the northern part of the basin, the Bonifacio anticyclone is resolved in a similar way in the ADT map and in the TYREM results, while the Bonifacio cyclone is wider in the latter. In August, the main vortices in the central and eastern TYS are present both in the ADT and in the model results, but the anticyclonic cell present in the latter is not evident in the altimeter data. The presence of such a cell is instead suggested by the August map of SLA of Figure 5.8. Finally, in October the model and the AVISO data are again in good

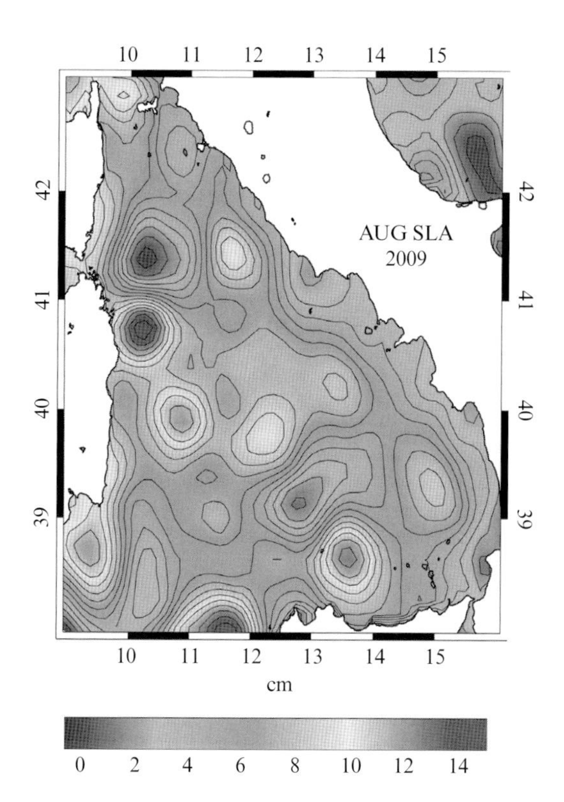

Figure 5.8 SLA of August 2009. For color detail, please see color plate section.

agreement over the whole basin; note, in particular, how the three-lobed structure of the Bonifacio system is reproduced in both.

The overall comparison is quite comforting, considering that: 1) we are comparing a geostrophic reconstruction of the flow with model outputs that contain an important wind-driven component; 2) the definition of the MDT may be problematic in a basin as the TYS, in which the circulation has a strong seasonal variability. Some of the differences noted above, and in particular those relative to the summer circulation, could therefore be due to unrealistic features in the MDT reconstruction, which locally distort the information present in the SLA (see *Iacono et al.* [2012] for a more detailed discussion of this issue).

5.3.2. The Intermediate Circulation

Traditional knowledge on the motion of the intermediate water masses (Levantine Intermediate Water, LIW hereafter) in the Mediterranean Sea is mostly based on hydrographic observations [*Millot*, 1999; *Millot and Taupier Letage*, 2005], with a few direct current measurements [*Astraldi et al.*, 1999], mainly made with moorings in the major channels and straits (Corsica, Sicily, Otranto, and Gibraltar). Only recently *Menna and Poulain* [2010] have proposed a description of the intermediate circulation based on the analysis of drifter data. These studies, however, have not been looking in detail at the TYS, and consequently have not added much to *Krivosheya*'s [1983] picture of a large-scale intermediate circulation with the same cyclonic structures of the surface layer in the western TYS (Bonifacio gyre and cyclonic region between Sardinia and Corsica), and a cyclonic cell, wider in winter, in the central and eastern parts of the basin. In a recent work by *Vetrano et al.* [2010], focusing on the 2004 TYS spring circulation, evidence is found of a global cyclonic cell, even though the dynamics of the intermediate layer appear dominated by mesoscale eddies, some of which related to corresponding surface structures.

Generally speaking, it can be said that the LIW patterns inside the basin are not well known, and that even less is known about how such patterns could vary through the year. For example, it is not clear whether the LIW exiting from the Sardinia Channel is LIW that has circulated around the whole basin, or a large part of it (this would seem to be the case for the spring situation analyzed in *Vetrano et al.* [2010]) or LIW that has reached the Channel through a more direct northwestern path, such as the one found in the simulations of *Pierini and Rubino* [2001].

The TYREM simulations provide interesting information about these issues. Figure 5.9, showing the average circulation at 400 m of depth for February (a), April (b), August (c), and September (d) allows us to depict four different scenarios. In the February map (Figure 5.9a), which is similar to that of January, the vigorous LIW jet outflowing from the Sicily Strait veers to the east and flows along the shelf all around the basin, forming a global cyclonic cell. Part of this stream exits through the Corsica Channel, while another branch borders the Sardinia coast and finally leaves through the Sardinia Channel. In the central part of the basin there are many eddies, some of which have a surface counterpart (1,2,3,4,5). It can be noted that before reaching the Aeolian islands the LIW inflow appears to bifurcate, with a minor stream that moves to the north, and then to the west, meandering around some of the vortices, and eventually reaching the southward stream along the Sardinia coast. This stream forms a smaller cyclonic cell, embedded in the global one, which is also present in March (not shown), when the global cyclonic cell breaks down.

In April (Figure 5.9b), a new situation settles that persists until June (not shown): there is no LIW stream on the Italian shelf, and most of the inflow goes northward, along 12°E, until reaching the southern border of the Bonifacio anticyclone, and then recirculates cyclonically toward the Sardinia Channel. A small part of the inflow appears to veer directly to the east and reaches the channel. In the August map (Figure 5.9c), the inflow appears weaker. To the north, the signature of the Bonifacio dipole is clearly seen, while in the eastern part of the basin an

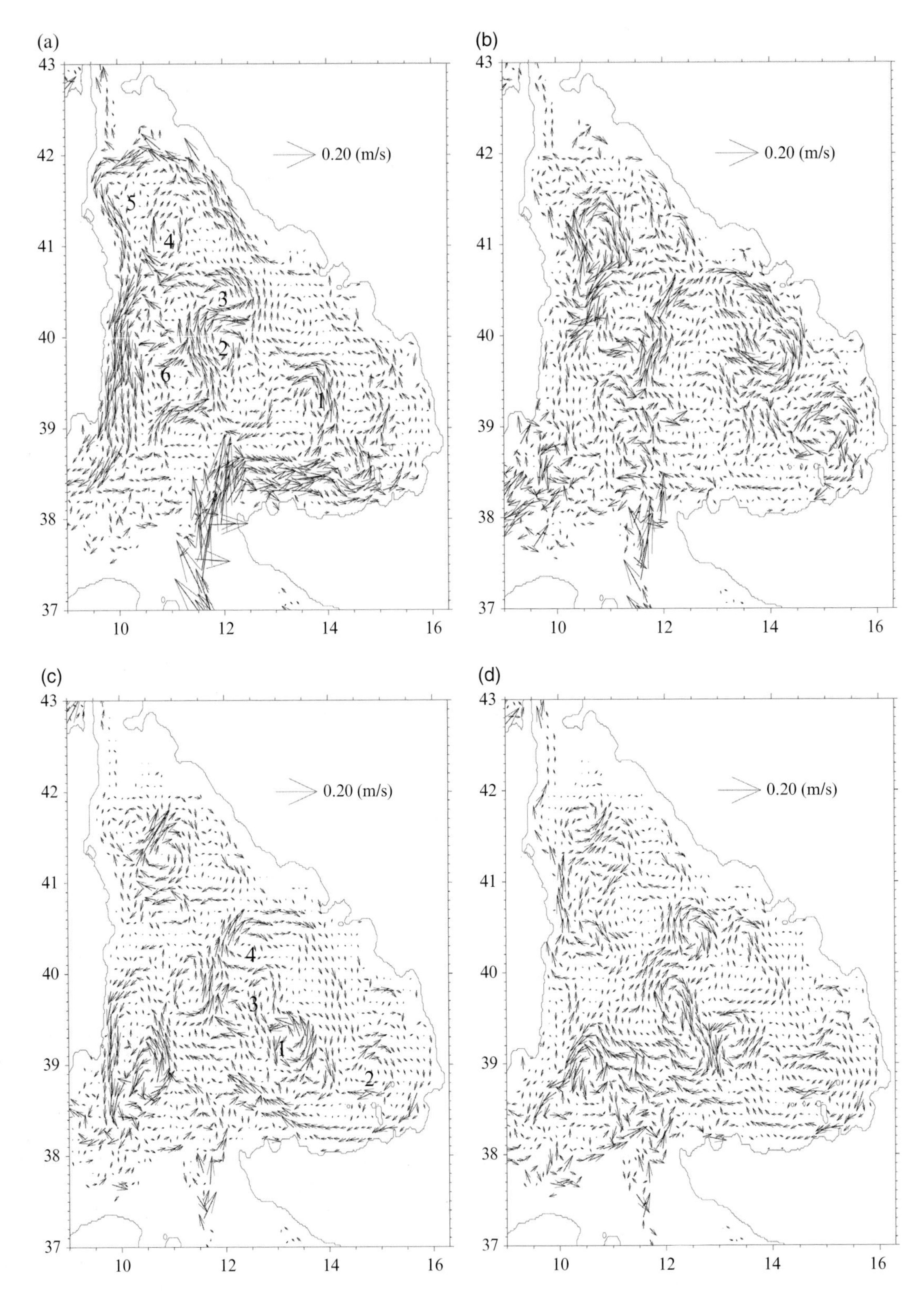

Figure 5.9 Velocity field at 400 m (arrows are drawn every 6 grid points) for (a) February, (b) April, (c) August, and (d) September.

anticyclonic circulation has formed, which mirrors the cell observed in the surface layer. The cell encompasses several eddies (1 to 4) that have their counterparts in the surface layer. The big anticyclonic cell is also present in the September map (Figure 5.9d). Now, however, the LIW inflow borders this cell in the northwestern direction, rounding a cyclone located at 39°N 10.6°E, and finally exiting through the Sardinia Channel. This makes a quite direct path of the basin, which is also found in October, after the breakdown of the anticyclonic cell. In November and December, the LIW inflow to the east forms again, and we recover the winter configuration.

Thus, in 2009, except perhaps in April, a "direct" westward branch of the LIW inflow is never really present. However, from midspring to the beginning of autumn there is a portion of the inflow that veers toward north-northwest, gets trapped in the cyclonic recirculations present between Sardinia and Sicily, and then finds its way out of the Sardinia Channel. This pattern becomes more direct from August to October, and further evidence of this will be given below.

The other point to be stressed is the presence of a strong barotropic component of the circulation. All the large-scale patterns present in the surface circulation—the winter global cyclonic cell, the summer anticyclonic cell in the eastern TYS, and the spring northward flow—are also found in the intermediate layer. Likewise, the main vortices found in the different seasons appear in both layers. To further illustrate this point, we show in Figure 5.10 the February barotropic stream function, in which the main patterns and structures are visible.

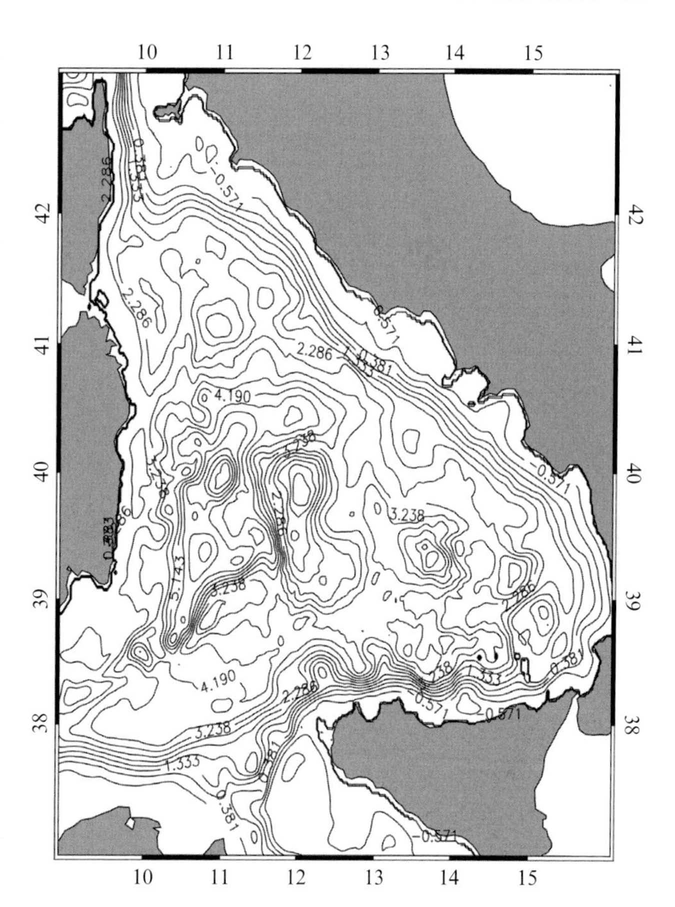

Figure 5.10 Barotropic stream function of February 2009.

5.4. WATER MASSES AND TRANSPORTS

We now look at some monthly maps of salinity minimum (S_{min}) and maximum (S_{max}), which provide useful information on the water masses distribution [for a given month, the average salinity field is computed, and then the S_{min} (S_{max}) map is constructed, by associating to each grid point of the horizontal domain the salinity minimum (maximum) along the water column]. The surface water mass of the TYS, characterized by low salinity values, is the result of the AW inflow through the Sardinia Channel. The February map of S_{min} (Figure 5.11a) is quite similar to the January/February 1958 map of *Astraldi and Gasparini* [1994]. The map clearly shows the inflowing AW stream, with the lowest values of S_{min}, forming a sharp zonal front above 38°N, in the region between Sardinia and Sicily. Inside the TYS, the values of S_{min} progressively increase along the Sicilian and Italian coasts, indicating the mixing of the AW with preexistent surface waters. High values of S_{min} are found in the western part of the basin, with maxima localized in the area of the Bonifacio cyclone, probably due to intense vertical mixing with underlying saltier waters. A meridional transect of salinity along 10.5°E (Figure 5.12) shows this mixing in the cyclonic area between Sardinia and Sicily, and, even stronger, in the area of the Bonifacio cyclone (around 41.5°N), where the mixed layer is deeper than 100 m.

The August S_{min} map (Figure 5.11b) is also similar to the corresponding map of *Astraldi and Gasparini* [1994]. In most of the eastern basin, S_{min} is low with values from 37.8 to 38, while maxima are found again in the western TYS, in correspondence of the main cyclonic structures. However, a pretty strong zonal gradient is present between the two regions, between 11°E and 12°E. This gradient, evidence of which is also found in *Astraldi and Gasparini* [1994], may be the result of the northward inflow of AW occurring in spring, which progressively fills the central part of the basin with fresher water, and of the formation of the summer anticyclonic cell, which then induces horizontal mixing and almost homogenization in the eastern part of the basin.

Figure 5.13 shows the map of S_{max}, in February, August, September, and October, respectively. In all maps, the LIW inflow is evident. In February (Figure 5.13a),

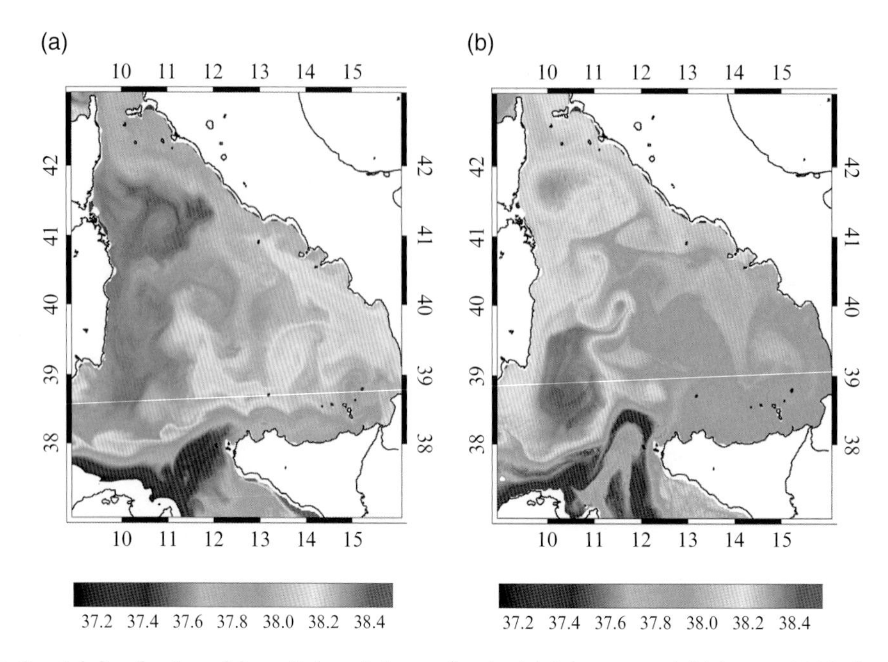

Figure 5.11 Spatial distribution of the salinity minimum S_{min} for (a) February and (b) August 2009. For color detail, please see color plate section.

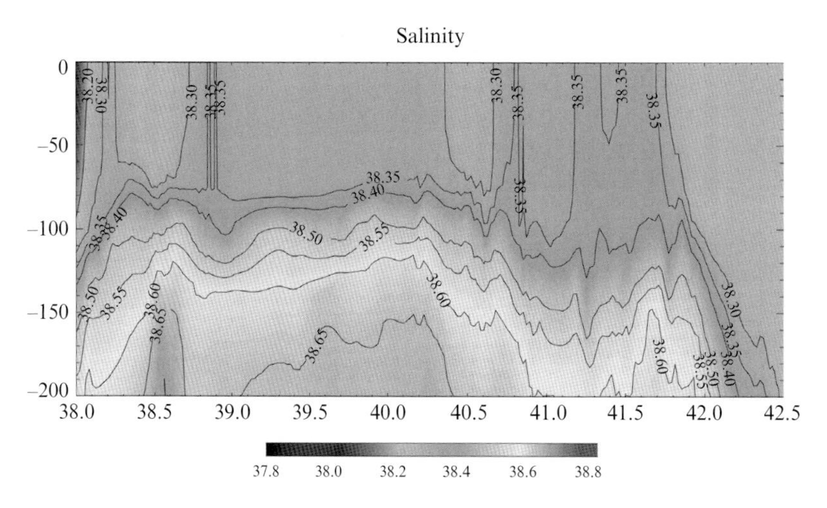

Figure 5.12 Meridional transect of salinity along 10.5°E for February 2009. For color detail, please see color plate section.

maxima of S_{max} are concentrated in the inflow area and to the east of it, along the Sicilian coast, while minima are found in the northern part of the basin. The summer situation (Figure 5.13b, 5.13c) is different. Maxima of S_{max} are always in the LIW inflow area, but now high values are spread over a wider area in the central and eastern parts of the basin. In the August and September maps, values around 38.8 form a tongue that extends toward northwest, and in October (Figure 5.13d), although slightly diluted, reaches the Sardinia Channel, consistent with the description of the LIW circulation previously given.

Finally, we discuss the water transports through the Corsica Channel. This region is a key place for the western Mediterranean, since it connects the TYS with the Liguro-Provencal basin, where in winter intense atmospheric processes induce deep-water formation. Analysis of the long series of current measurements made since mid1980 [*Astraldi et al.*, 1999; *Gasparini et al.*, 2008] shows that the flow through the channel is almost always northward and involves both AW and LIW moving together coherently. *Astraldi and Gasparini* [1994] estimated that this flow consists of about 80% of AW and about 20% of LIW. The total transport is stronger from autumn to spring, with winter maxima that may exceed 2 Sv, and much weaker in summer.

Figure 5.14 shows the transports of AW and LIW from the model, calculated assuming that the two water masses in the channel are separated by the σ = 28.9 reference

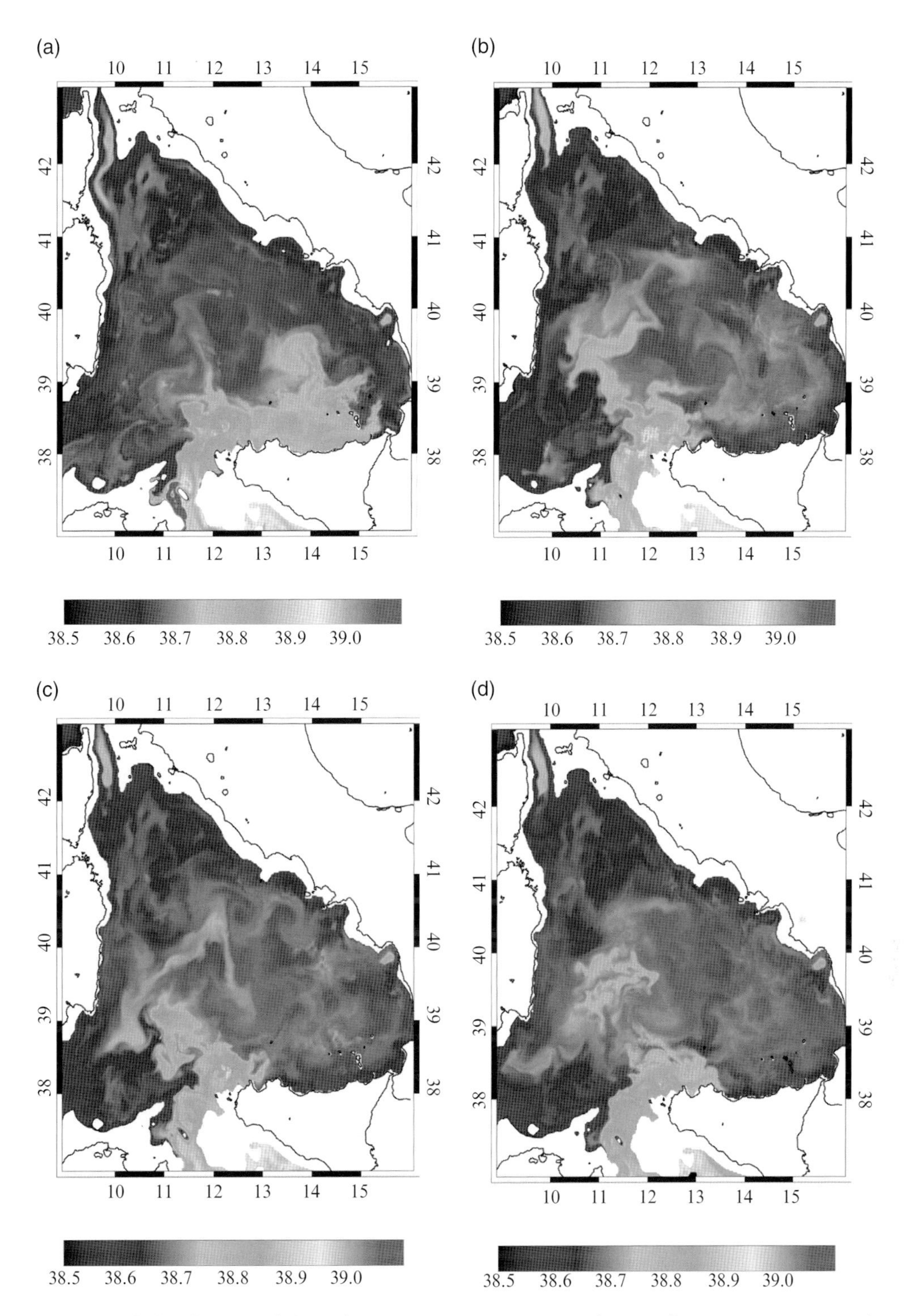

Figure 5.13 Spatial distribution of the salinity maximum S_{max}: (a) February, (b) August, (c) September, and (d) October. The last three panels illustrate the formation of a LIW tongue that moves toward northwest, gets trapped in the cyclonic circulation present in the area, and finally reaches the Sardinia Channel. For color detail, please see color plate section.

isopycnal surface [see *Vetrano et al.*, 2010], which roughly lies at 200 m of depth. For this specific year, the maximum transports of AW (1.54 Sv) and LIW (0.6 Sv) occur in February and December, respectively, while minimum values are found in August (0.33 Sv for the AW, and 0.21 Sv for the LIW). The figure also shows that the ratio between the LIW and AW transports varies throughout the year. This ratio is about 1/3 in winter, but increases in

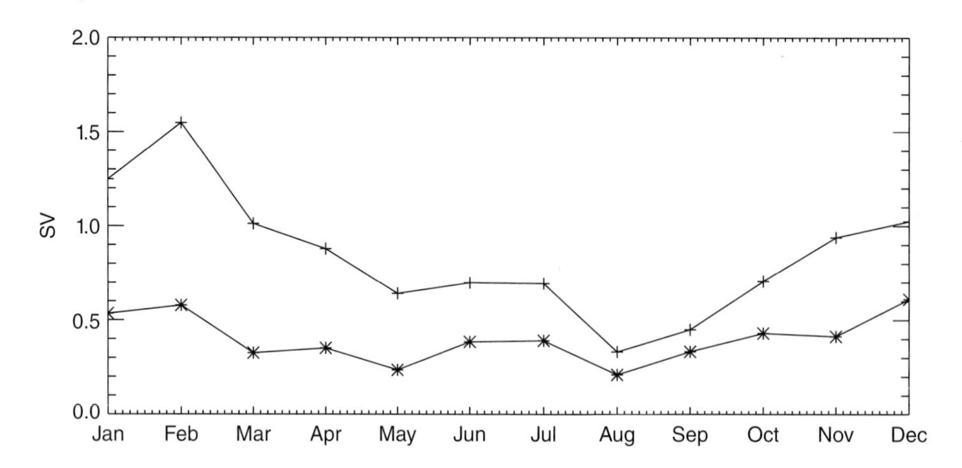

Figure 5.14 Transports at the Corsica Channel, (+) AW, (*) LIW.

summer, and has even larger values in September and October, the months in which the AW outflow is minimum.

5.5. SUMMARY

An assessment of the TYS 2009 surface and intermediate circulation has been presented, based on the results of the high-resolution operational ocean model TYREM. The main results can be summarized as follows.

1. The analysis of the model outputs shows that the main features of the surface circulation are similar to those highlighted in the recent work by *Iacono et al.* [2012], where the seasonal variability of the average circulation has been assessed through the analysis of altimeter data. In this respect, the year 2009 appears to be quite typical. One of the main features of the winter-spring surface circulation evidenced in *Iacono et al.* [2012], that is, the existence of a global pattern formed by a meandering AW stream with on the shore (offshore) side anticyclones (cyclones) nested in the anticyclonic (cyclonic) meanders is also evident in the February-March circulation maps of Figure 5.6. However, the figure also shows that the number of meanders, and of associated structures, may be larger than that suggested by the analysis of the altimeter dataset; for example, the anticyclonic regions to the north of Sicily and off Naples appear each to be formed by two distinct cores. As noted in *Iacono et al.* [2012], the winter-spring pattern suggests an instability of the AW stream, which fully develops toward the end of the winter, when the cyclonic wind-stress cell weakens over the basin. The model outputs support this picture, but further dedicated numerical work will be needed to investigate the nature of this instability.
2. In spring, the global cyclonic cell breaks down, but the AW stream coming from the Sardinia Channel now deviates northward when entering the TYS, forming a narrower cell in the western part of the basin. In April and May, this yields a direct path for the AW from the south to the Corsica Channel, which has apparently not been discussed before in the literature.

   In summer, consistent with what is known, there is no clear inflow from the south, and little outflow from the Corsica Channel. The AW appears mostly to recirculate in the wide cyclonic area between Sardinia and Sicily, while the eastern TYS is occupied by a wide anticyclonic cell, formed by a stream that meanders around several well-defined eddies, some of which result from the evolution of structures present in winter and spring. This anticyclonic cell is mostly driven by the interaction between the local wind stress (that is typically anticyclonic in summer) and the basin topography. However, since the western boundary of the anticyclonic cell is adjacent to the eastward boundary of the cyclonic area, there might be exchanges of momentum and vorticity between them.
3. The intermediate circulation has also been found to display marked seasonal variability. It appears closely related to that of the upper layer, with the main streams and structures present in both layers in all seasons. The numerical simulations have also shed some light on the long-standing issue concerning the existence of a direct path of the LIW from the Sicily Strait to the Sardinia Channel. The results (see, for example, Figure 5.13) indicate that although a direct westward path is not present, in summer and autumn the reinforcement of the cyclonic cell between Sardinia an Sicily favors the removal of the LIW resident in the area to the north of the Sicily Strait, which is entrained into a relatively short cyclonic path toward the Sardinia Channel and, ultimately, toward the Algero-Provencal basin.

APPENDIX

Methods Used to Derive Surface Heat Fluxes

The net surface heat flux at the air-sea interface has been estimated using a combination of ECMWF analysis (sea level atmospheric pressure, air temperature at 2 m, dew point temperature, cloud cover, zonal and meridional wind components) and MFS system data. Each component of the net surface heat flux (Q_{tot}) has been estimated using bulk formulae based on the knowledge available through meteorological parameters. More in detail:

Reed's [1977] relation has been used to estimate mean daily insolation including cloud attenuation. Then the shortwave radiation budget including the effect of the sea surface albedo is given by

$$Q_s = \alpha * Q_0 * (1 - 0.637 * C + 0.0019 * h) * (1 - A),$$

where Q_0 is the clear sky radiation, C the mean daily cloud cover, h the solar elevation at noon, A the sea surface albedo, and α the mean atmospheric transmittance.

The latent heat flux has been computed using

$$Q_e = \rho * C_e * w * (q - q_s) * L,$$

where ρ is the air density (corrected by the actual value of the atmospheric pressure), $C_e = 1.14 \times 10^{-3}$ the Dalton number [*Gilman and Garret*, 1994], w the wind intensity, q the specific humidity, q_s the saturation humidity, and $L = 2.456*10^6$ J Kg^{-1} the latent heat of water. The specific humidity and saturation humidity have been computed from water vapor pressure and saturated water vapor pressure [*Gill*, 1982, pp. 40–41] that in turn have been computed from the dew point temperature (derived from ECMWF analysis) and sea surface temperature, respectively, including atmospheric pressure correction and salt correction [*Gill*, 1982, p. 606].

The sensible heat flux has been computed using

$$Q_h = \rho * C_p * C_h * (T_s - T_a) * w,$$

where $C_p = 1005$ JKg^{-1}K^{-1} is the specific heat of air, T_s the sea surface temperature, and T_a the air temperature. The Stanton number C_h has been estimated as a function of the wind intensity

$$C_h = \{0.720 + [0.0175 * w * (T_s - T_a)]\} * 10^{-3} \quad \text{if } w < 8\,\text{ms}^{-1}$$

$$C_h = \{1.0 + [0.0015 * w * (T_s - T_a)]\} * 10^{-3} \quad \text{if } w \geq 8\,\text{ms}^{-1}$$

Finally, the infrared radiation budget has been computed using the formula of *Bignami et al.* [1995]:

$$Q_e = \in * \sigma * T_s^4 - [\sigma * T_a^4 * (0.653 + 0.00535 * e)] * (1 + 0.1762 * C),$$

where *e* is the water vapor pressure, $\in$ the surface emissivity, and σ the Stefan-Boltzman constant.

REFERENCES

Artale, V., M. Astraldi, G. Buffoni, and G. P. Gasparini (1994), Seasonal variability of gyre-scale circulation in the northern Tyrrhenian Sea, *J. of Geoph. Res.*, *99*, 14127–37.

Astraldi, M., and G. P. Gasparini (1994), The seasonal characteristics of the circulation in the Tyrrhenian Sea, in *Seasonal and Interannual Variability of the Western Mediterranean Sea, Coastal and Estuarine Studies*, Vol. *46*, 115–134, American Geophysical Union (AGU).

Astraldi, M., S. Balopoulos, J. Candela, J. Font, M. Gacic, G. P. Gasparini, B. Manca, A. Theocaris, and J. Tintorè (1999), The role of straits and channels in understanding the characteristics of Mediterranean circulation, *Progress in Oceanography*, *44*, 65–108.

Bignami, F, S. Marullo, R. Santoleri, and M. E. Schiano (1995), Long-wave radiation budget in the Mediterranean Sea, *J. of Geoph. Res*, *100*, 2501–2514.

Blumberg, A., and G. L. Mellor (1987), A description of a three-dimensional coastal ocean circulation model, in *Three Dimensional Coastal Ocean Models, Coastal and Estuarine Sciences*, 4, 1–16, American Geophysical Union (AGU), Washington, D.C.

Budillon, G, G. P. Gasparini, and K. Shroeder (2009), Persistence of an eddy signature in the central Tyrrhenian basin, *Deep Sea Res. II*, *56*, 713–724.

Flather, R. A. (1976), A tidal model of the northwest European continental shelf, *Memo. Soc. Roy. Sci. Liege*, *6* (10), 141–164.

Gasparini, G., K. Schroeder, and S. Sparnocchia (2008), Straits and channels as key regions of an integrated marine observatory of the Mediterranean: Our experience on their long-term monitoring, in *Toward an Integrated System of Mediterranean Marine Observatories*, Vol. *34*, 75–79, of CIESM Workshop Monographs, CIESM.

Gill, A. (1982), *Atmosphere–ocean Dynamics*, Academic Press, San Diego, Calif.

Gilman, C., and C. Garret (1994), Heat flux parameterization for the Mediterranean Sea: The role of atmospheric aerosol and constraints from the water budget, *J. of Geoph. Res*, *99*, 5119–5134.

Haney, R. L. (1991), On the pressure gradient force over steep topography in sigma coordinate ocean models, *J. of Phys. Oceanogr.*, *21*, 4, 610–619.

Herbaut, C., L. Mortier, and M. Crepon (1996), A sensitivity study of the general circulation of the western Mediterranean Sea, I: The response to density forcing through the strait, *J. of Phys. Oceanogr.*, *96* (1), 65–84.

Iacono, R, E. Napolitano, S. Marullo, V. Artale, and A. Vetrano (2012), Seasonal variability of the Tyrrhenian Sea surface circulation as assessed by altimeter data, *Submitted.*

Kantha, L. H., and C. Clayson (1994), An improved mixed layer model for geophysical applications, *J. of Geoph. Res*, *99*, doi:10.1029/94JC02257.

Korres, G., N. Pinardi, and A. Lascaratos (2000), The ocean response to low-frequency interannual atmospheric variability in the Mediterranean Sea, Part I: Sensitivity experiments and energy analysis, *J. of Climate*, *13*, 705–773.

Korres, G. N., and A. Lascaratos (2003), A one-way nested eddy resolving model of the Aegean and Levantine basin: Implementation and climatological runs, *Annales Geoph.*, *21*, 205–220.

Krivosheya, V. G. (1983), Water circulation and structure in the Tyrrhenian Sea, *Oceanology*, *23*, 166–171.

Lermusiaux, P. F. J., and A. R. Robinson (2001), Features of dominant mesoscale variability, circulation patterns and dynamics in the Strait of Sicily, *Deep Sea Research Part I*, *48*, 1953–1997.

Marchesiello, P., J. C. Mc Williams, and A. Shchepetkin (2001), Open boundary conditions for long-term integration of regional oceanic models, *Oce. Modell.*, *3* (1–2), 1–20.

Marullo, S., R. Santoleri, and F. Bignami (1994), The surface characteristics of the Tyrrhenian Sea: Historical satellite data analysis, in *Seasonal and Interannual Variability of the Western Mediterranean Sea, Coastal and Estuarine Stud.*, *46*, 135–154, ed. P. E. La Violette, AGU, Washington, D.C.

Mellor, G. L. (2004), User's guide for a three-dimensional, primitive equation numerical ocean model, *Int. Rep. Program in Atmos. Ocean. Sci.*, 11–35, Princeton Univ., Princeton N.J.

Mellor, G. L., and T. Yamada (1982), Development of turbulent closure model for geophysical fluid problem, *Rew. in Geoph. and Space Phys.*, *20*, 4, 851–875.

Mellor, G. L., T. Ezer, and L.-Y. Oey (1994), The pressure gradient conundrum of sigma coordinate ocean models, *J. of Atmosph. and Oce. Tech.*, 1126–1134.

Menna, M., and P. M. Poulain (2010), Mediterranean intermediate circulation estimated from Argo data in 2003–2010, *Ocean Science*, *6*, 331–343.

Millot, C. (1999), Circulation in the western Mediterranean Sea, *J. of Mar. Sys.*, *20*, 423–442.

Millot C., and I. Taupier Letage (2005), Additional evidence of LIW entrainment across the Algerian subbasin by mesoscale eddies and not by a permanent westward flow, *Progr. in Oceanogr.*, *66* (2–4), 231–250.

Oddo, P., N. Pinardi, M. Zavatarelli, and A. Coluccelli (2006), The Adriatic Basin forecasting system, *Acta Adriatica*, *47*, 169–184.

Pierini, S., and A. Rubino (2001), Modelling the oceanic circulation in the area of the Strait of Sicily: The remotely forced dynamics, *J. of Phys. Oceanogr.*, *31*, 6, 1397–1412.

Pierini, S., and A. Simioli (1998), A wind-driven circulation model for the Tyrrhenian Sea area, *J. of Mar. Sys.*, *18*, 811–825.

Reed, R. K. (1977), On estimating insolation over the ocean, *J. of Phys. Oceanogr.*, *7*, 482–485.

Rinaldi, E., B. Buongiorno Nardelli, E. Zambianchi, R. Santoleri, and P. M. Poulain (2010), Lagrangian and Eulerian observations of the surface circulation in the Tyrrhenian Sea, *J. of Geoph. Res.*, *115*, C04024.

Rio, M. H., P. M. Poulain, A. Pasqual, E. Mauri, G. Larnicol, and R. Santoleri (2007), A mean dynamic topography of the Mediterranean Sea computed from altimetric data, in-situ measurements and a general circulation model, *J. of Mar. Sys.*, *65*, 484–508.

Roussenov, V., E. Stanev, V. Artale, and N. Pinardi (1995), A seasonal model of the Mediterranean Sea general circulation, *J. of Geoph. Res.*, *100*, 13-515–13-538.

Samgorinsky, J. (1963), General circulation experiments with primitive equations, I: The basic experiment, *Mon. Weath. Rev.*, *91*, 99–164.

Tonani, M., N. Pinardi, S. Dobricic, I. Pujol, and C. Fratianni (2008), A high-resolution free surface model of the Mediterranean Sea, *Ocean Science*, *4*, 1–14.

Vetrano, A., E. Napolitano, R. Iacono, K. Shroeder, and G. P. Gasparini (2010), Tyrrhenian Sea circulation and water mass fluxes in spring 2004, observations and model results, *J. of Geoph. Res.*, *115*, C06023.

Vetrano, A., G. P. Gasparini, R. Molcard, and M. Astraldi (2004), Water flux estimates in the central Mediterranean Sea from an inverse box model, *J. of Geoph. Res*, *109*, doi: 10.1029/2003JC001903.

Zavatarelli, M., and N. Pinardi (2003), The Adriatic Sea modeling system: A nested approach, *Ann. Geophys.*, *21*: 345–364.

Zavatarelli, M., and G. L. Mellor (1995), A numerical study of the Mediterranean Sea circulation, *J. of Phys. Oceanogr.*, *25*, 1384–1414.

6

The Eastern Mediterranean Transient: Evidence for Similar Events Previously?

Wolfgang Roether[1], Birgit Klein[2], and Dagmar Hainbucher[3]

6.1. INTRODUCTION

Of the two potential source regions of Eastern Mediterranean (EMed) deep waters, that is, the Adriatic and the Aegean Seas, the former was in the past generally identified as the dominant one [*Pollack*, 1951; *Wüst*, 1961; *Hopkins*, 1978; *Malanotte-Rizzoli and Hecht*, 1988]. However in the 1990s, the role of the two regions became reversed. The Aegean Sea, which in 1987 had only formed a shallower water mass (Cretan Intermediate Water, CIW) [*Schlitzer et al.*, 1991; *Roether et al.*, 1999], began to discharge unusually dense waters [*Theocharis et al.*, 1992, 1999; *Roether et al.*, 1996], inducing what became known as the Eastern Mediterranean Transient (EMT). There is now a consensus that the EMT was caused by a combination of high-salinity waters intruding into the Aegean and the two particularly strong winters of 1991–1992 and 1992–1993 [*Josey et al.*, 2003; *Gačić et al.*, 2013].

As already noted by *Nielsen* [1912], the Aegean Sea waters are characterized by temperature and salinity exceeding the values of the EMed at large, which makes any Aegean signatures well apparent. The actual density increase in the Aegean, ongoing since the late 1980s [*Theocharis et al.*, 1999; *Zervakis et al.*, 2000; *Beuvier et al.* 2010], was primarily a matter of enhanced salinity [*Roether et al.*, 2007], and the event produced profound changes in the EMed's hydrography, specifically a salinity increase throughout the deep waters and pronounced temperature-salinity (T-S) inversions [*Roether et al.*, 2007]. After peaking in 1993, the Aegean source strength gradually declined, and by about 2002, the Adriatic took over again [*Rubino and Hainbucher*, 2007]. But the changes in the deep-water hydrography persisted even though strongly modified by circulation and mixing.

It has been argued that extensive Aegean dense-water production might be recurrent in time [*Josey*, 2003; *Beuvier*, 2010; *Gačić et al.*, 2011, among others], although *Gačić et al.* [2013] did not find firm evidence of that for the past 60 years. To resolve the issue is the topic of the present work.

The approach is to search for specific signatures in the historic hydrographic observations, which date back to 1910. To deal with the problem that up into the 1950s the data not only are of limited precision but also have gaps of about 20 years, we take advantage of the fact that the evolution of the actual EMT is rather well documented over a similar time span. This allows us to deduce plausible EMT-type signatures at the location of a historic station should such an event have occurred. The following begins with outlining the characteristics of the current EMT. Thereafter we select suitable hydrographic observations among the available historic data and compare these with signatures expected from the evolution of the actual EMT.

6.2. LESSONS FROM THE ACTUAL EMT

Figure 6.1 presents the history of dense-water release from the Aegean Sea. It is based on salinity excesses in 1995 [*Roether et al.*, 1996] and in 1991 relative to 1987 and on the excess in the tracer chlorofluorocarbon 12 (CFC-12) in 2001 relative to 1995, below 1000 or 1200 m depth. The onset of dense Aegean outflow in early 1990,

[1] *Institut für Umweltphysik, Univ. Bremen, Bremen, Germany*
[2] *Bundesanstalt für Seeschifffahrt und Hydrographie, Hamburg, Germany*
[3] *Institut für Meereskunde, ZMAW, Universität Hamburg, Hamburg, Germany*

The Mediterranean Sea: Temporal Variability and Spatial Patterns, Geophysical Monograph 202. First Edition.
Edited by Gian Luca Eusebi Borzelli, Miroslav Gačić, Piero Lionello, and Paola Malanotte-Rizzoli.

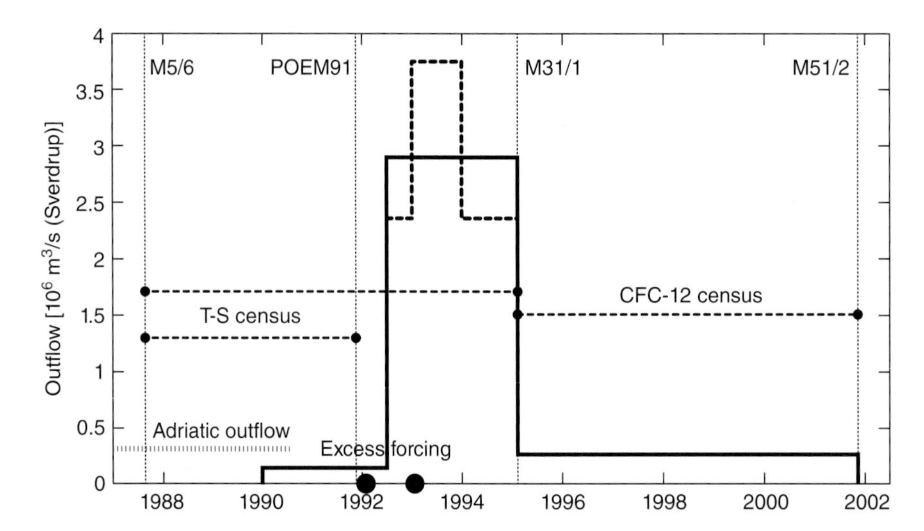

Figure 6.1 Histogram of dense-water outflow rates (Sverdrups, Sv) from the Aegean Sea, 1987–2001. The values are based on differential T-S and CFC-12 inventories (dashed lines) for the hydrographic surveys shown [*Roether et al.*, 2007]. The average pre-EMT outflow of the Adriatic (dotted; *Roether and Schlitzer* [1991]) and the periods of the two excessively strong winters (bullets on the time axis; *Josey* [2003]) are indicated. M5/6, M31/1, and M51/2 are cruises of the F.S. *Meteor*, and POEM91 is a multiship survey [*Malanotte-Rizzoli et al.*, 1999]. Year marks are 1 January of the year.

the beginning of its peak period in mid-1992, and the outflow maximum in 1993 (dashed) are based on hydrographic information [*Roether et al.*, 2007; see also *Gertman et al.*, 2006]. Only about 3% of the total was released before 1992. The bulk of the output occurred within 2.5 years (mid-1992 to late 1994). Whereas the Adriatic supply was rather constant in time, indicating a quasi-steady situation, that of the Aegean was peak-shaped, with the maximum (1993) exceeding the Adriatic supply about 10-fold. Integrating over the histogram, the total Aegean discharge becomes twice the complete volume of the Aegean Sea. Since even in 1993 large parts of the Aegean Sea had insufficient density to contribute to the deep outflow [e.g., *Velaoras and Lascaratos*, 2010], it is clear that the Aegean dense-water formation must have incorporated waters from outside the Aegean. Note that after 1995 the Aegean outflow began to be delivered into layers less than about 2500 m deep [*Theocharis et al.*, 2002]. Using an eddy-permitting model, *Beuvier et al.* [2010] obtain a temporal distribution of the outflow similar to that of Figure 6.1, but their rates and densities of the outflow are lower than those of *Roether et al.* [2007].

Figure 6.2 characterizes the related hydrographic changes. Presented are T-S diagrams in 1995, that is, immediately after the Aegean peak outflow ended, in 2001, and in 2011, 19 years after the onset of the peak outflow. Typical diagrams are shown for the central Levantine Sea (marked by triangles), the Hellenic Trench southeast (pentagrams) and west (diamonds) of Crete, and in the central Ionian (circles), together with an Ionian pre-EMT reference diagram (squares). Note that the latter diagram, which displays steady declines of temperature and salinity with depth, closely characterizes the entire EMed (see *Roether et al.* [2007].

In 1995 (Figure 6.2, left), the effect was still largely confined to the Levantine and the Hellenic Trench, where all stations showed pronounced T-S inversions (T-S minima at roughly 1500 m depth). These and the distinctly higher salinities and higher minimum temperatures make the EMT impact very prominently apparent. The central Ionian station, in contrast, only showed comparatively moderate increases in salinity.

By 2001 (Figure 6.2, center), the salinities in the Ionian had approached those of the other regions, and a moderate inversion was present. The diagrams of the eastern Hellenic Trench and the Levantine had become virtually identical. In 2011 (Figure 6.2, right), the Ionian Sea minimum temperature was about 0.07 °C higher than in 2001, the previous inversion was lost, but a new inversion toward the bottom had developed. The diagrams of the eastern Hellenic Trench and the Levantine have an additional inversion in the deepest waters. In the former region this inversion is strong and the original one has become less prominent. In the Levantine, in contrast, the original, upper inversion is still strong and the lower one quite moderate. The minimum temperature exceeds the pre-EMT temperature by about 0.3 °C, the T-S is much reduced, and the T-S relationship is exceptionally smooth.

In Figure 6.2 one notes a steady evolution of the T-S relationship between 1995 and 2011. A notable feature is that the temperature and salinity averages in the Levantine

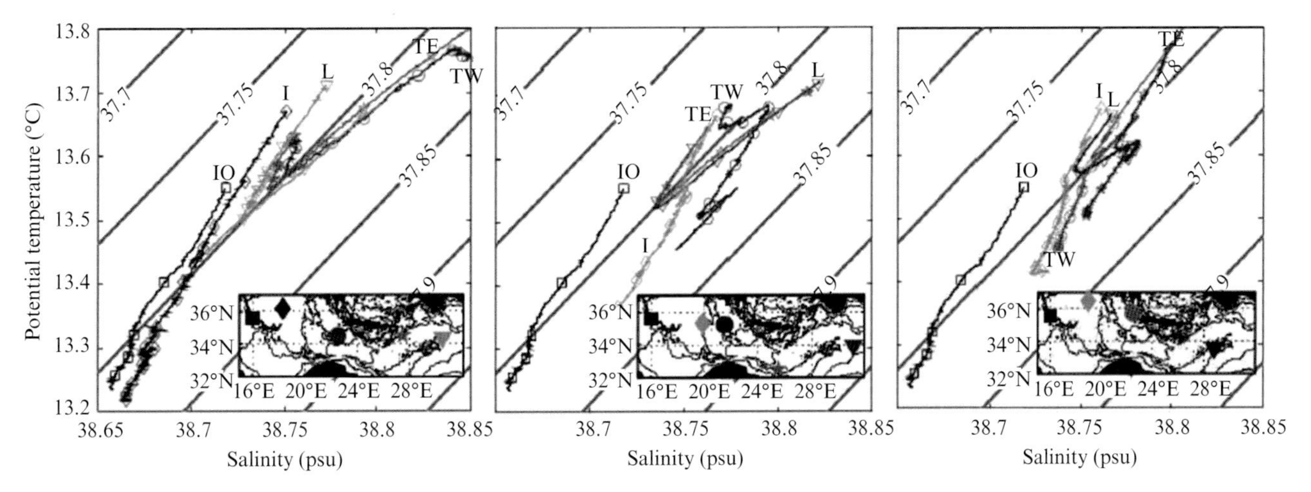

Figure 6.2 T-S diagrams below 800 m depth in 1995 (left), 2001 (center), and 2011 (right), both with a reference diagram for 1987 (M5/6, Sta. 779). The individual diagrams are denoted as IO = Ionian 1987 (square marks), I = central Ionian (circles), L = Levantine (triangles), TW = Hellenic Trench west (diamonds), and TE = Hellenic Trench east (pentagrams). For station positions see inset maps. Density isolines are σ_2 (= potential density anomaly referenced to 2000 dbar pressure, kg/m^3). The 2011 data were obtained on cruise M84/3 of F.S. *Meteor* in April 2011 [*Tanhua*, 2007].

below the upper inversion hardly changed in time (averages roughly 13.6 °C and 38.76). Our interpretation of this finding is that the temporal evolution was governed by vertical mixing, although the moderate return toward the bottom in 2011 is a result of influx from the Ionian [*Roether et al.*, 2007]. The near-bottom inversion at the central Ionian station is produced by recent Adriatic outflow below the inversion. That water has a maximum density of σ_2 = 37.82 kg/m^3 (σ_0 = 29.20), which exceeds the density observed in 1987 by 0.02 units and that in 1995 by 0.01 units. The high density in 2011 results primarily from the increased salinity.

A prerequisite for the strong Aegean outflow (Figure 6.1) were elevated densities in the deep waters of the southeastern Aegean (the eastern Cretan Sea). The data collection of *Roether et al.* [2007, see their Figures 9 and 7] and also *Theocharis et al.* [1999] in fact shows densities at the Kasos Strait sill depth (1000 m, see below) increasing from σ_2 = 37.76 kg/m^3 in 1987 to 37.93 kg/m^3 (σ_0 = 29.34) through the peak outflow period. The density difference across Kasos Strait in 1992, when the peak outflow began, was about 0.15 units. By 1998, the density outside had increased by about 0.025 units and that in the Cretan Sea decreased by 0.05 units (see Figure 6.5 below), so that the density difference across Kasos Strait was reduced to approximately 0.075 units, that is, half the maximum value. According to Figure 6.1, the 0.15 units density difference in 1992 supported outflow at about 2.5 Sv (1992) but the 0.075 units in 1998 supported far less than half the rate (a rough guess would be the 1995–2001 average, 0.3 Sv). It thus appears that the relationship between the density difference across the strait and the outflow is highly nonlinear.

Many of the features in Figure 6.2 are a consequence of the EMed's bathymetry (Figure 6.3). Dense waters formed in the Aegean Sea mainly exit through Kasos Strait, east of Crete (denoted K in the figure, ~1000 m sill) and a shallower exit is Antikithira Strait (A, ~560 m sill). Both exits connect to the Hellenic Trench (HT), which extends all along the Cretan Island Arc. To the south and west, the trench is bounded by the East Mediterranean Ridge (EMR). Over much of the Levantine Sea, the EMR represents a well-defined barrier close to 2500 m depth (the white isobath in Figure 6.3 is 2525 m). In the Cretan Passage the EMR is distinctly shallower; here also the trench is shallower, forming a sill 2560 m deep close to 24.5°E. Further along the Cretan Arc, the EMR gradually sinks to 3000 m depth, finally opening to the Ionian Sea fully near to 37°N. For deep waters, the Hellenic Trench forms the principal pathway between the Levantine and Ionian seas. The Herodotus Trough (Herodotus, ~2100 m sill) is a further deep connection between the Ionian and Levantine seas. The general flow direction was toward the Ionian Sea while the Aegean dense water source was dominant (Figure 6.1) and it was reverse in the classical situation with the Adriatic dominating, but this does not mean that the direction was uniform with depth [*Roether et al.*, 2007]. Otranto Strait (OS, ~800 m sill) connects to the Adriatic Sea.

Aegean dense water accumulating in the Hellenic Trench has two possibilities to expand further. One is southward overflow of the EMR, the other, to follow,

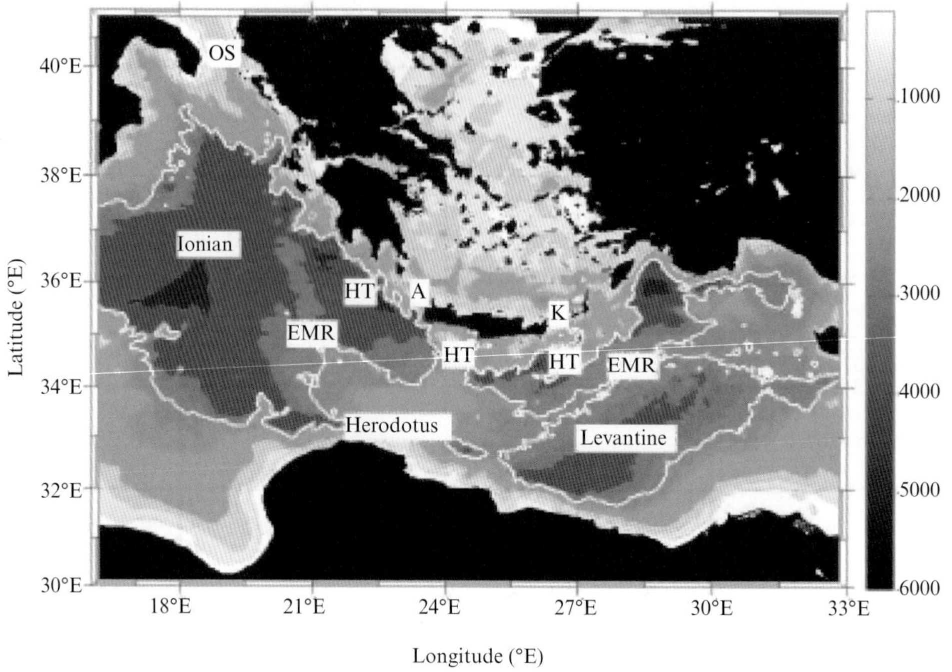

Figure 6.3 Bathymetry of the EMed. K = Kasos Strait, A = Antikithera Strait, HT = Hellenic Trench, extending all along the Cretan Island Arc, EMR = East Mediterranean Ridge, Herodotus = Herodotus Trough, OS = Otranto Strait. The white line shows the 2525 m isobath. The Ionian Sea is west of the Cretan Passage, the Levantine Sea is east of it.

steered by geostrophy, the Cretan Arc slopes toward the Ionian Sea. *Roether et al.* [2007] demonstrated absence of EMR overflow for the 1991 POEM-BC survey, in contrast to the subsequent peak outflow period (Figure 6.1) during which such overflow was most prominent. One must, thus, distinguish between excessive and more moderate rates of outflow, with only the former ones affecting the Levantine south of the EMR directly. For cases of more moderate Aegean outflow rates, the important consequence is that the Aegean-derived deep waters have to follow the Hellenic Trench to the Ionian before they can reach the deep waters of the Levantine Sea south of the EMR, so that there the impact will be slow.

The lessons for the evaluation of the Mediterranean's T-S history are the following: The most easily distinguishable signatures of an Aegean dense-water outflow event are to be found in the Levantine Sea south of the EMR, in the Hellenic Trench, and in the eastern Cretan Sea. Those in the Levantine are particularly long-lived, apparently for more than 20 years (Figure 6.2). However, an impact will be missing if events are incapable of inducing overflow of the EMR. Both the southeastern Cretan Sea and the Hellenic Trench will naturally record all events. Trench stations show a T-S inversion early on, while later only elevated minimum temperatures remain. Intruding Ionian waters toward the bottom limit that effect, naturally more so west of Crete than south of it. The situation in the Ionian Sea is more complex. This sea receives a delayed signature, as is evident in Figure 6.2 and also from the finding of *Roether et al.* [2007] that the Ionian salinity in 2001 had nearly completed its approach toward that of the Levantine Sea, but not yet in 1999 (data of cruise M44/4 in May 1999, not shown). Furthermore, deep-water addition from the Adriatic can vary in its T-S characteristics and in the depths to which it sinks, which complicates the interpretation of T-S diagrams in the Ionian.

6.3. HISTORIC T-S SIGNATURES IN COMPARISON WITH POTENTIAL EFFECTS OF EMT-TYPE EVENTS

Historic T and S data were extracted from the MEDAR data set [*MEDATLAS*, 2002] selecting deep hydrographic stations only. We excluded stations that showed excessive scatter or questionable salinity values. Some recent data from other sources (e.g., cruises of F. S. *Meteor* and recent Aegean cruises) were added. For the Levantine Sea south of the EMR, we selected stations typically about 5 years apart. However, prior to 1950, only stations about 20 years apart are available, but even such a time gap will, as was demonstrated above, allow one to detect an event.

We found that the data situation for the Hellenic Trench does not allow constructing a useful time series, while for the eastern Cretan Sea since 1948 a comparatively dense series of observations exists. The Levantine series begins with *Nielsen* [1912], who in the summer of

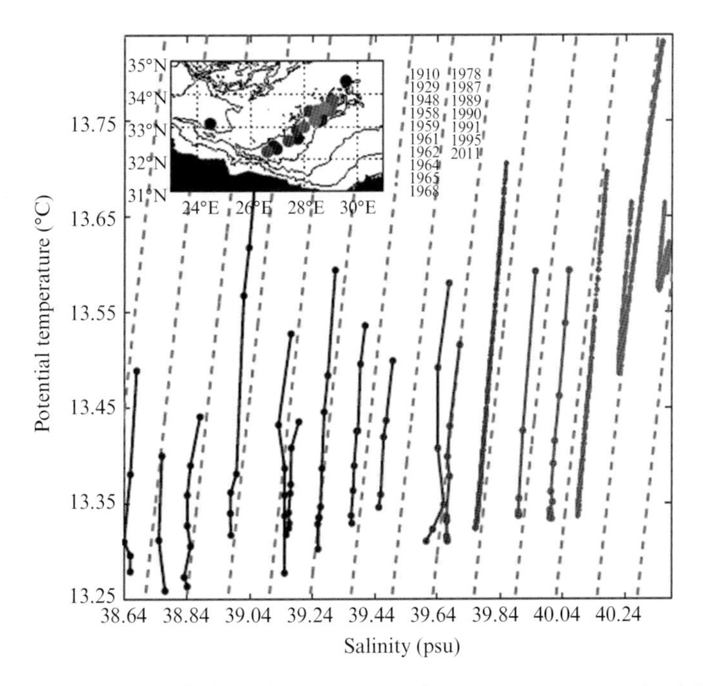

Figure 6.4 Family plot of T-S diagrams below about 800 m in the Levantine Sea south of the EMR, 1910–2011. Each diagram is shifted by 0.1 psu from the previous one. Density isolines are σ_2. For station positions, see inset map. The *Dana* ran only one station in the Levantine Sea, which is located somewhat away from the rest but still well within the region in question. The diagrams for 1995 and 2011 are those of Figure 6.2.

1910 made observations of salinity and temperature in the EMed. His and the other early data, which used chlorine titrations for salinity determination, are discussed by *Wüst* [1960]. He notes salinity uncertainties of $\Delta S = 0.03$ units for Nielsen and 0.02 units for the cruises of R.V. *Dana* (1928–1930), *Atlantis* (1948), and *Calypso* (1955). All later data sets have salinity precisions improved roughly 10-fold, but accuracy is sometimes uncertain. For temperature measurement, *Nielsen* [1912], using reversing thermometers, reports $\Delta T = 0.02\,°C$, which should be valid also for the other early cruises. It follows that detection of changes in temperature is far more sensitive and reliable than of changes in salinity. *In situ* measurement of temperature and salinity began in 1964. It achieved modern precision ($\Delta S = \pm\ 0.003$) when the Neil Brown CTDs entered the market in the early 1970s [*Brown*, 1991].

Figure 6.4 shows the selected T-S diagrams, 1910–2011, in the central and southern Levantine Sea in the form of a family plot. In the diagrams of 1995 and 2011, which are those of Figure 6.2, the actual EMT is very evident. All the earlier ones, in contrast, miss distinctly raised minimum temperatures and T-S inversions, which were identified above as the principal criteria to detect addition of Aegean dense outflow. The result is that we can exclude direct overflow of Aegean-derived dense waters across the EMR. Due to the long lifetime of the related signature, the exclusion dates back beyond 1910, at least to the year 1900. One notes that an EMT signature is not apparent up to 1991. The salinity gaps between the individual diagrams in Figure 6.4 vary, but we ascribe this to salinity measurement uncertainties rather than to true variations. The data show a slow increase in time of the minimum temperatures, in keeping with other work. Because the stations reach to different depths, the increase is subject to scatter.

Figure 6.5 presents a density versus time curve in the eastern Cretan Sea, averaged over 800–1200 m depth, bracketing the Kasos Strait sill depth (1000 m), 1948–2011. Density was chosen, because it is this quantity that governs the rate of outflow. If more than one station is available in any year, the standard deviation between the values for each station is shown to indicate the degree of uncertainty. For years with only one station, no error is given, but of course similar uncertainties apply. The uncertainty ranges seem to be maximal in the 1970s, when the largest error bar touches the values during the EMT peak outflow period 1992–1994. One notes however that the density rise during 1988–1992 toward the outflow maximum and the subsequent decline (cf. section 2) extend over several years. No such systematic structure is apparent in the 1970 data. The reason why Figure 6.5 is less decisive than Figure 6.4 in excluding events is that the densities in Figure 6.5 largely depend on salinity, measurement of which, as mentioned, has sometimes been problematic. In Figure 6.4, on the other

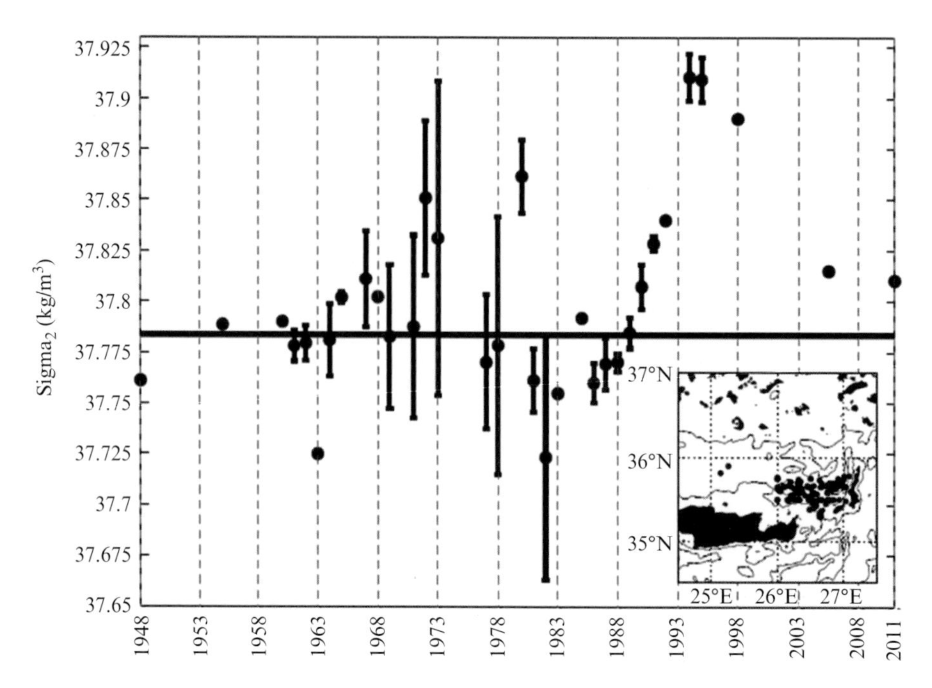

Figure 6.5 Density (σ_2, kg/m^3) versus time in the eastern Cretan Sea, 1948–2011. The values are averages, 800–1200 m, averaged over the year if more than one station is available. In that case (only) the standard deviation between the stations is shown as an error bar. The horizontal line shows the 1948–1988 density average. The values for 1998 and 2005–2006 originate from *Roether et al.* [2007] and *Vervatis et al.* [2011].

hand, one searches for the minimum temperature and the qualitative feature of a T-S inversion, which are far less prone to error.

6.4. DISCUSSION AND CONCLUSION

Our analysis allows us to exclude Aegean outflow events capable of inducing EMR overflow to fill the central and southern Levantine Sea since about 1900 (Figure 6.4). The situation for events incapable to induce such overflow to a significant degree (Figure 6.5) is somewhat more uncertain in view of the considerable error bars, even if a consistent evolution like observed for the EMT period (after 1988) is at no time apparent.

To assist our argumentation, we introduce the following additional observations: First, *Pollak* [1951, his Figure 3] reports T and S profiles in the Levantine and Ionian Seas for the 1948 *Atlantis* cruise, which are virtually identical below 800 m and show the steady decline of T_{pot} and S with depth that is typical for the Adriatic source (cf. Figure 6.2). This observation excludes events back to at least 1945. To bridge the data gap in Figure 6.5 between 1948 and 1955, Figure 6.6a shows T-S diagrams of R.V. *Calypso* in 1955 in the Hellenic Trench west of Crete. There is no evidence of T-S inversions and of an elevated minimum temperature; slightly higher minimum temperatures at two of the stations are ascribed to their lesser maximum depths (see caption). To deal with the period of the highest uncertainties, that is, the 1970s, Figure 6.6b shows T-S diagrams for 1978 [*Roether et al.*, 1992], with the station locations being similar to those of Figure 6.2. The diagrams virtually agree, except for a small west-to-east increase of bottom temperatures, which is to be ascribed to downward mixing of shallower waters on the way from the Adriatic source to the Levantine Sea. Figure 6.6b thus excludes any significant contribution of Aegean-derived deep waters since about 1970. We further mention the observation of a virtual absence of deep-water fine structure in the CTD recordings of the pre-EMT *Meteor* cruise (M5/6, 1987), in contrast to recordings in the EMT era. This observation excludes Aegean events for quite a number of years prior to 1987. In total, it follows that we can also exclude more moderate Aegean outflow events, that is, events incapable to produce noticeable EMR overflow, since at least 1945.

Aegean events producing EMR overflow would undoubtedly be considered as EMT-like events, but the limit between large and more moderate according to that criterion and the magnitude of the "moderate" events remain uncertain. To allow a rough guess, we estimated the Aegean-derived fractions residing in the deep waters (i.e., below 1000 m) of the Hellenic Trench and the Levantine and the Ionian seas (both excluding the trench part), right after the peak outflow period in early 1995.

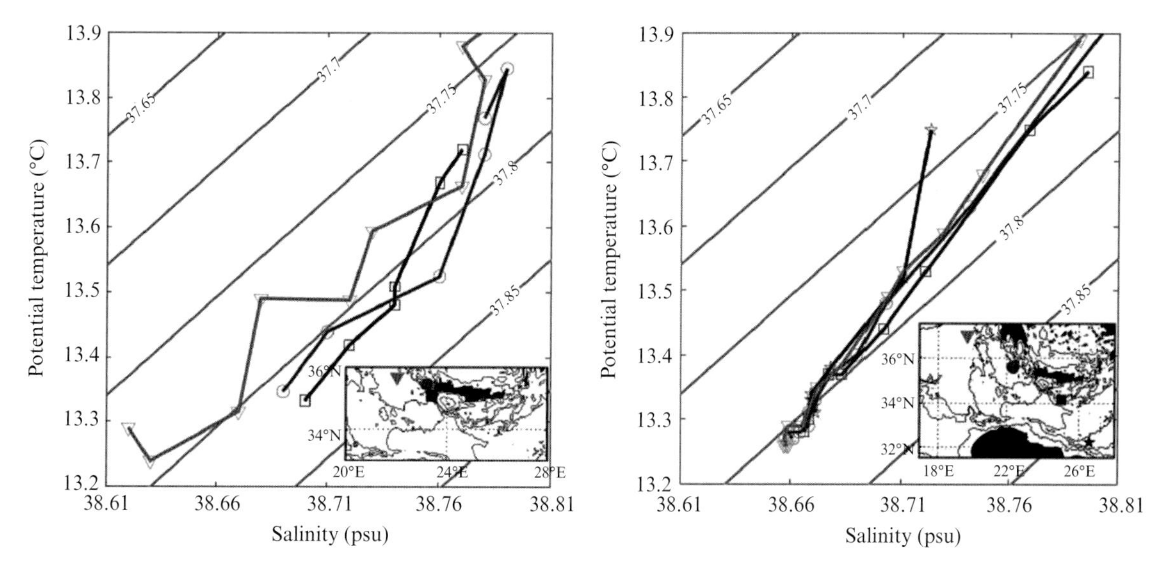

Figure 6.6 T-S diagrams of 1955 R.V. *Calypso* (left) and of 1978 *Meteor* cruise M50/3 (right). Shown are bottle data for M50/3 temperatures from a CTD controlled by reversing thermometers and salinities measured using an Autosal instrument. For positions see inset map. The *Calypso* stations, marked by open circles and squares, reach less than 3000 m deep while those in triangles are more than 4000 m deep. Density isolines as in Figure 6.2.

The relative volumes of the three parts were determined from estimated area and mean vertical extent. Using reported outflow properties [*Roether et al.*, 2007, Figure 9], T-S diagrams from the same work (Figure 7) and of Figure 6.2 above, we find average volume fractions of Aegean dense-water outflow of about 33% in the trench, 27% in the Levantine Sea, and 7% in the Ionian Sea (the latter two with the trench area excluded). The product of Aegean fractions and volumes amounts to more than 50% of the total in the central and southern Levantine, roughly 25% in the Hellenic Trench, and at most that much in the Ionian west of the trench. These estimates receive backing from the fact that the so determined total comes close to that found by the detailed T-S census up to 1995 that is indicated in Figure 6.1.

Considering that a smaller event would not reduce the Levantine part only, we expect that an event a factor of two smaller than the current EMT would still produce sufficient overflow to be detectable in the Levantine. The time curve in Figure 6.5, in combination with the supporting information presented above, should lower the limit for detecting an event by another factor of two, that is, to a magnitude of 25% of the current EMT. That magnitude should be well detectable, considering that the Aegean dense-water outflow up to late 1991, which was roughly one magnitude smaller (less than 3% of the current EMT total, Figure 6.1), was clearly recognizable in the Hellenic Trench [*Roether et al.*, 2007, Figure 6].

Our findings contradict recent suggestions that an EMT could be a recurrent phenomenon. *Bozec et al.* [2008] argued that EMT-like events can be reproduced partly by random atmospheric forcing. *Gačić et al.* [2011] noted that recurrent maxima in Levantine near-surface salinities could act as a precursor, and *Josey* [2003] interpreted his finding that there were severe winters not only in 1991–1993 (cf. Figure 6.1) but also in earlier years as another potential preconditioning. It appears that, in reality, a coincidence of the mentioned phenomena, and possibly additional ones, such as Levantine Intermediate Water being diverted into the Aegean Sea [*Malanotte-Rizzoli et al.*, 1999; *Samuel et al.*, 1999], is needed to produce an event.

We conclude that between about 1900 and the onset of the current EMT no such coincidence occurred. A finding supporting our conclusion is the increasing trend of upper-ocean salinities 1945–2001 [cf. *Mariotti*, 2010], with the highest values ever in the central Levantine between 1990 and 1995 [*Gačić et al.*, 2013]. The latter authors report another, somewhat smaller salinity peak in the 1970s. For that same period, *Josey* [2003] reported enhanced winter cooling, and *Beuvier et al.* [2010] found formation of unusually dense water in the Aegean Sea, noting however that the event would probably not have produced significant signatures outside of that sea. The 1970s, thus, are a somewhat critical period, but Figure 6.6b does exclude any significant event. We caution here that we only deal with Aegean waters sufficiently dense to enter the deep waters of the EMed and do not address the intermediate water depth range.

In summary, we exclude EMT-like events back to about the year 1900, which should be valid also for events a factor of about two smaller than the actual EMT. Beginning in about 1945, we can exclude events still smaller by probably an additional factor of two, which should cover all events strong enough to produce significant Aegean-derived water signatures throughout the EMed. The present work adds to the observational basis for modeling efforts addressing the occurrence of EMT-like events in the EMed, such as in the recent work of *Beuvier et al.* [2010].

Acknowledgments. This study relies on the high-quality hydrographic work that has been carried out in the EMed since the early twentieth century and on the effort to collect and quality-control the data for the MEDAR data depository. Special thanks for assistance go to T. Tanhua, Kiel, chief scientist of *Meteor* cruise M84/3 in 2011. The *Meteor* cruises were funded by the Deutsche Forschungsgemeinschaft, Bonn-Bad Godesberg, Germany; that for M84/3 within its core program METEOR/MERIAN.

REFERENCES

Beuvier, J., F. Sevault, M. Herrmann, H. Kontoyiannis, W. Ludwig, M. Rixen, E. Stanev, K. Béranger, and S. Somot (2010), Modeling the Mediterranean Sea interannual variability during 1961–2000: Focus on the Eastern Mediterranean Transient, *J. Geophys. Res.*, *115*, C08017, doi:10.1029/2009JC005950.

Bozec, A., P. Bouret-Aubertot, D. Iudicone, and M. Crépon (2008), Impact of penetrative solar radiation on the diagnosis of water mass transformation in the Mediterranean Sea, *J. Geophys. Res.*, *113*, *C06012*, doi:10.1029/2007JC004606.

Brown, N., (1991) The History of Salinometers and CTD Sensor Systems, http://www.biodiverstiylibrary.org/blibliography/item/17421.

Gačić, M., G. Civitarese, G. L. Eusebi Borcelli, V. Kovačević, P.-M. Poulain, A. Theocharis, M. Menna, A. Catucci, and N. Zarokanellos (2011), On the relationship between the decadal oscillations of the northern Ionian Sea and the salinity distributions in the eastern Mediterranean, *J. Geophys. Res.*, *116*, C12002, doi:10.1029/2011JC007280.

Gačić, M., K. Schroeder, G. Civitarese, S. Cosoli, A. Vetrano, and G. L. Eusebi Borzelli (2013), Salinity in the Sicily Channel corroborates the role of the Adriatic-Ionian Bimodal Oscillating System (BiOS) in shaping the decadal variability of the Mediterranean overturning circulation, *Ocean Sci.*, *9*, 83–90, doi:10.5194/os-9-83-2013.

Gertman, I., N. Pinardi, Y. Popov, and A. Hecht (2006), Aegean Sea watermasses during the early stages of the Eastern Mediterranean Climatic Transient (1988–90), *J. Phys. Oceanogr.*, *36*, 1841–1859.

Hopkins, T., S. (1978), *Physical Processes in the Mediterranean Basins*, in The Belle W. Baruch Library in Marine Science, Columbia, SC, No. *7*, 269–310.

Josey, A. S. (2003), Changes in the heat and freshwater forcing of the Eastern Mediterranean and their influence on deep water formation, *J. Geophys. Res.*, *108* (C7), 3237, doi:10.1029/2003JC001778.

Malanotte-Rizzoli, P., and A. Hecht (1988), Large-scale properties of the Eastern Mediterranean: A review, *Oceanolog. Acta*, *11*, 4, 323–335.

Malanotte-Rizzoli., P., B. B. Manca, M. Ribera d'Alcala, A. Theocharis, S. Brenner, G. Budillon, and E. Ozsoy (1999), The Eastern Mediterranean in the 80s and in the 90s: The big transition in the intermediate and deep circulations, *Dynamics Atmosph. Oceans*, *29* (2–4): 365–395.

Mariotti, A. (2010), Recent changes in the Mediterranean water cycle: A pathway toward long-term regional hydroclimatic change?*J. Climate*,*23*,1513–1525,doi:10.1175/2009JCLI3251.1.

MEDATLAS (2002), available from http://www.ifremer.fr/medar.

Nielsen, J. N. (1912), Hydrography of the Mediterranean and adjacent waters, in *Report of the Danish Oceanographic Expedition 1908–1910 to the Mediterranean and Adjacent Waters*, Copenhagen, *1*, 72–191.

Pollack, M., J. (1951), The sources of the deep water of the eastern Mediterranean Sea, *J. Marine Res.*, *10*, 128–152.

Roether, W., and R. Schlitzer (1991), Eastern Mediterranean deep water renewal on the basis of chlorofluoromethane and tritium data, *Dynamics Atmosph. Oceans*, *15*, 333–354.

Roether, W., B. B. Manca, B. Klein, D. Bregant, D. Georgopoulos, V. Beitzel, V. Kovačević, and A. Lucchetta (1996), Recent changes in Eastern Mediterranean deep waters, *Science*, *271*, 333–335.

Roether, W., B. Klein, B. B. Manca, A. Theocharis, and S. Kioroglou (2007), Transient eastern Mediterranean deep waters in response to the massive dense-water output of the Aegean Sea in the 1990s, *Prog. Oceanogr.*, *74*, 540–571, doi:10.1016/j.pocean.2007.03.001, 2007.

Roether W., P. Schlosser, R. Kuntz, and W. Weiss (1992), Transient-tracer studies of the thermohaline circulation of the Mediterranean, in *Winds and Currents of the Mediterranean Basin*, Proc. NATO Workshop, Atmospheric and Oceanic Circulations in the Mediterranean Basin, 7–14 Sept. 1983, Santa Teresa, Italy, Vol. *II*, 291–317, ed. H. Charnock, Harvard University, Cambridge, MA.

Roether, W., V. Beitzel, J. Sültenfuß, and A. Putzka (1999), The eastern Mediterranean tritium distribution in 1987, *J. Marine Syst.*, *20*, 49–61.

Rubino, A., and D. Hainbucher (2007), A large abrupt change in the abyssal water masses of the eastern Mediterranean, *Geophys. Res. Let.*, *34*, L23607, doi:10.1029/2007GL031737.

Samuel, S., K. Haines, S. Josey, and P. G. Myers (1999), Response of the Mediterranean Sea thermohaline circulation to observed changes in the winter wind stress field in the period 1980–1993, *J. Geophys. Res.*,*104*, 7771–7784.

Schlitzer, R., W. Roether, M. Hausmann, H. G. Junghans, H. Oster, H. Johannsen, and A. Michelato (1991), Chlorofluoromethane and oxygen in the Eastern Mediterranean, *Deep Sea Res.*, *38* (12), 1531–1551.

Tanhua, T. (2011), Short Cruise Report of RV *Meteor* Cruise M84/3, Istanbul (Turkey) —Vigo (Spain), April 5–28, 2011, http://www.ifm.zmaw.de/fileadmin/files/leitstelle/meteor/M84/M84-3-SCR.pdf.

Theocharis, A., B. Klein, K. Nittis, and W. Roether (2002), Evolution and status of the Eastern Mediterranean Transient (1997–1999), *J. Marine Syst.*, *33–34*, 91–116.

Theocharis, A., D. Georgopoulos, P. Karagevrekis, A. Iona, L. Perivoliotis, and N. Charalambidis (1992), Aegean influence in the deep layers of the Eastern Ionian Sea, in *Rapport de la Commission international pour l'Exploration Scientifique de la Mer Méditerranée*, *33*, 235.

Theocharis, A., K. Nittis, H. Kontoyiannis, E. Papageorgiou, and E. Balopoulos (1999), Climatic changes in the Aegean Sea influence the Eastern Mediterranean thermohaline circulation (1986–1997), *Geophys. Res. Lett.*, *26* (11), 1617–1620.

Velaoras, D., and A. Lascaratos (2010), North-Central Aegean Sea surface and intermediate water masses and their role in triggering the Eastern Mediterranean Transient, *J. Marine Syst.*, *83*, 58–66.

Vervatis, V. D., S. S. Sofianos, and A. Theocharis (2011), Distribution of the thermohaline characteristics in the Aegean Sea related to water mass formation processes (2005–2006 winter surveys), *J. Geophys. Res.*, *116*, C09034, doi:10.1029/2010JC006868.

Wüst, G. (1960), Die Tiefenzirkulation des Mittelländischen Meeres in den Kernschichten des Zwischen- und Tiefenwassers, *Deutsche Hydrogr. Z.*, *13*, 106–131 (with 9 separate plates).

Wüst, G. (1961), On the vertical circulation of the Mediterranean Sea, *J. Geophys. Res.*, *66*, 3261–3271.

Zervakis, V., D. Georgopoulos, P. G. Drakopoulos (2000), The role of the north Aegean in triggering the recent Eastern Mediterranean Transient, *J. Geophys. Res.*, *105* (C11), 26103–26116.

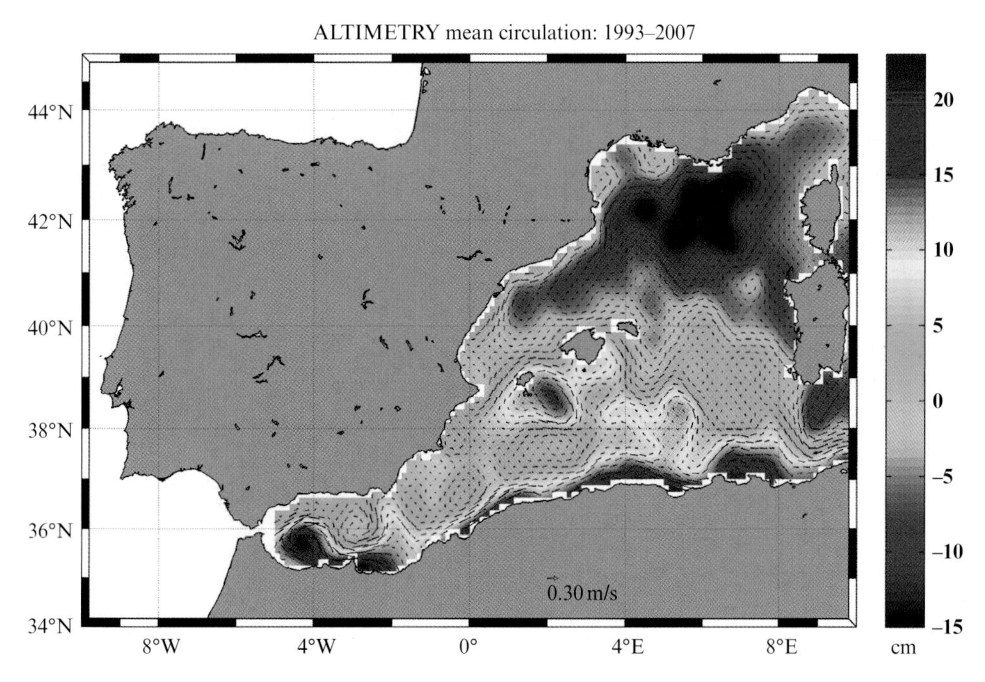

Figure 2.1. Altimetry mean geostrophic circulation in the WMED derived from Absolute Dynamic Topography for the 1993–2007 period.

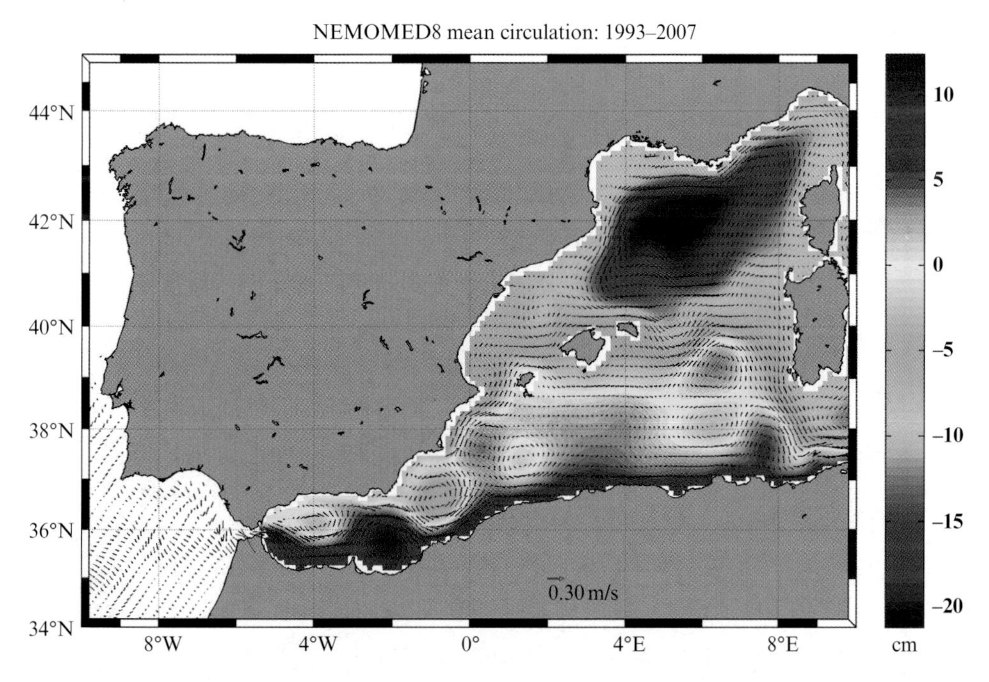

Figure 2.2. NM8 mean geostrophic circulation in the WMED derived from Sea Surface Height for the 1993–2007 period.

The Mediterranean Sea: Temporal Variability and Spatial Patterns, Geophysical Monograph 202. First Edition.
Edited by Gian Luca Eusebi Borzelli, Miroslav Gačić, Piero Lionello, and Paola Malanotte-Rizzoli.

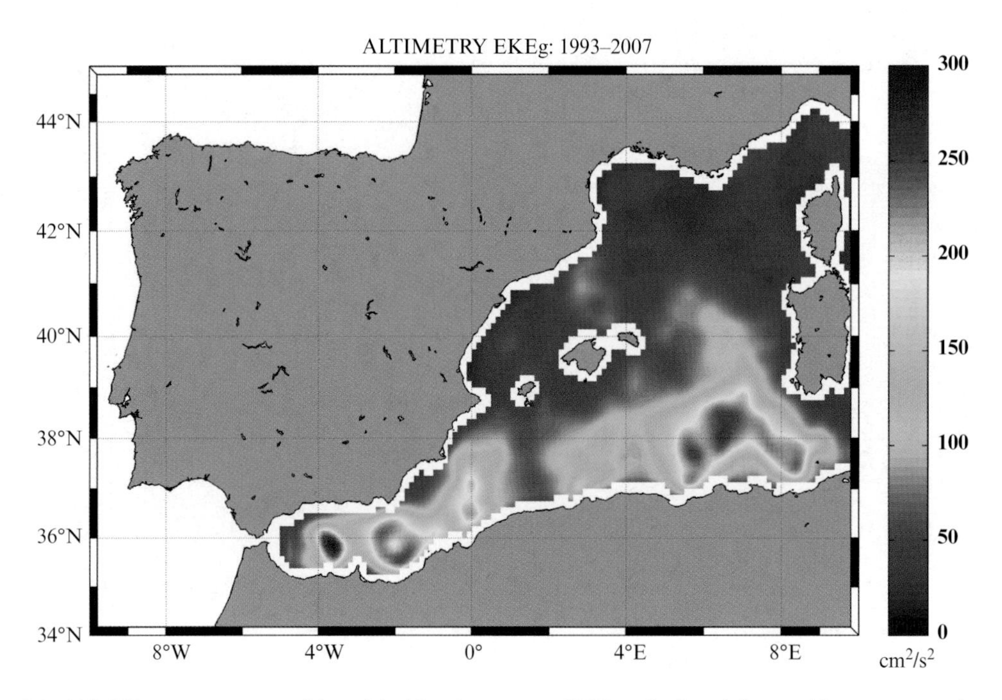

Figure 2.3. WMED mean geostrophic eddy kinetic energy (EKE) calculated from altimetry over the period 1993–2007.

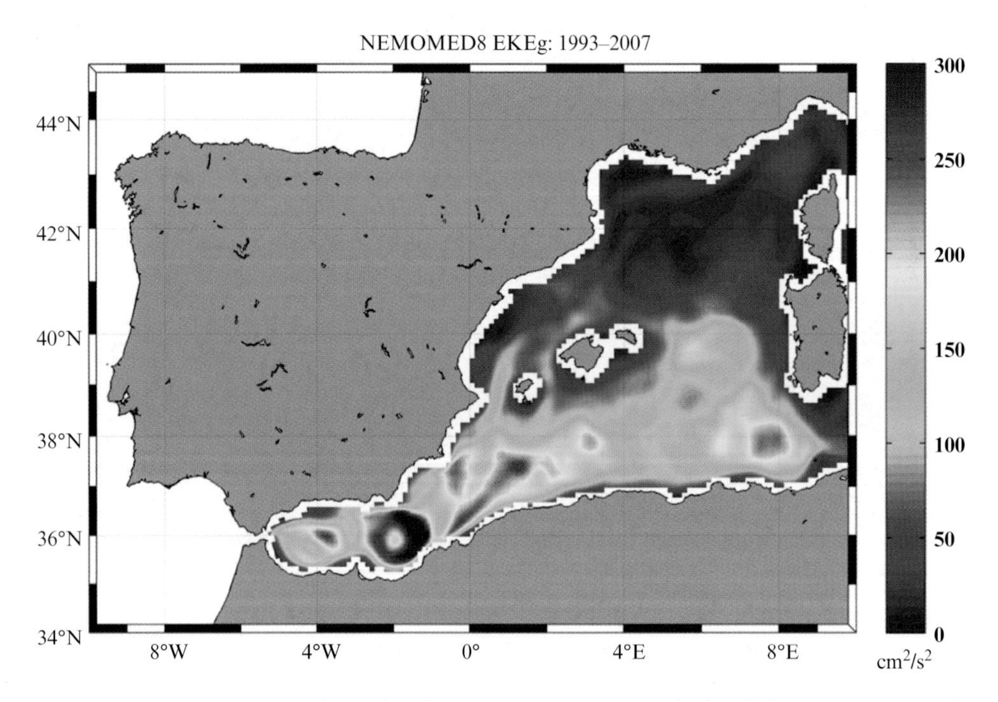

Figure 2.5. WMED mean geostrophic eddy kinetic energy (EKE) calculated from NM8 over the period 1993–2007.

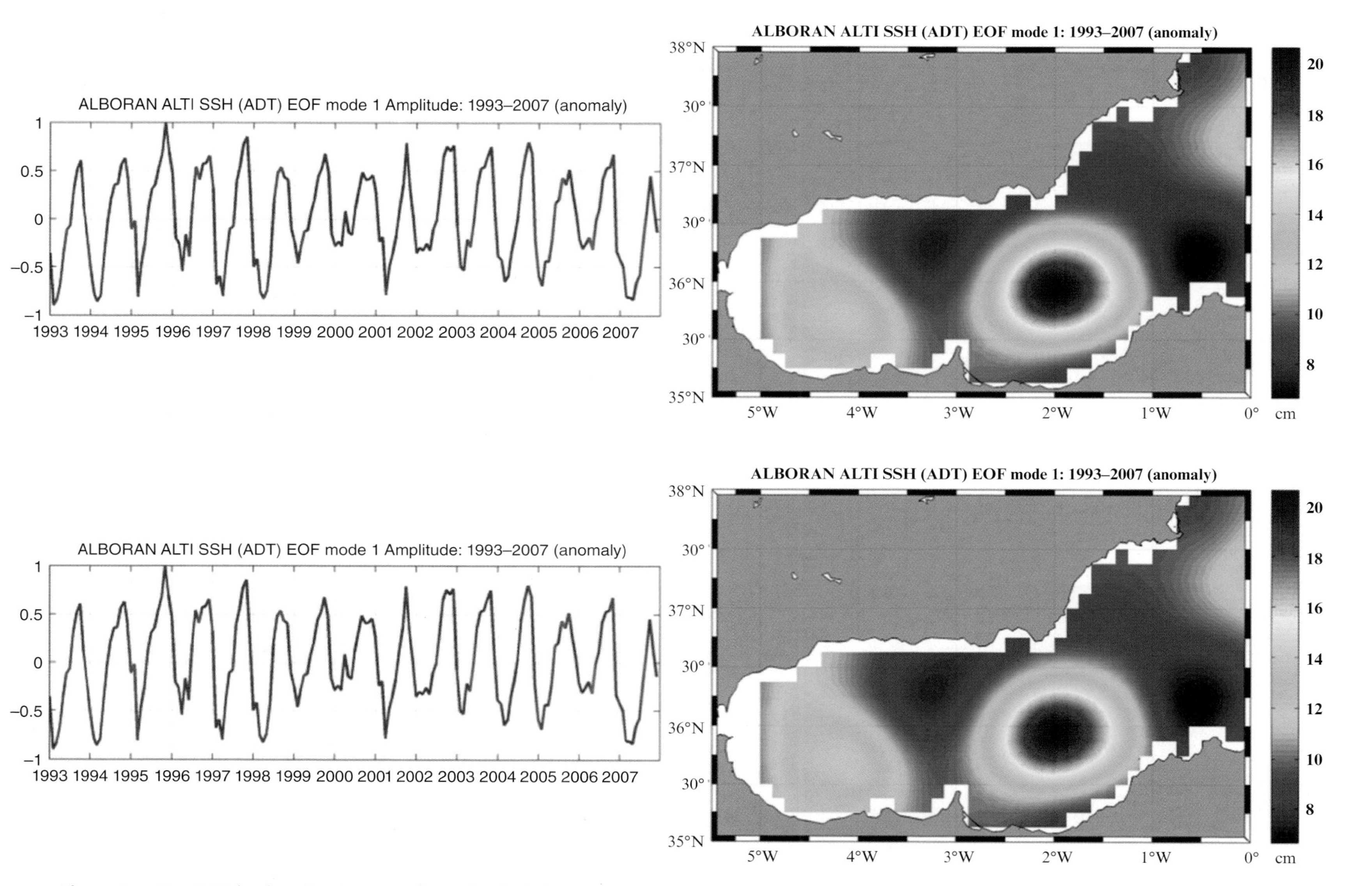

Figure 2.7. First EOF for the Alborán Sea with amplitude (left) and pattern (right); altimetry ADT (top), NM8 SSH (bottom).

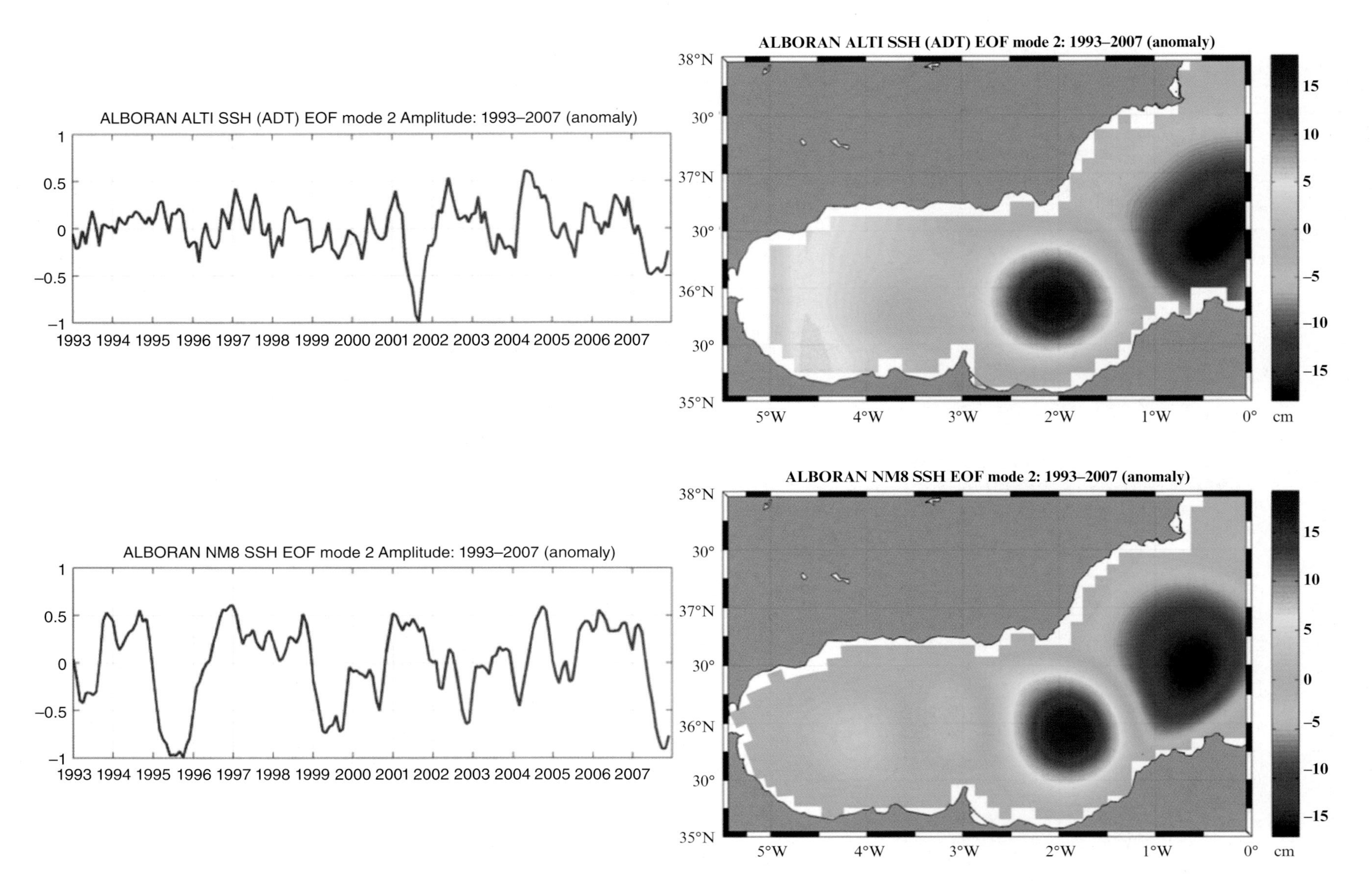

Figure 2.8. Second EOF for the Alborán Sea with amplitude (left) and pattern (right); altimetry ADT (top), NM8 SSH (bottom).

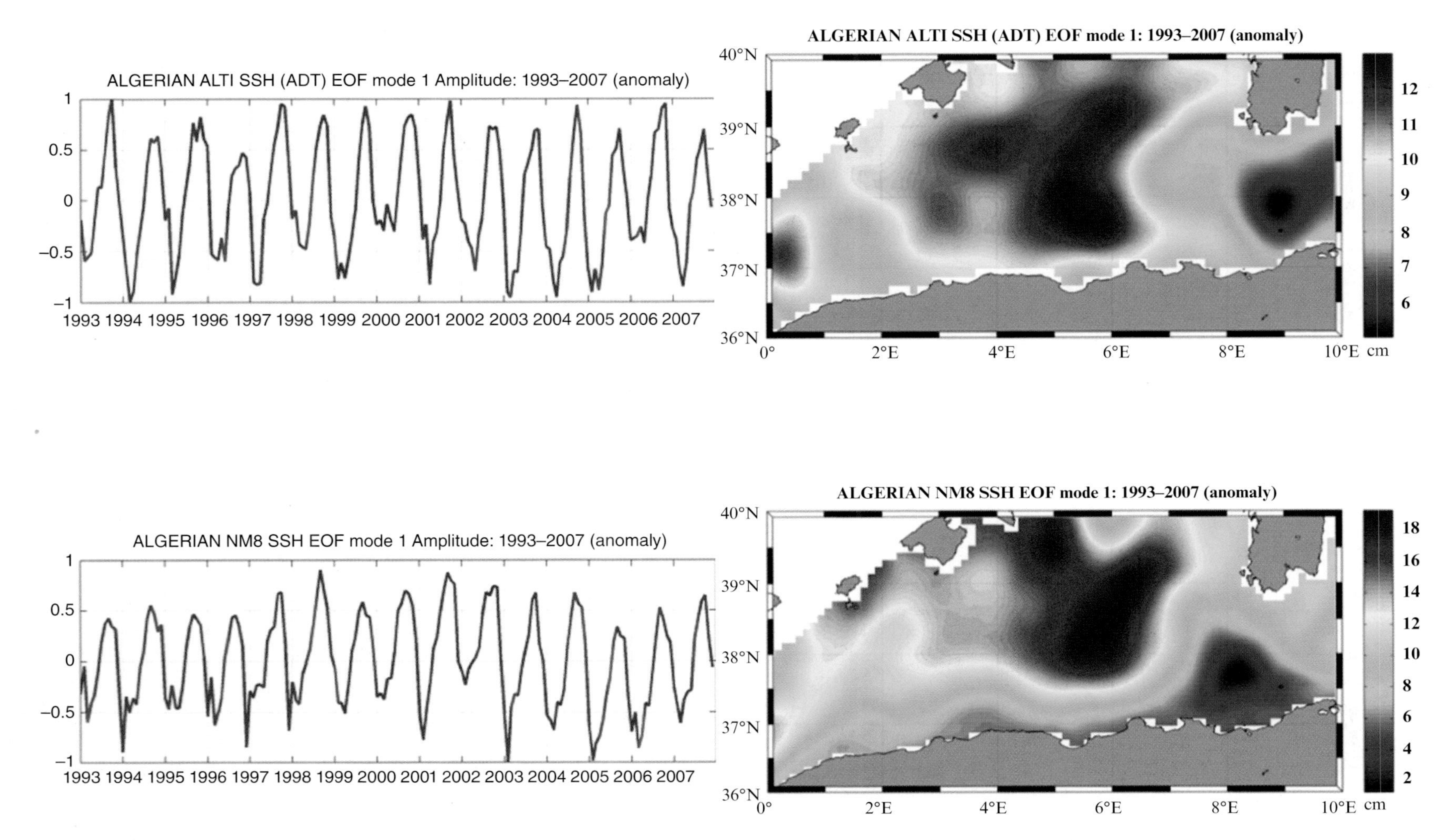

Figure 2.9. First EOF for the Algerian Basin with amplitude (left) and pattern (right); altimetry ADT (top), NM8 SSH (bottom).

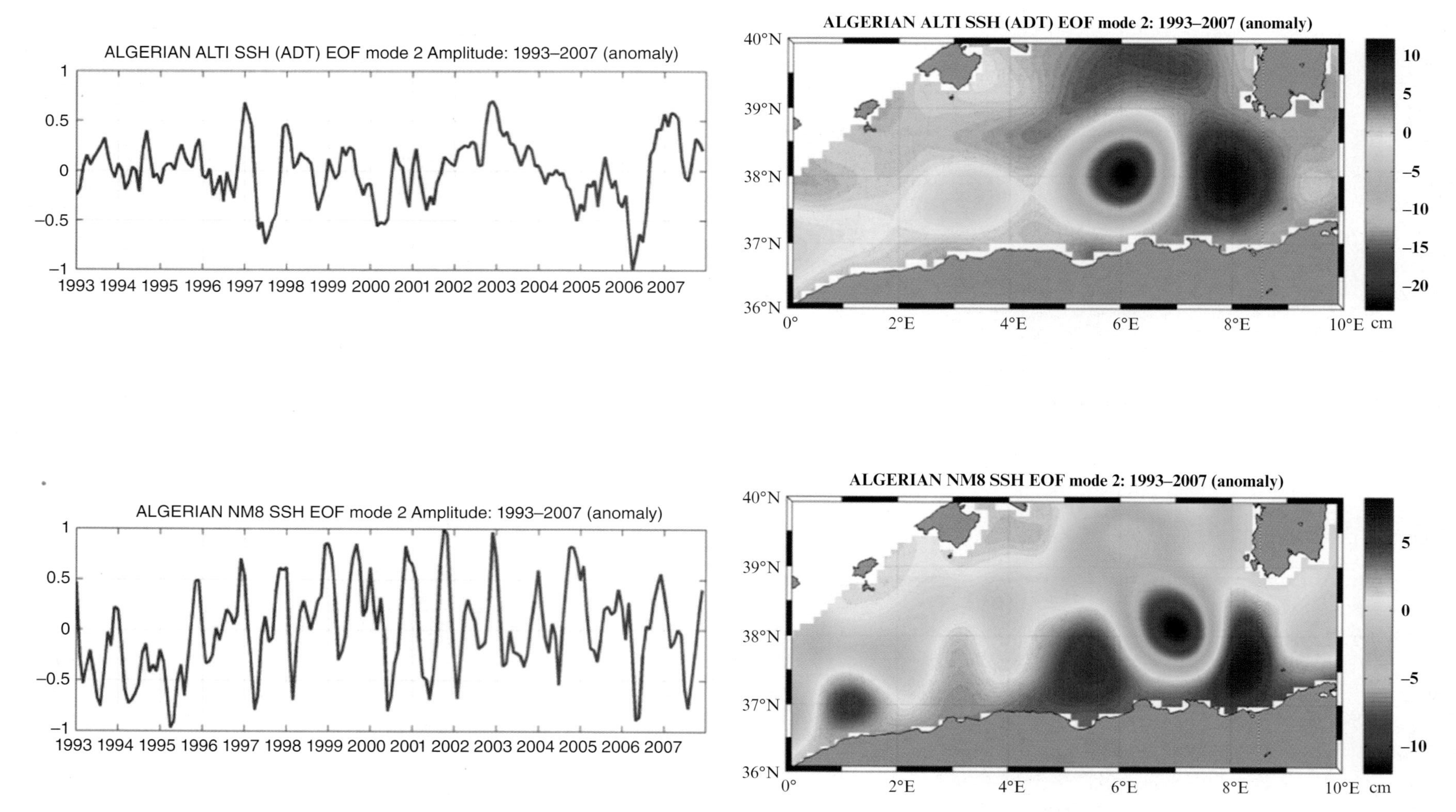

Figure 2.10. Second EOF for the Algerian Basin with amplitude (left) and pattern (right); altimetry ADT (top), NM8 SSH (bottom).

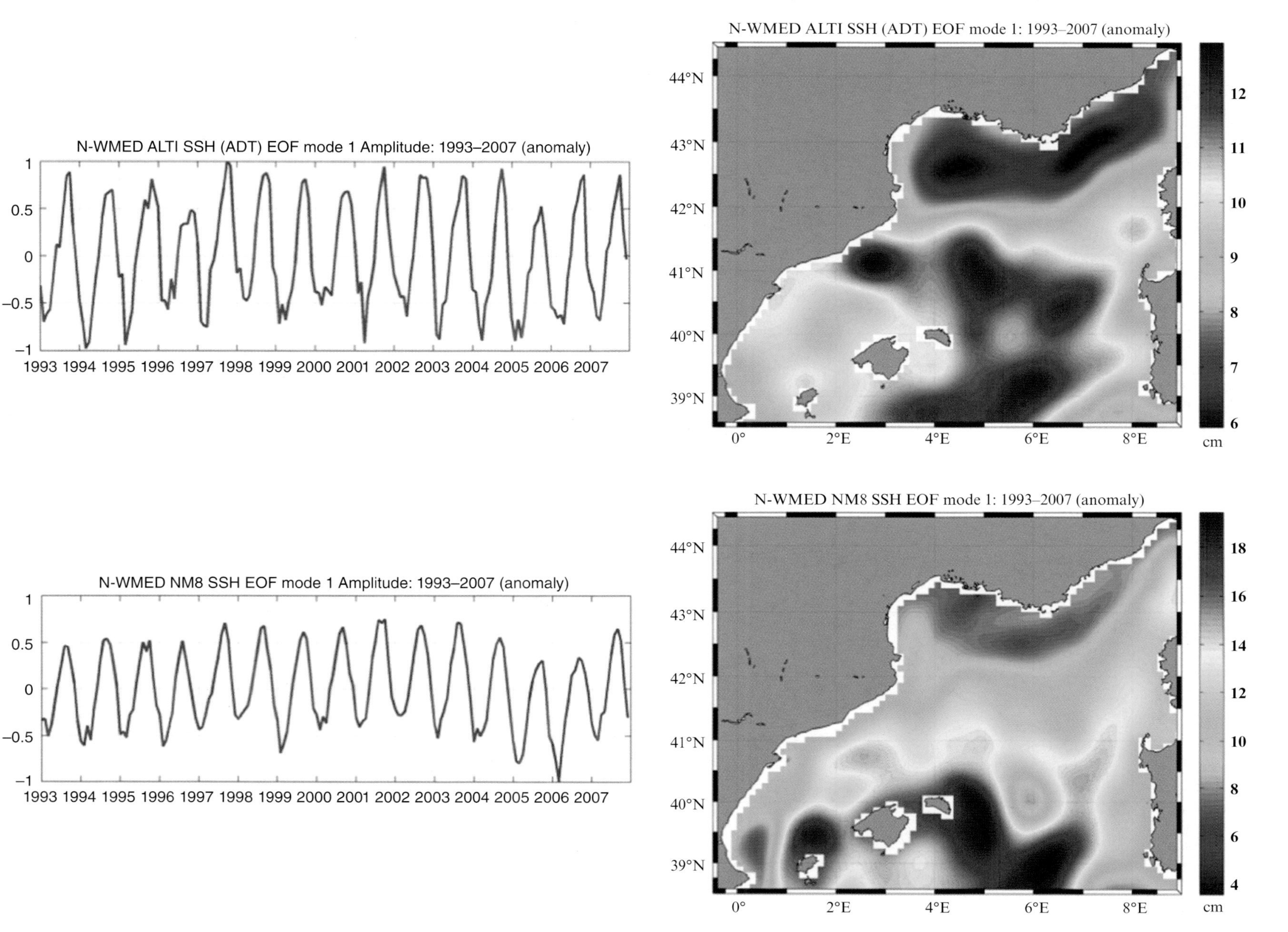

Figure 2.11. First EOF for the NWMED with amplitude (left) and pattern (right); altimetry ADT (top), NM8 SSH (bottom).

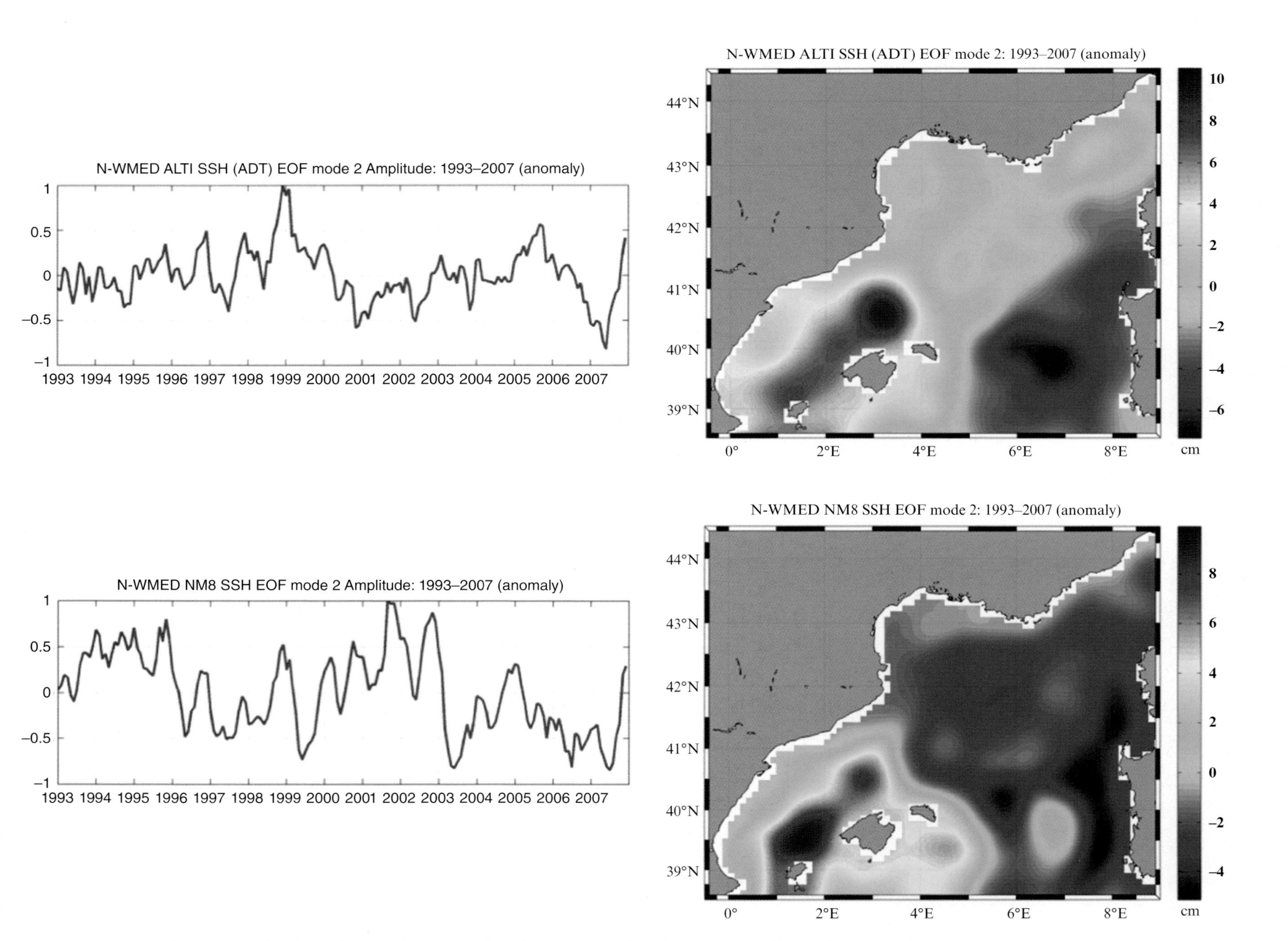

Figure 2.12. Second EOF for the NWMED with amplitude (left) and pattern (right); altimetry ADT (top), NM8 SSH (bottom).

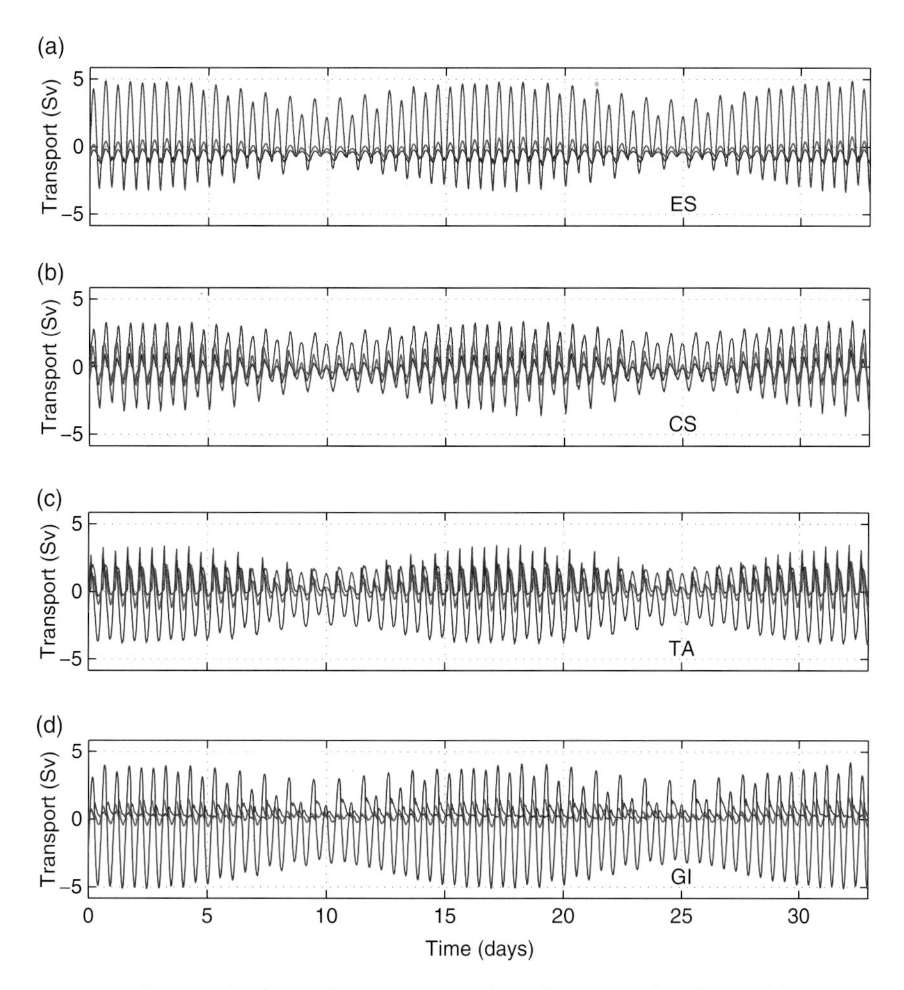

Figure 3.9. (a) Time evolution of Atlantic layer transport (blue line), interfacial-mixed layer (green line), and Mediterranean layer transport (red line) at Espartel section; (b) Same as (a) for Camarinal Sill section; (c) Same as (a) for Tarifa section; (d) Same as (a) for Gibraltar section.

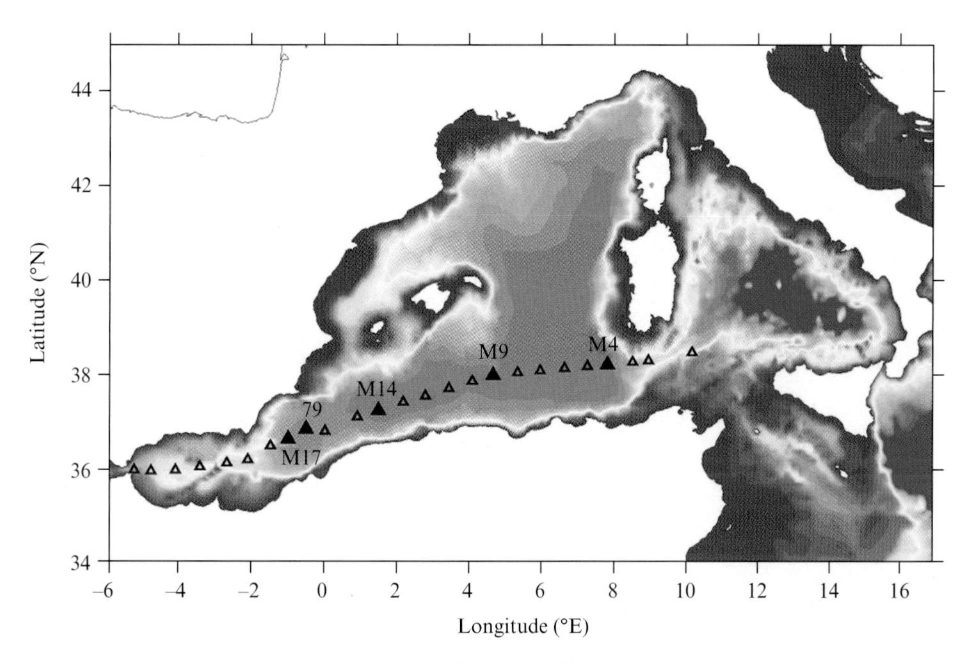

Figure 4.1. Map of CTD stations that were occupied by Urania during 2004, 2005, 2006, 2008, and 2010. These stations are part of regular repeat surveys of the western Mediterranean Sea since 2004 with support from Consiglio Nazionale delle Ricerche (CNR). Here we focus on stations M4–M17. Station 79 was used by *Borghini et al.* (2012) to estimate the overall changes in temperature and salinity of the Mediterranean waters as they are about to exit the Mediterranean.

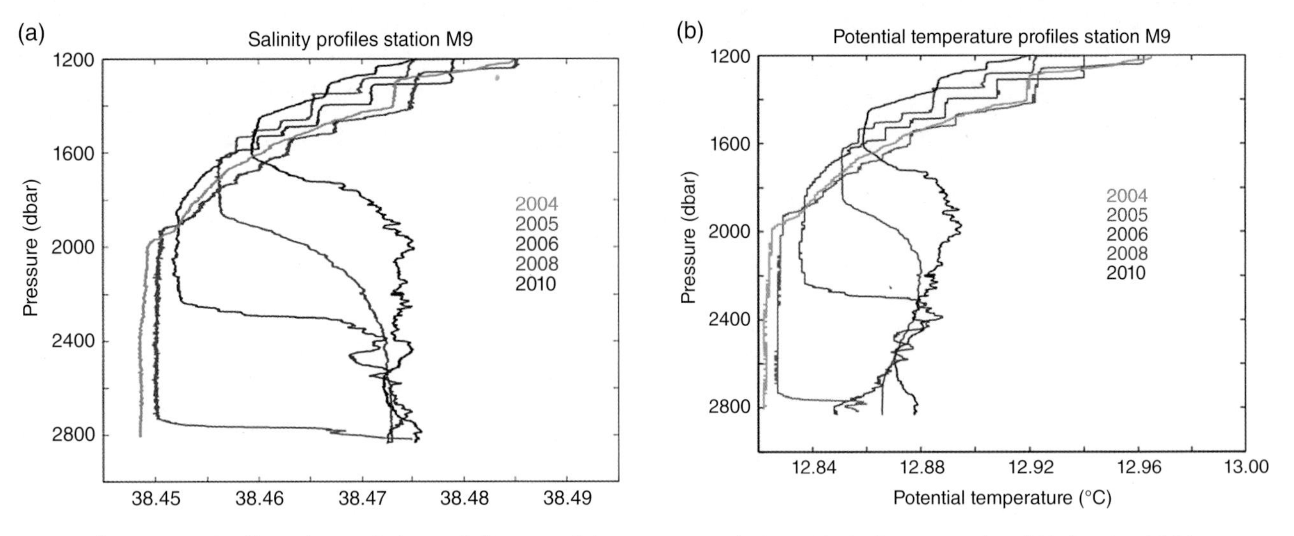

Figure 4.2. Profiles of (a) salinity and (b) potential temperature below 1200 dbar at station M9 for the 2004, 2005, 2006, 2008, and 2010 surveys.

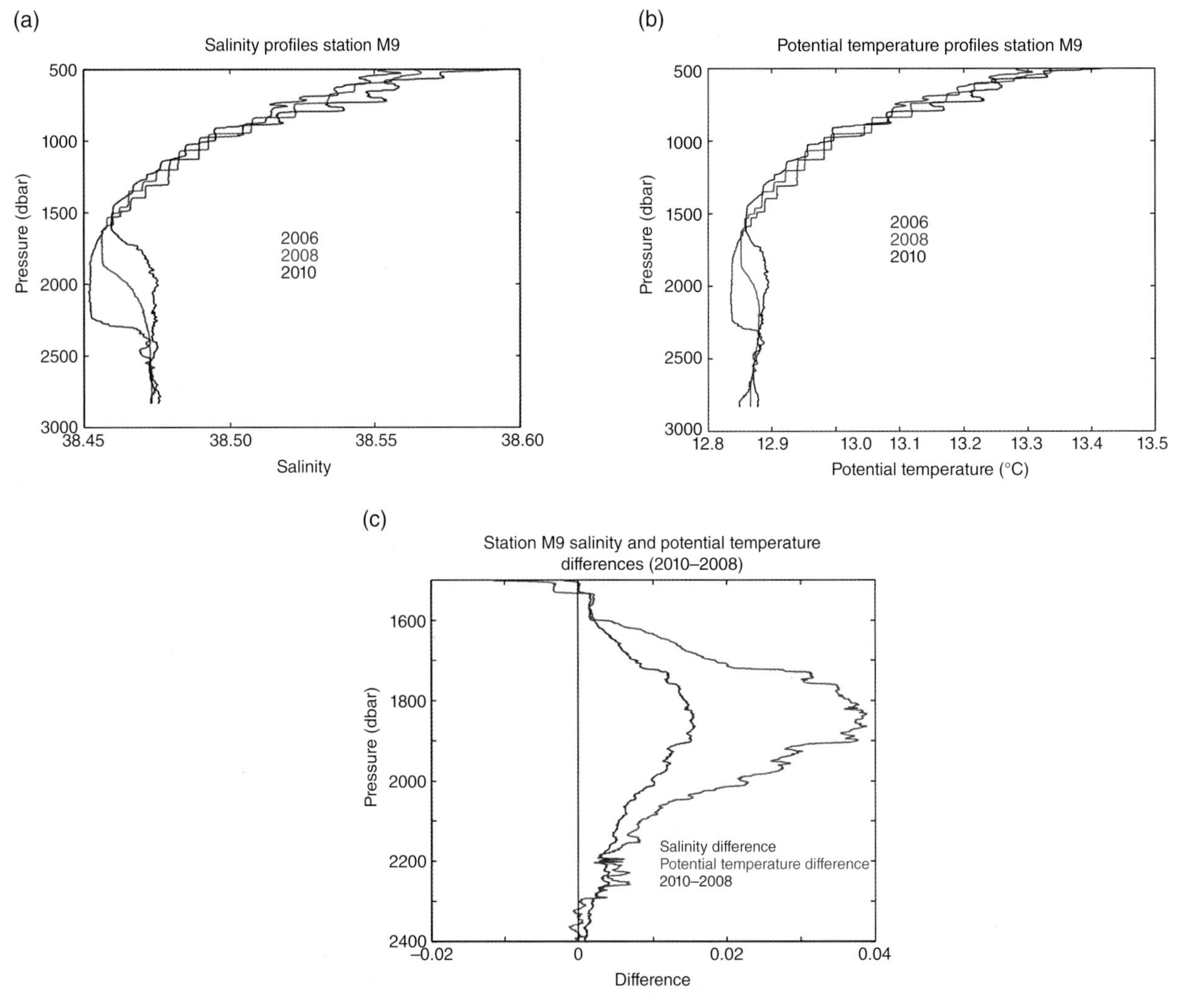

Figure 4.3. Profiles of (a) salinity and (b) potential temperature below 500 dbar at station M9 for 2006, 2008, and 2010 surveys. The halocline-thermocline from 500 dbar to 1500 dbar exhibits steplike features that are usually associated with salt finger processes; the deep waters below 2300 dbar exhibit similar characteristics of new deep water formed in 2005–2006. The only part of the water column where changes are immediately evident is the transition zone between the bottom of the halocline-thermocline and the top of the new deep water. Here the transition waters become progressively warmer and saltier from 2006 to 2008 to 2010. (c) Profiles of the 2010 minus 2008 differences in salinity and potential temperature in the transition zone between the bottom of the halocline-thermocline and the top of the new deep water at station M9 are used to quantify the amount of salt and heat flux convergences into this transition zone, tabulated in Table 4.1.

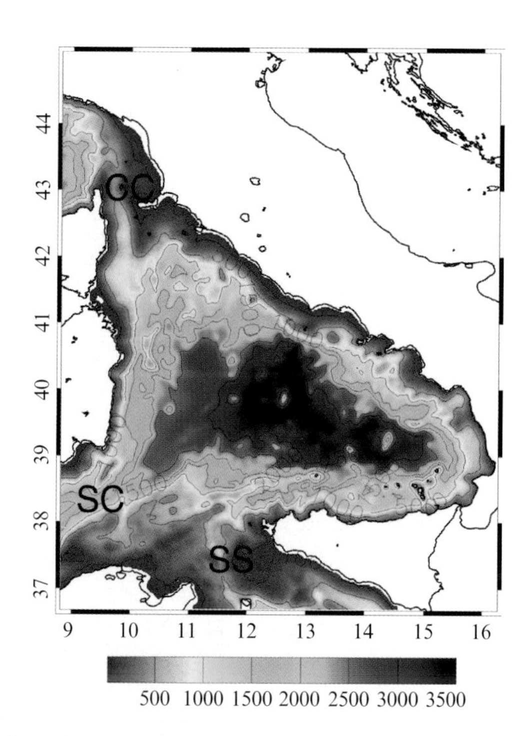

Figure 5.1. The Tyrrhenian Sea, with its three openings: the Corsica Channel (CC), the Sardinia Channel (SC), and the Sicily Strait (SS). Bathymetry is indicated by colors. The domain represented exactly coincides with the computational domain of the operational model TYREM, whose outputs are discussed in the text.

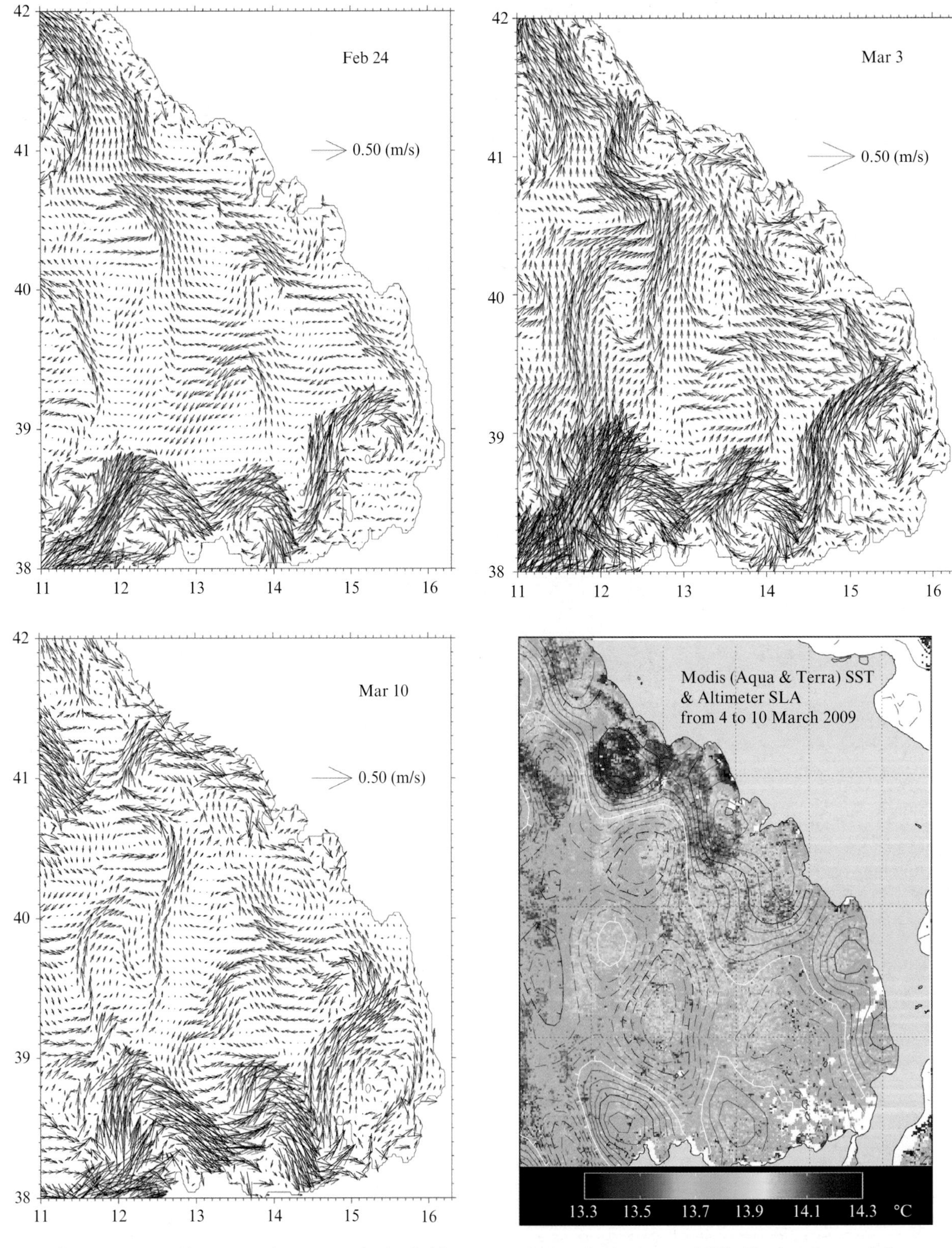

Figure 5.6. Weekly averaged TYREM velocity fields at 10 m (zooms in the eastern TYS). The fourth panel shows the SLA for the week 4–10 March (contours), with the SST of the same week superposed (colors).

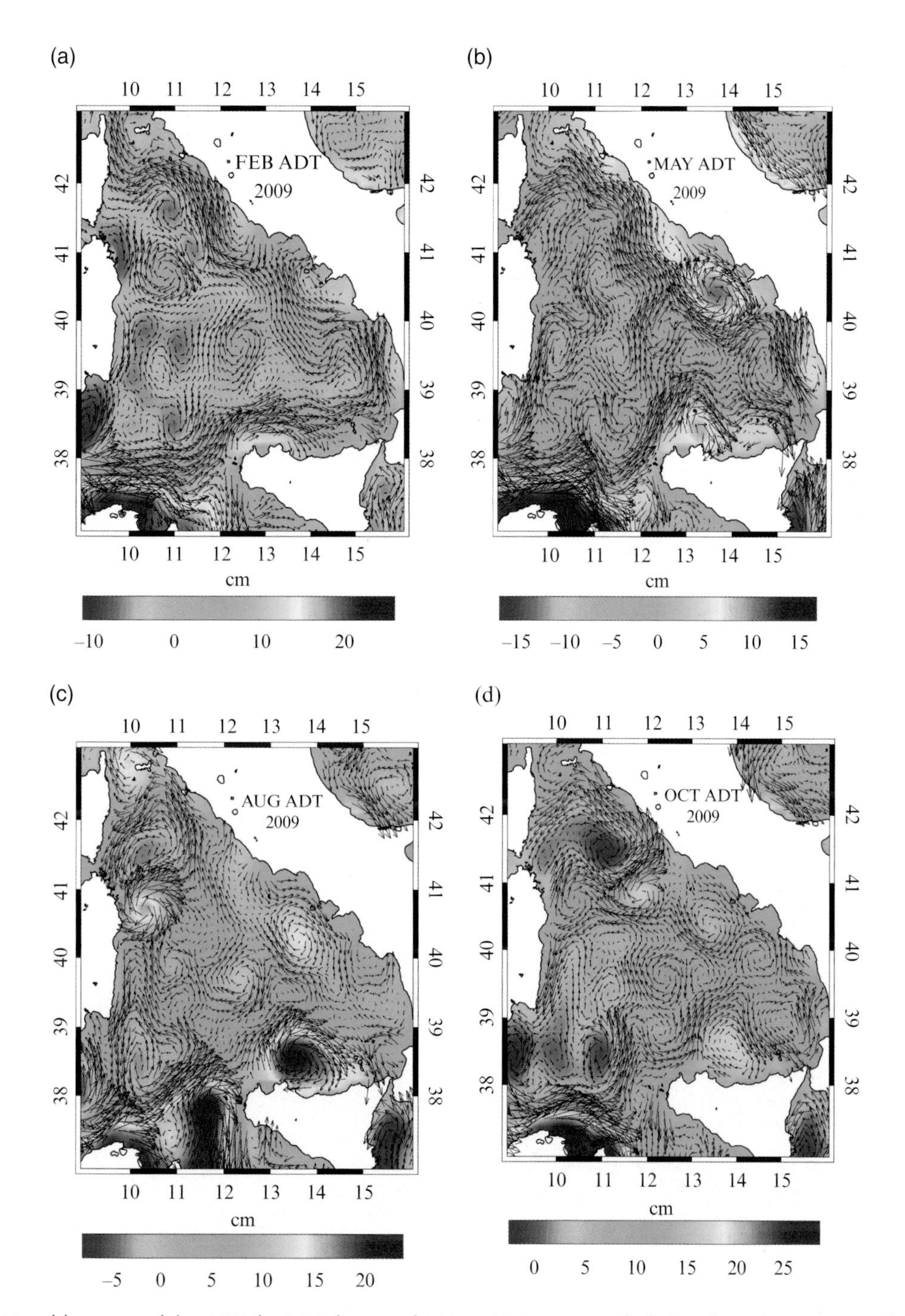

Figure 5.7. Monthly mean of the ADT for (a) February, (b) May, (c) August, and (d) October. The corresponding average geostrophic circulations are also shown.

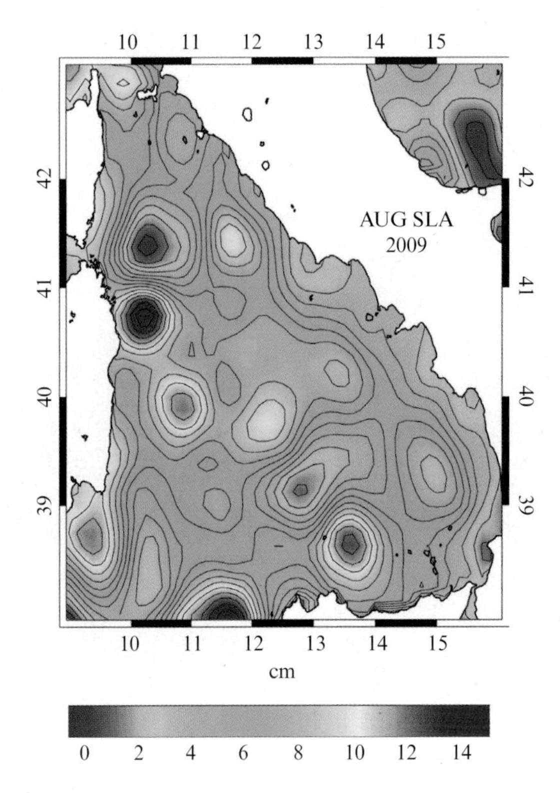

Figure 5.8. SLA of August 2009.

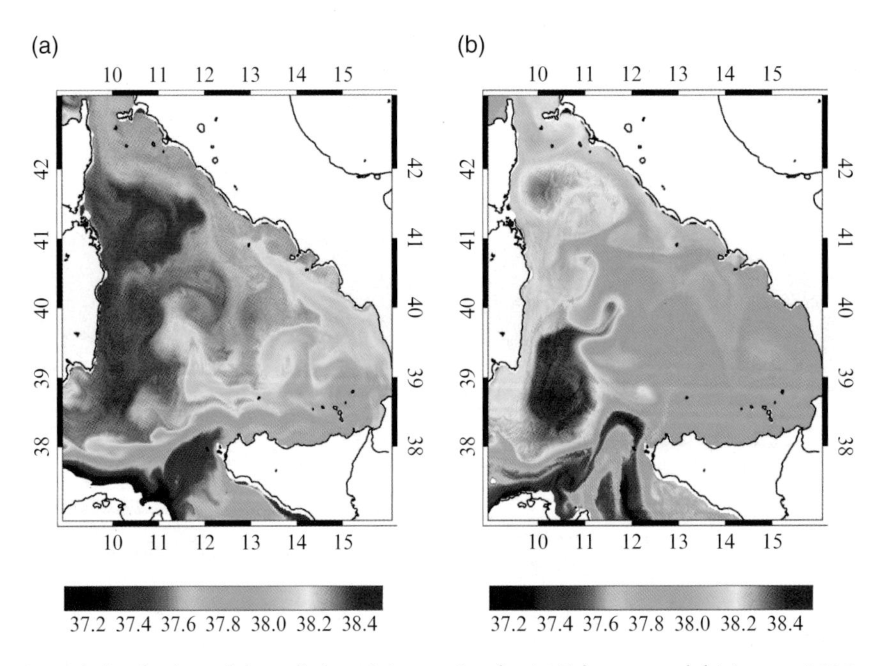

Figure 5.11. Spatial distribution of the salinity minimum S_{min} for (a) February and (b) August 2009.

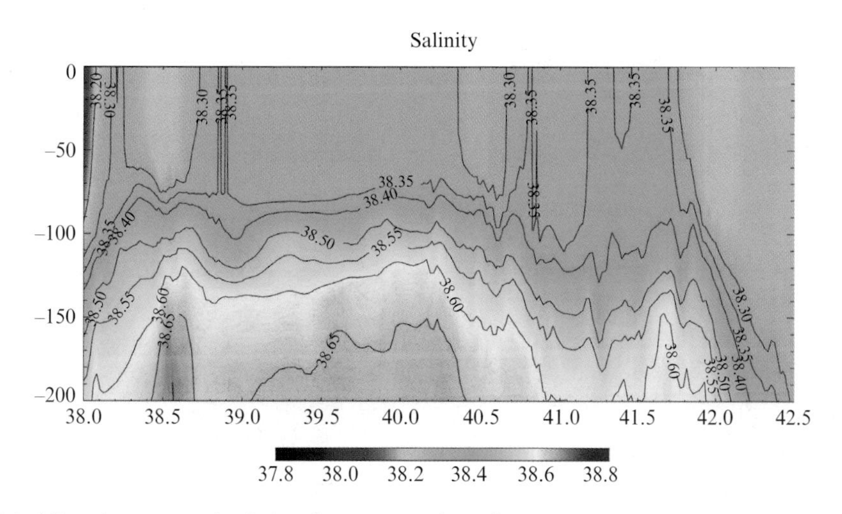

Figure 5.12. Meridional transect of salinity along 10.5°E for February 2009.

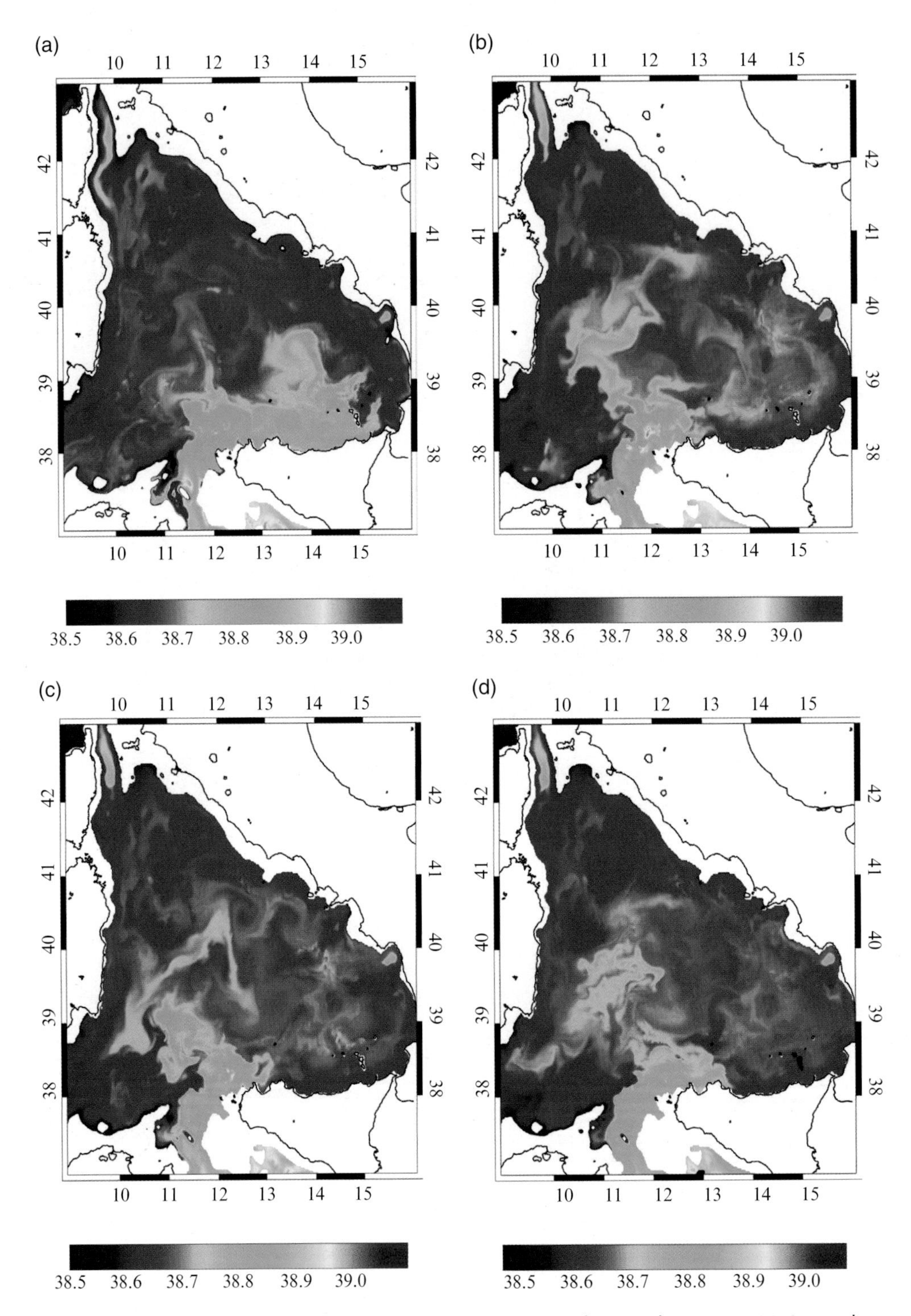

Figure 5.13. Spatial distribution of the salinity maximum S_{max}: (a) February, (b) August, (c) September, and (d) October. The last three panels illustrate the formation of a LIW tongue that moves toward northwest, gets trapped in the cyclonic circulation present in the area, and finally reaches the Sardinia Channel.

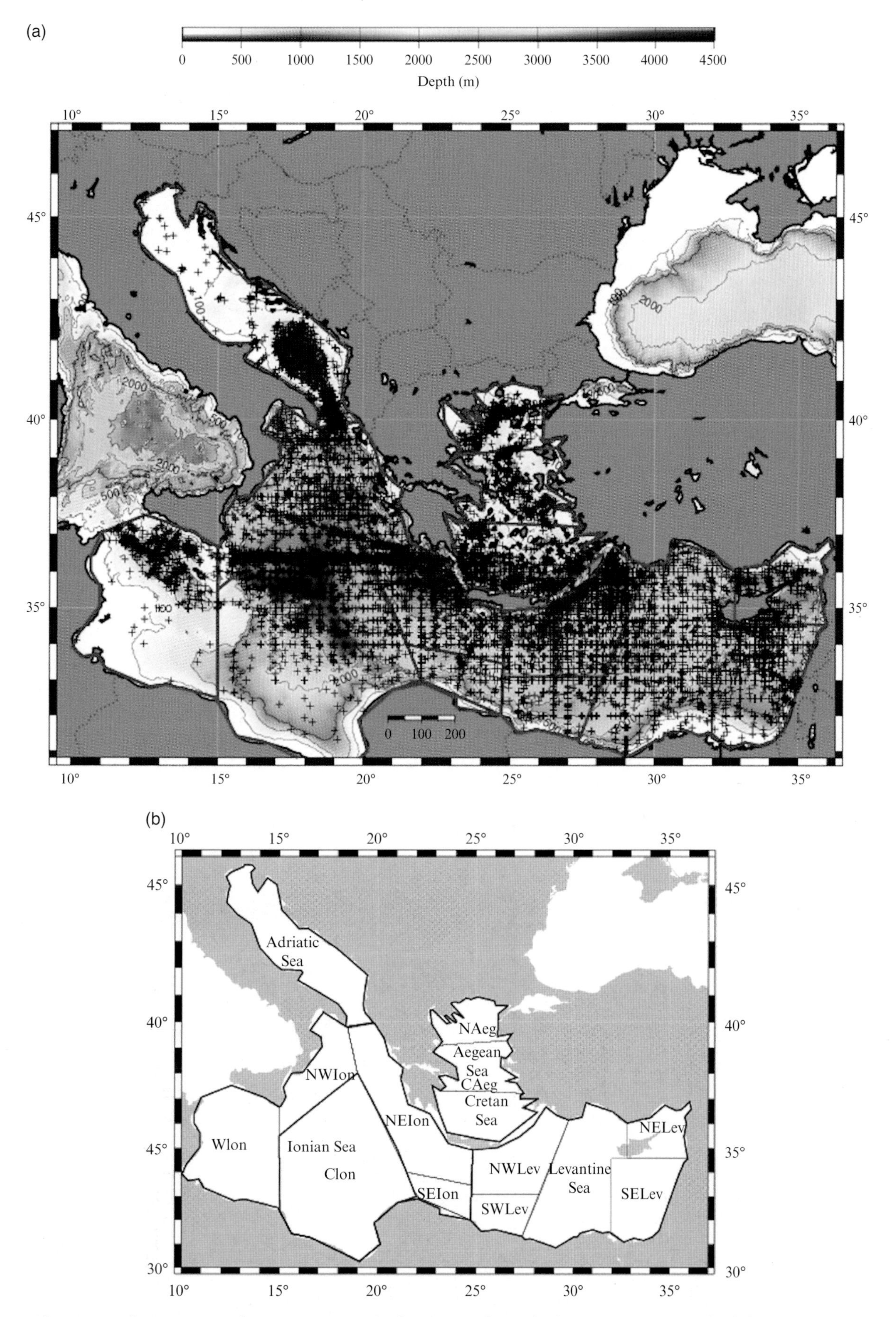

Figure 7.1. The Eastern Mediterranean Sea (a) bathymetry with overlaid station positions and subdomain boundaries, (b) subdomains of the study area.

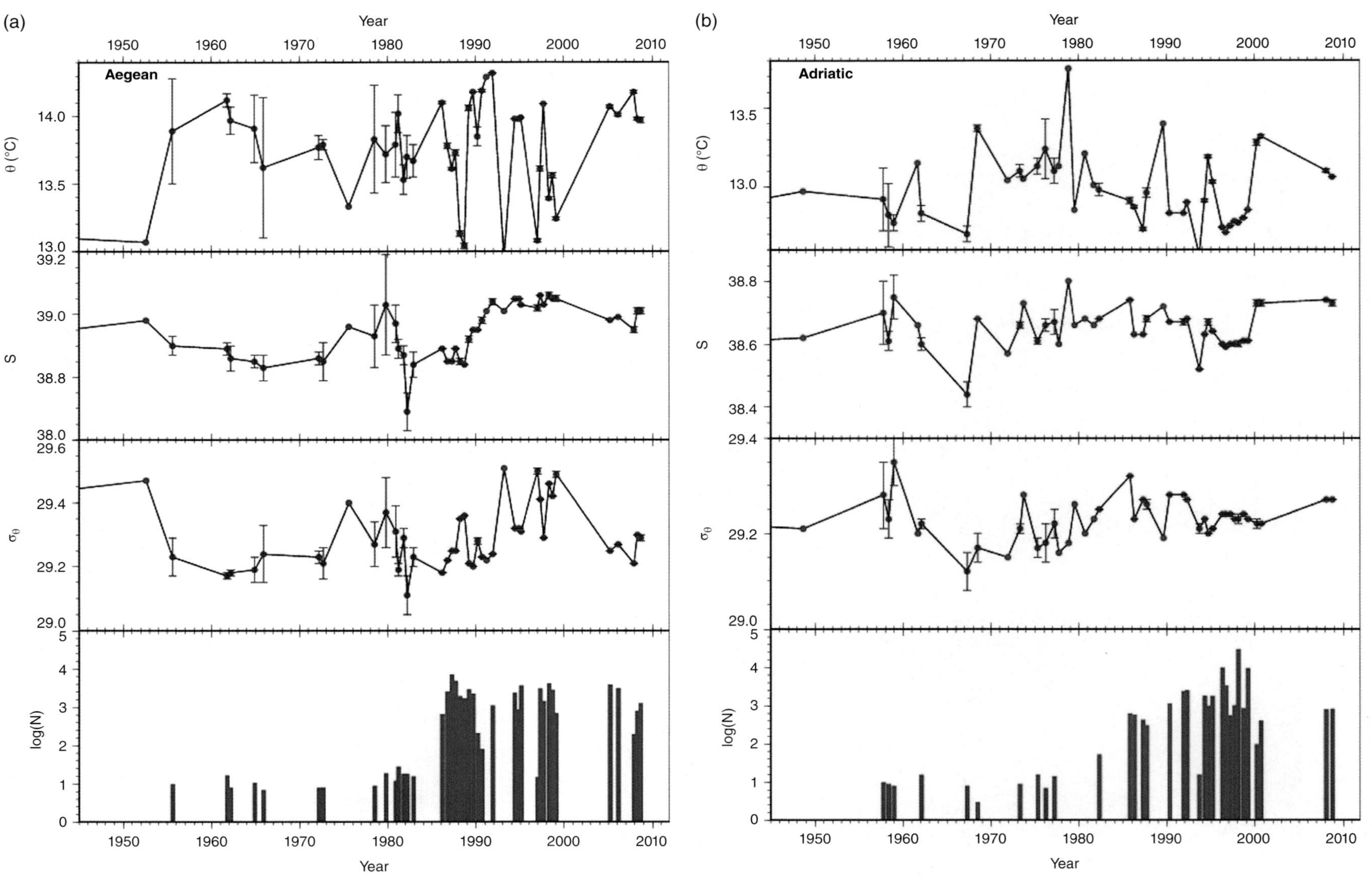

Figure 7.2. (a) Time series of average potential temperature, salinity, potential density, and log(N), the logarithm of the number of data points N in the 900–1100-m depth level, binned at half-year intervals in the Aegean Sea (vertical bars with end caps denote 95% confidence limits for data in each bin); (b) time series of average potential temperature, salinity, potential density, and log(N), the logarithm of the number of data points N in the 900–1100-m depth level, binned at half-year intervals in the Adriatic Sea (vertical bars with end caps denote 95% confidence limits for data in each bin).

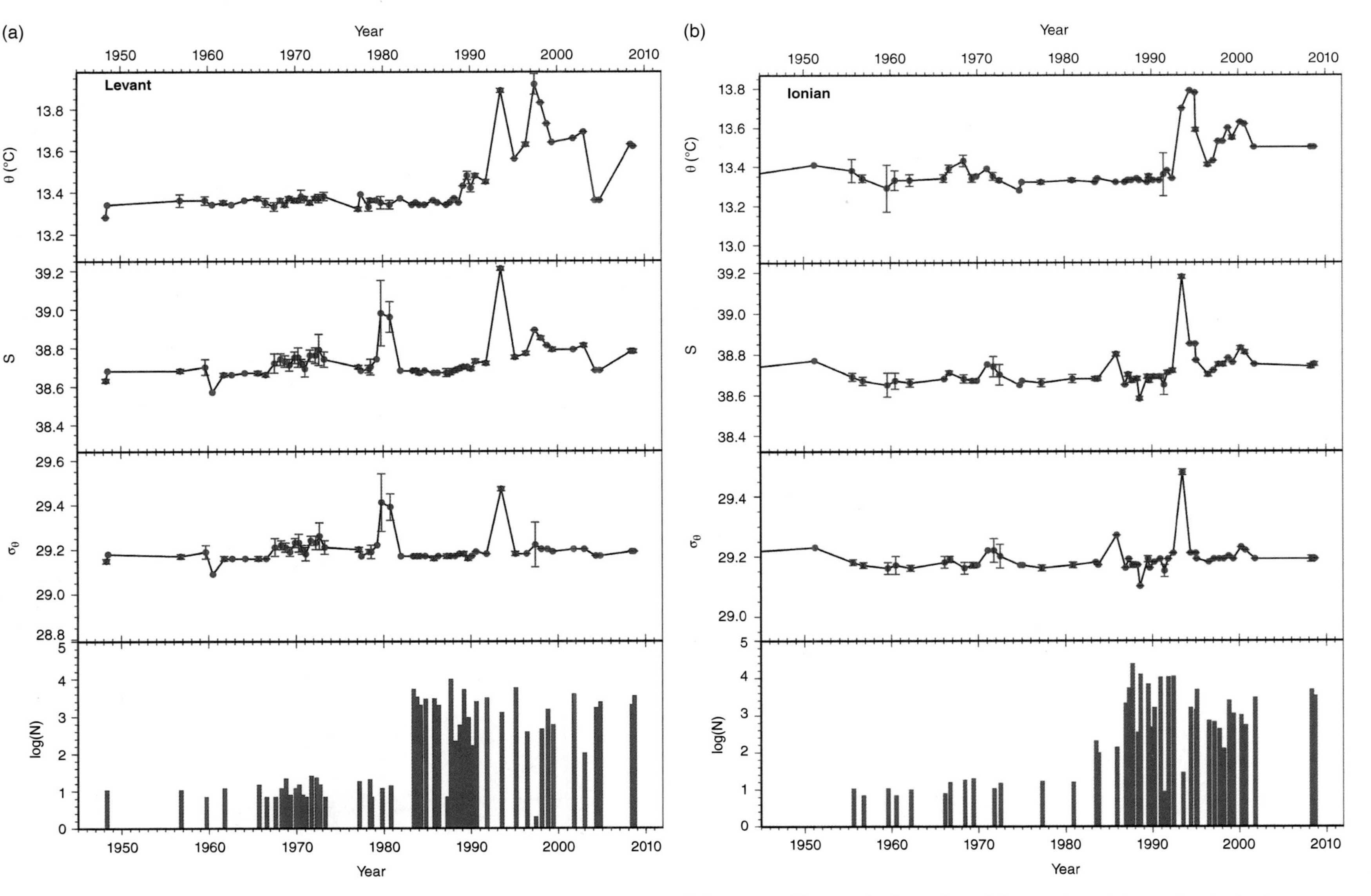

Figure 7.3. (a) Time series of average potential temperature, salinity, potential density, and log(N), the logarithm of the number of data points N, in the 900–1100-m depth level, binned at half-year intervals in the Levantine Sea (vertical bars with end caps denote 95% confidence limits for data in each bin); (b) time series of average potential temperature, salinity, potential density, and log(N), the logarithm of the number of data points N in the 900–1100-m depth level, binned at half-year intervals in the Ionian Sea (vertical bars with end caps denote 95% confidence limits for data in each bin).

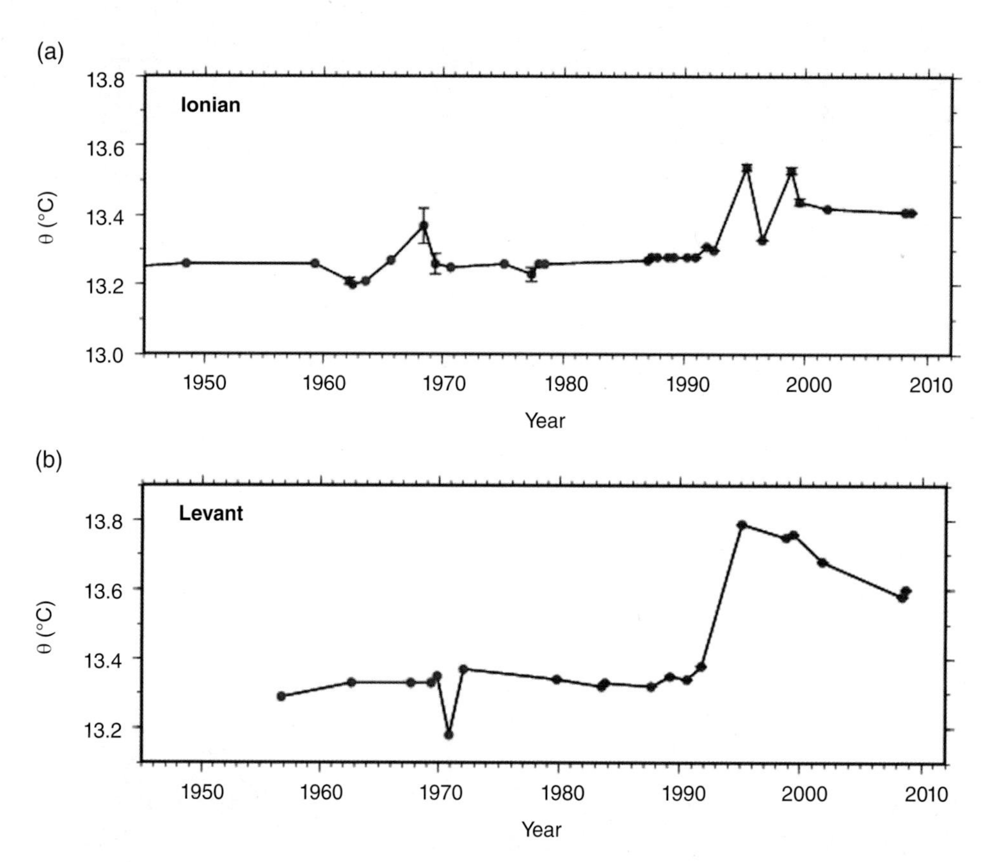

Figure 7.4. Time series of average temperature at depths of 2900–3100 m binned at half-year intervals in the (a) Ionian and (b) Levantine Seas.

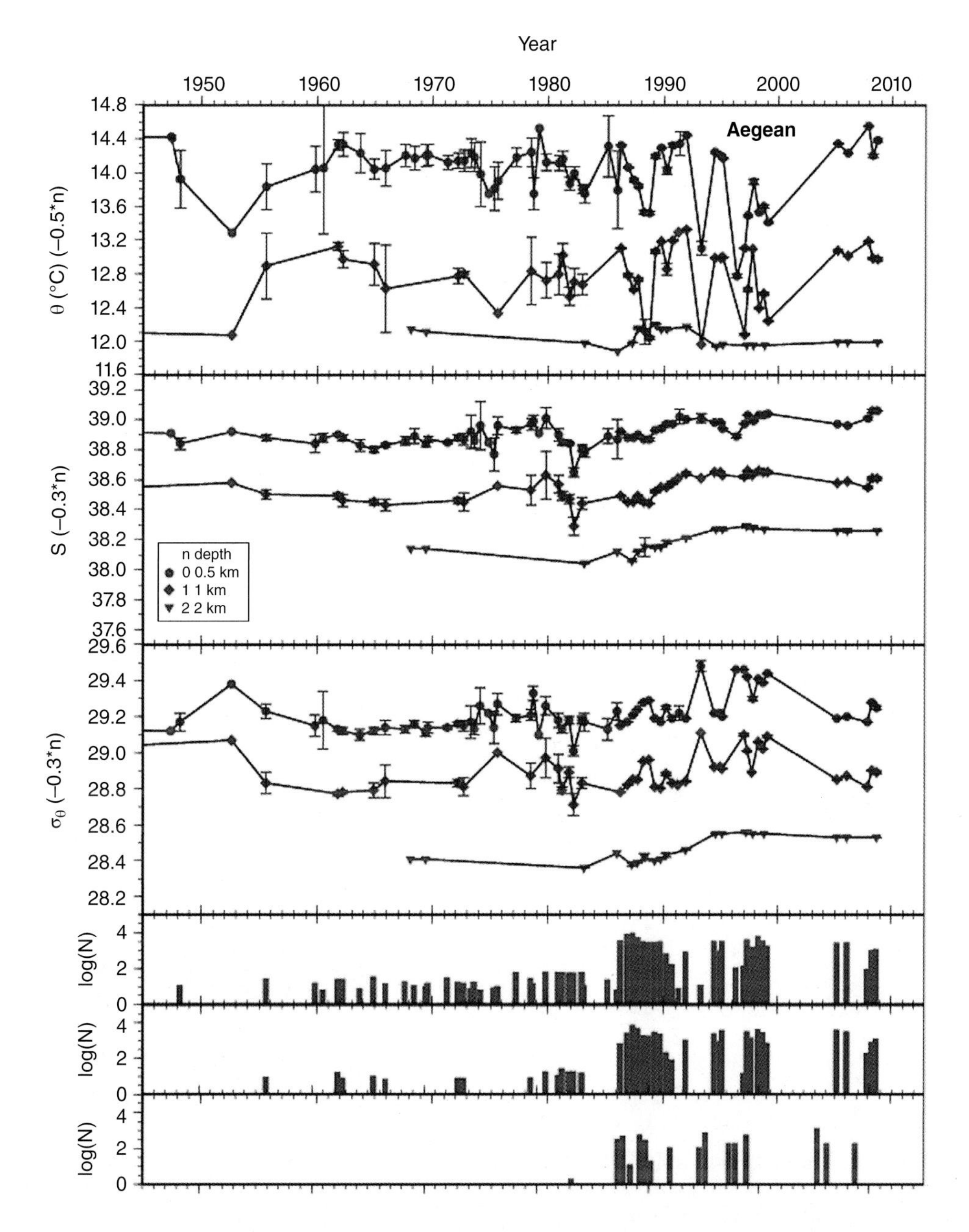

Figure 7.5. Aegean Sea potential temperature θ, salinity, and σ_θ density at depth layers of 450–550 m (n = 0), 900–1100 m (n = 1), and 1900–2100 m (n = 2) (vertically shifted down by a constant times n) and log(N), the logarithm of the number of data points N averaged for each half-year bin interval. (Vertical bars with end caps denote 95% confidence limits for data in each bin.)

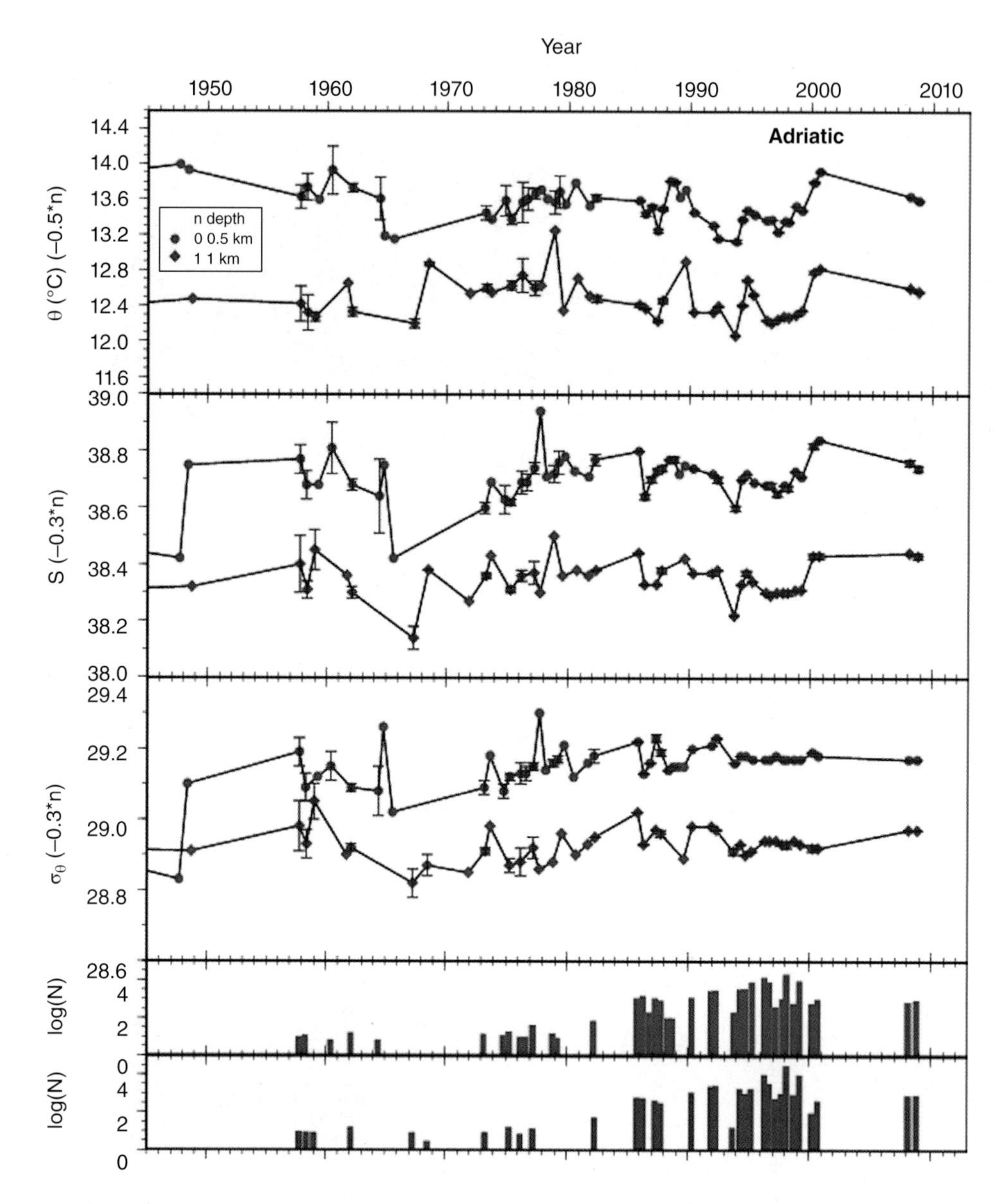

Figure 7.6. Adriatic Sea potential temperature θ, salinity, and σ_θ density at depth layers of 450–550 m (n = 0) and 900–1100 m (n = 1) (vertically shifted down by a constant times n), and log(N), the logarithm of the number of data points N averaged for each half-year bin interval.

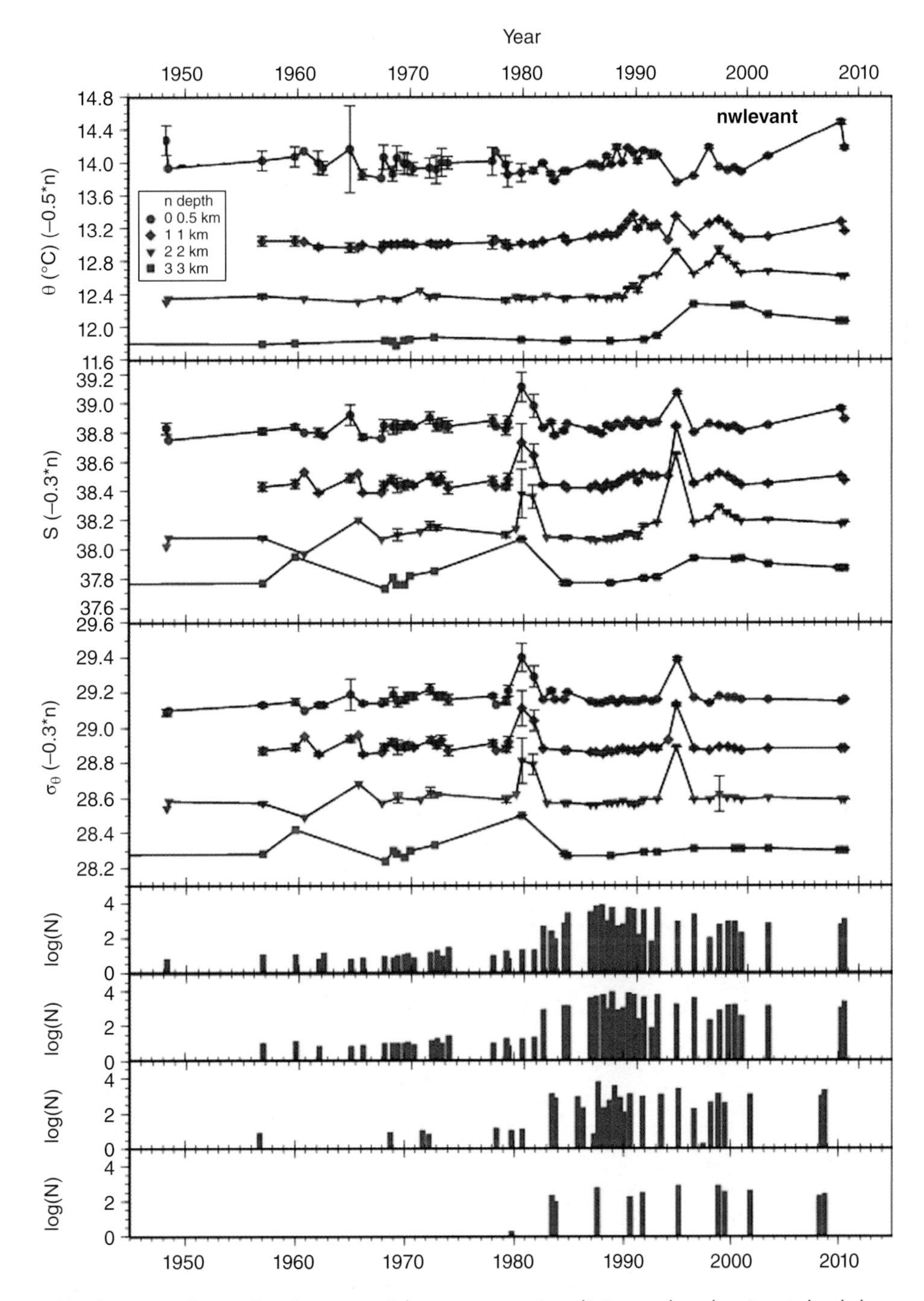

Figure 7.7. Northwestern Levantine Sea potential temperature θ, salinity, and σ_0 density at depth layers of 450–550 m (n = 0), 900–1100 m (n = 1), 1900–2100 m (n = 2), and 2900–3100 m (n = 3) (vertically shifted down by a constant times n) and log(N), the logarithm of the number of data points N averaged for each half-year bin interval.

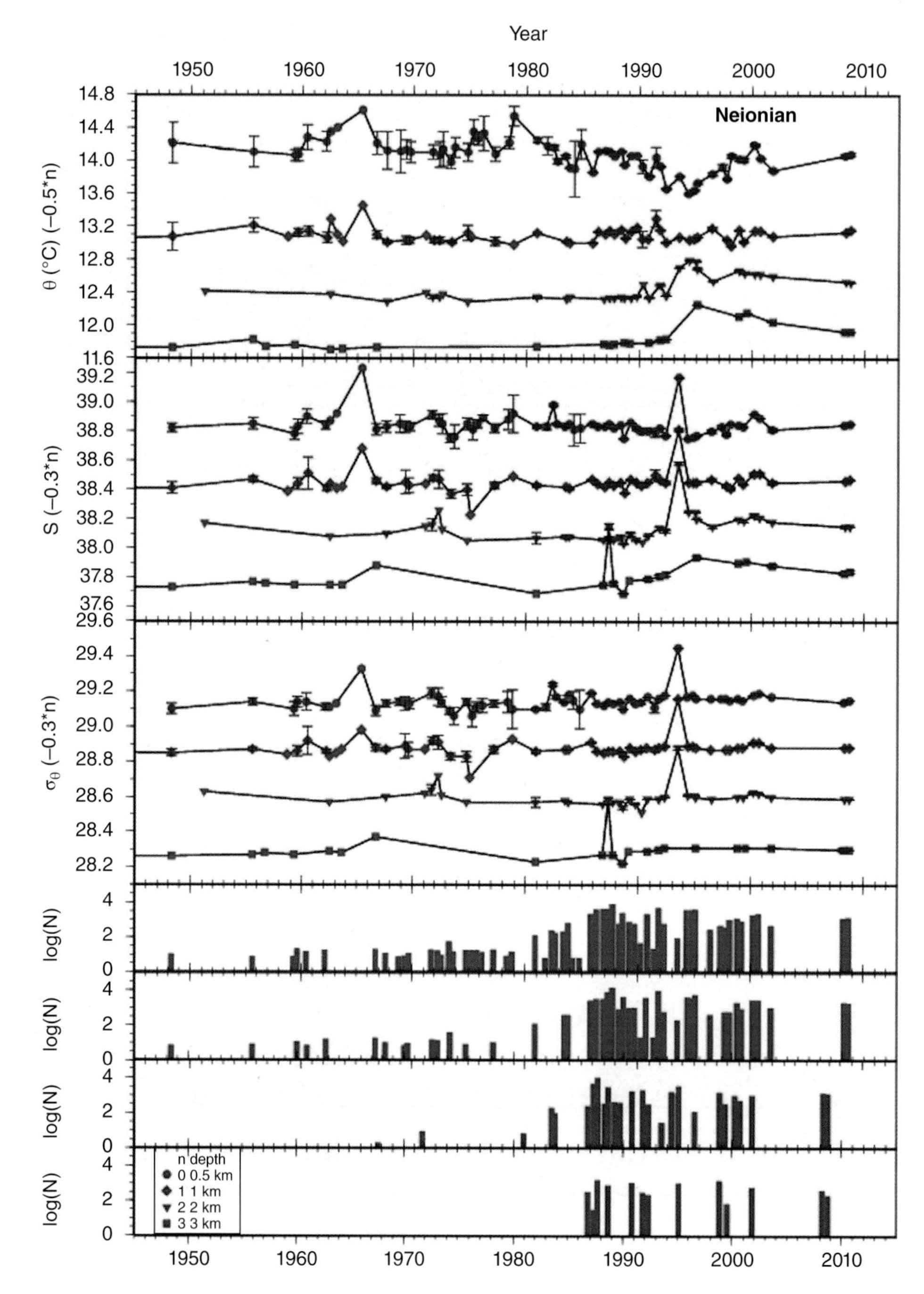

Figure 7.8. Northeastern Ionian Sea potential temperature θ, salinity, and σ_θ density at depth layers of 450–550 m (n = 0), 900–1100 m (n = 1), 1900–2100 m (n = 2), and 2900–3100 m (n = 3) (vertically shifted down by a constant times n), and log(N), the logarithm of the number of data points N averaged for each half-year bin interval.

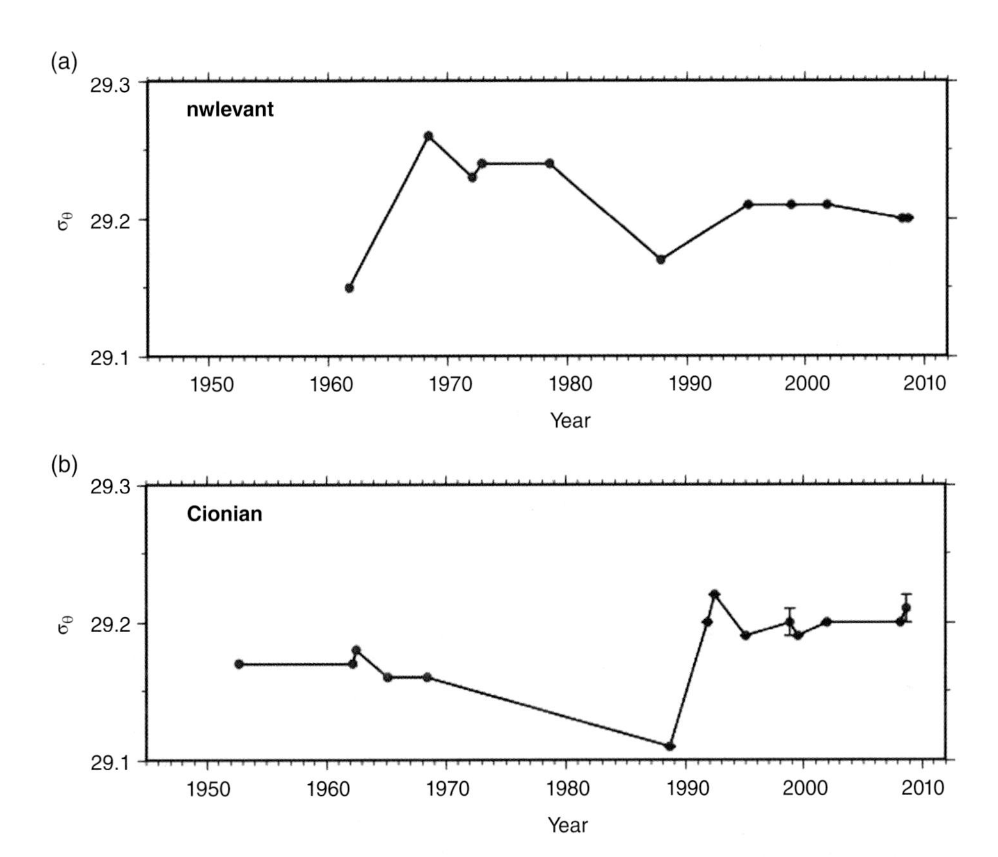

Figure 7.9. Potential density in the 3900–4100-m depth interval in the (a) northwestern Levantine and (b) central Ionian subdomains representing the deepest areas of the Levantine and Ionian seas.

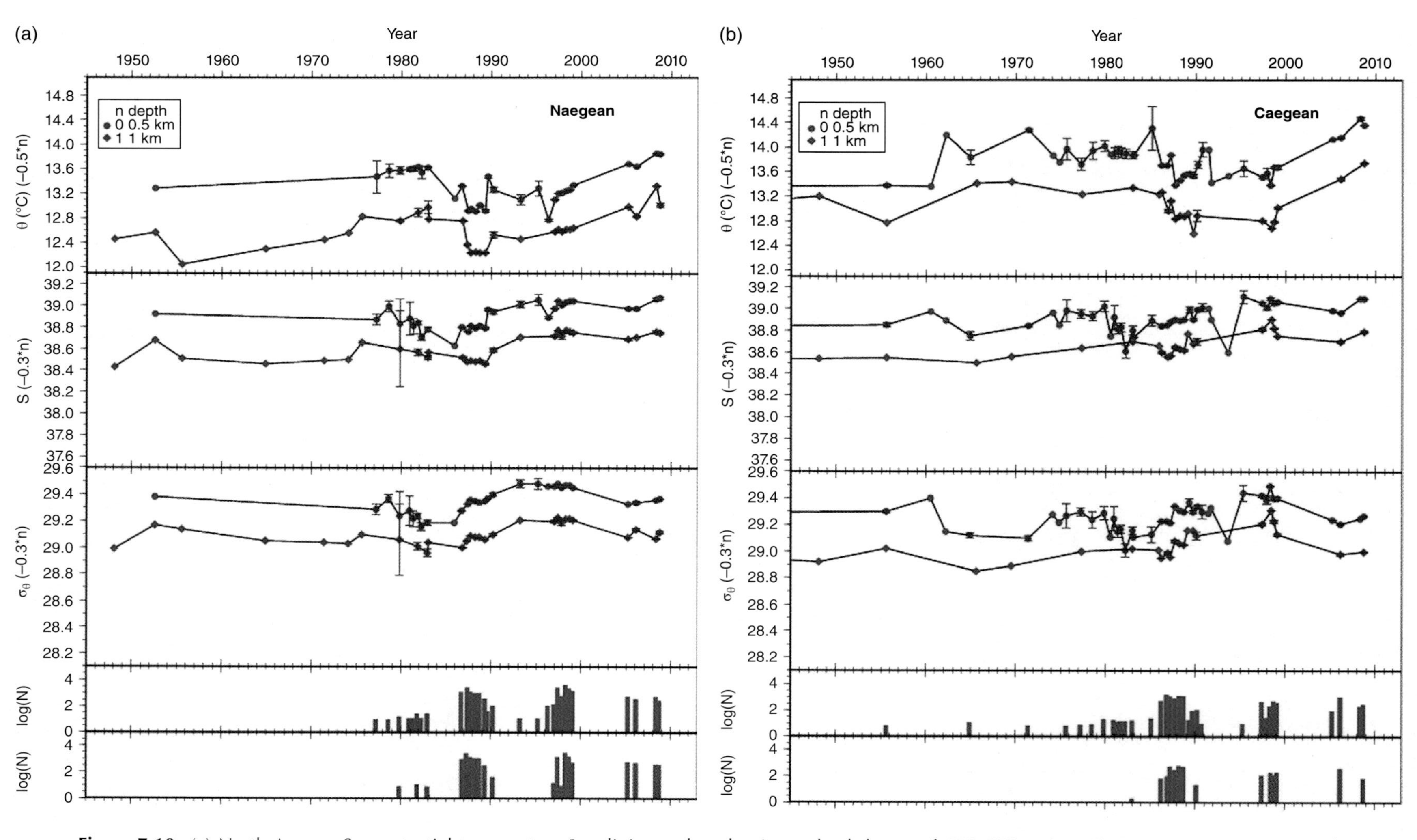

Figure 7.10. (a) North Aegean Sea potential temperature θ, salinity, and σ_θ density at depth layers of 450–550 m (n = 0), 900–1100 m (n = 1), and 1900–2100 m (n = 2) (vertically shifted down by a constant times n) and log(N), the logarithm of the number of data points N averaged for each half-year bin interval; (b) Central Aegean Sea potential temperature θ, salinity, and σ_θ density at depth layers of 450–550 m (n = 0), 900–1100 m (n = 1), and 1900–2100 m (n = 2) (vertically shifted down by a constant times n) and log(N), the logarithm of the number of data points N averaged for each half-year bin interval;

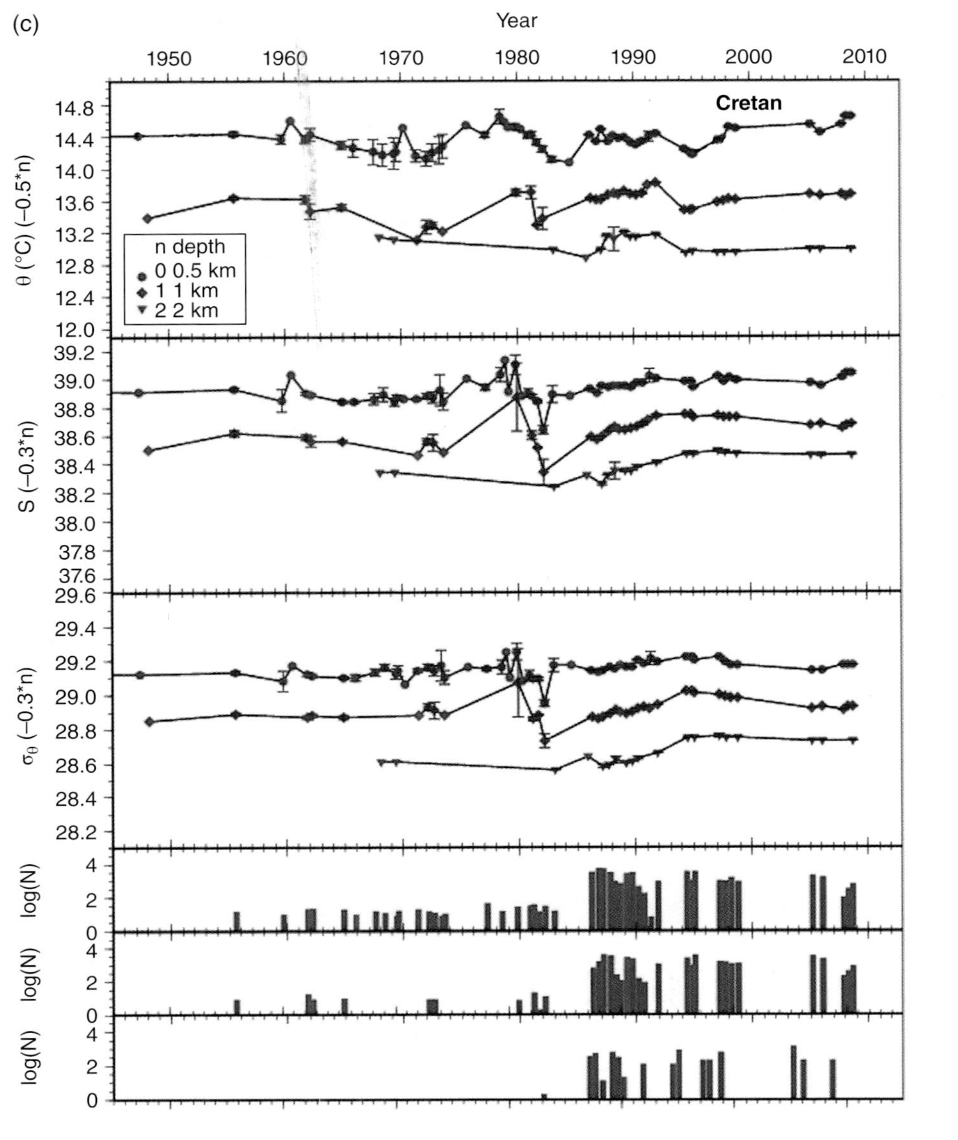

Figure 7.10. (Continued) (c) Southern Aegean Sea (Cretan Sea) potential temperature θ, salinity, and σ_θ density at depth layers of 450–550 m (n = 0), 900–1100 m (n = 1), and 1900–2100 m (n = 2) (vertically shifted down by a constant times n), and log(N), the logarithm of the number of data points N averaged for each half-year bin interval.

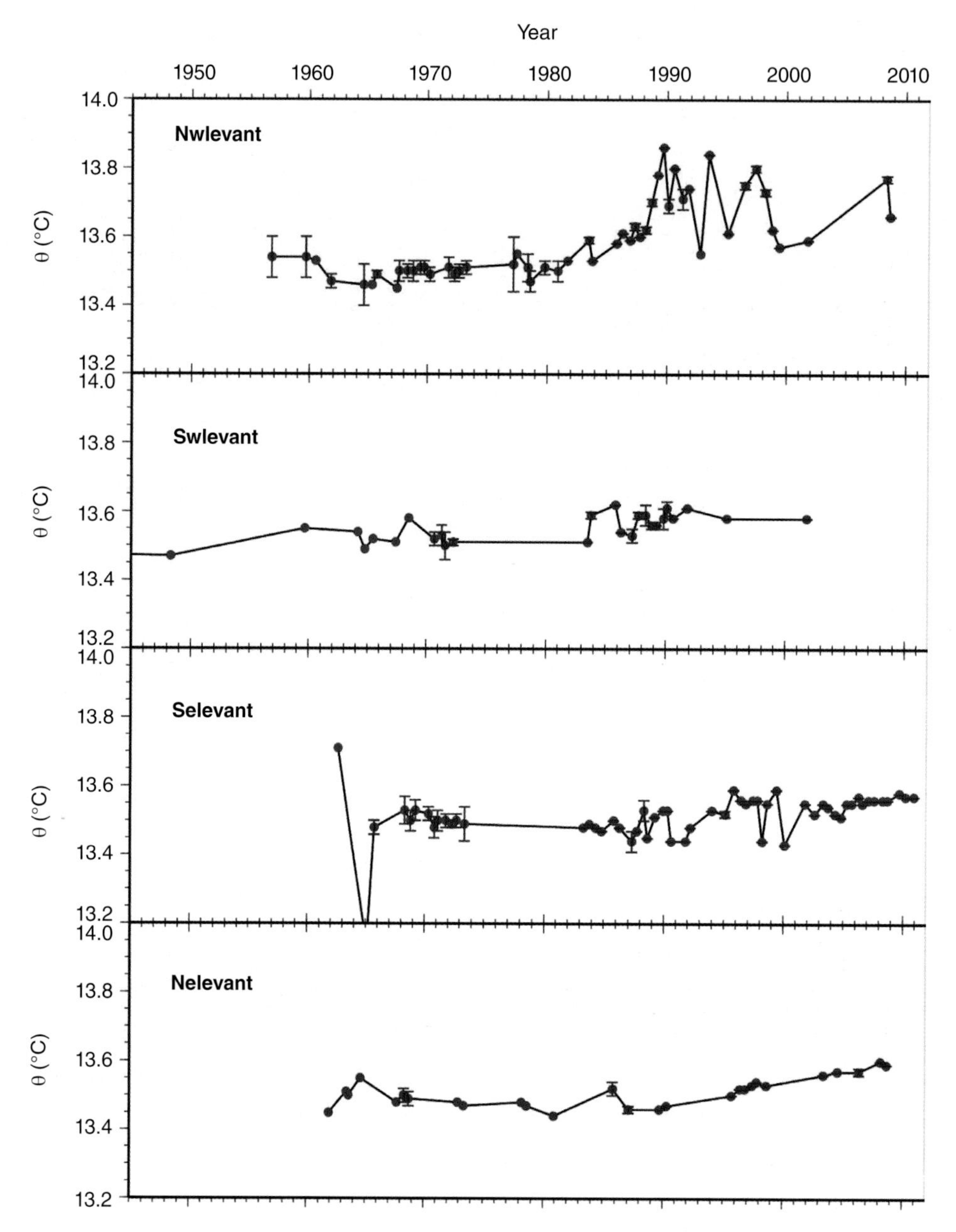

Figure 7.11. Potential temperature θ in the 900–1100-m depth level in the (a) northwestern, (b) southwestern, (c) southeastern, and (d) northeastern Levantine subdomains.

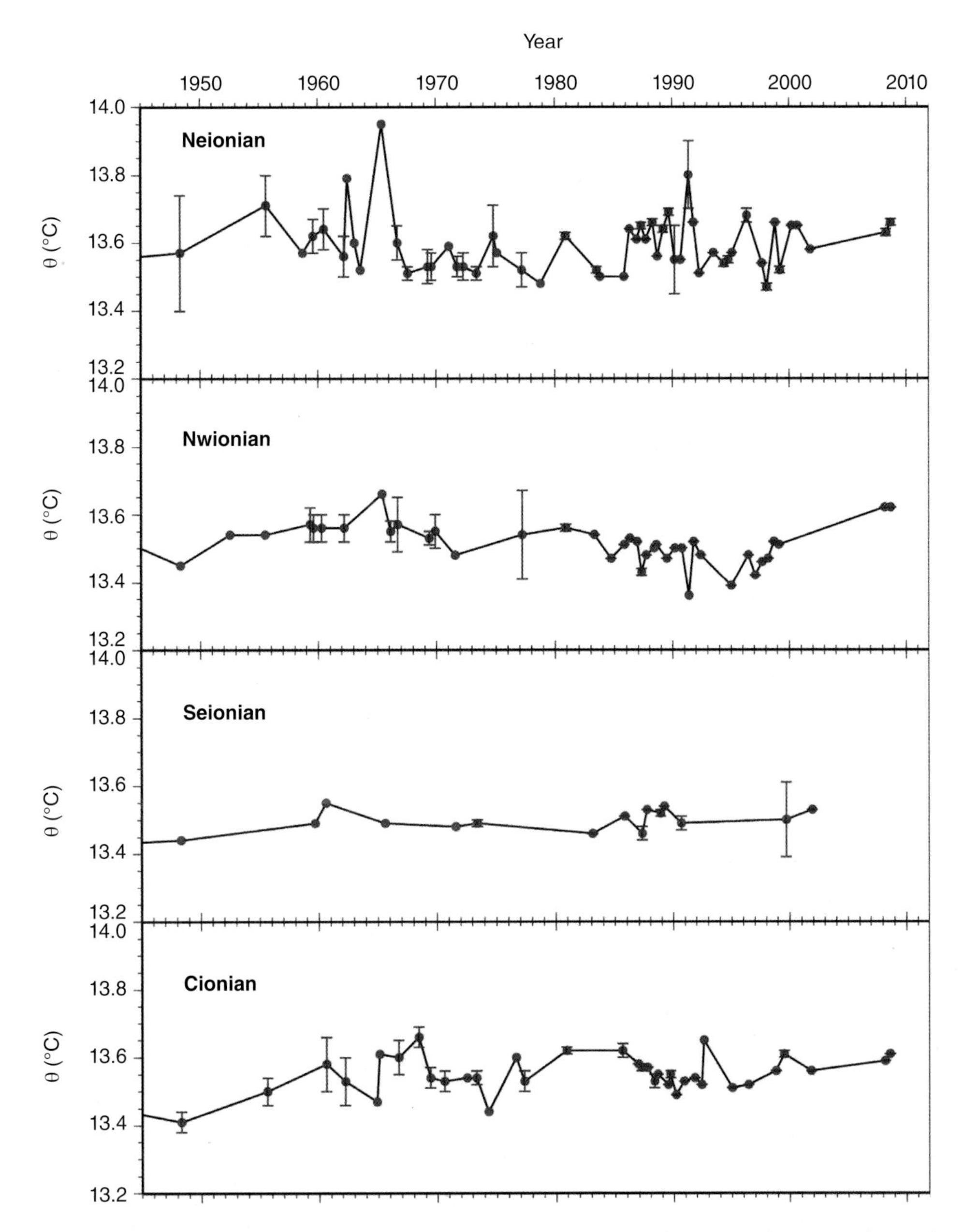

Figure 7.12. Potential temperature θ in the 900–1100-m depth level in the (a) northeastern, (b) northwestern, (c) southeastern, and (d) central Ionian subdomains.

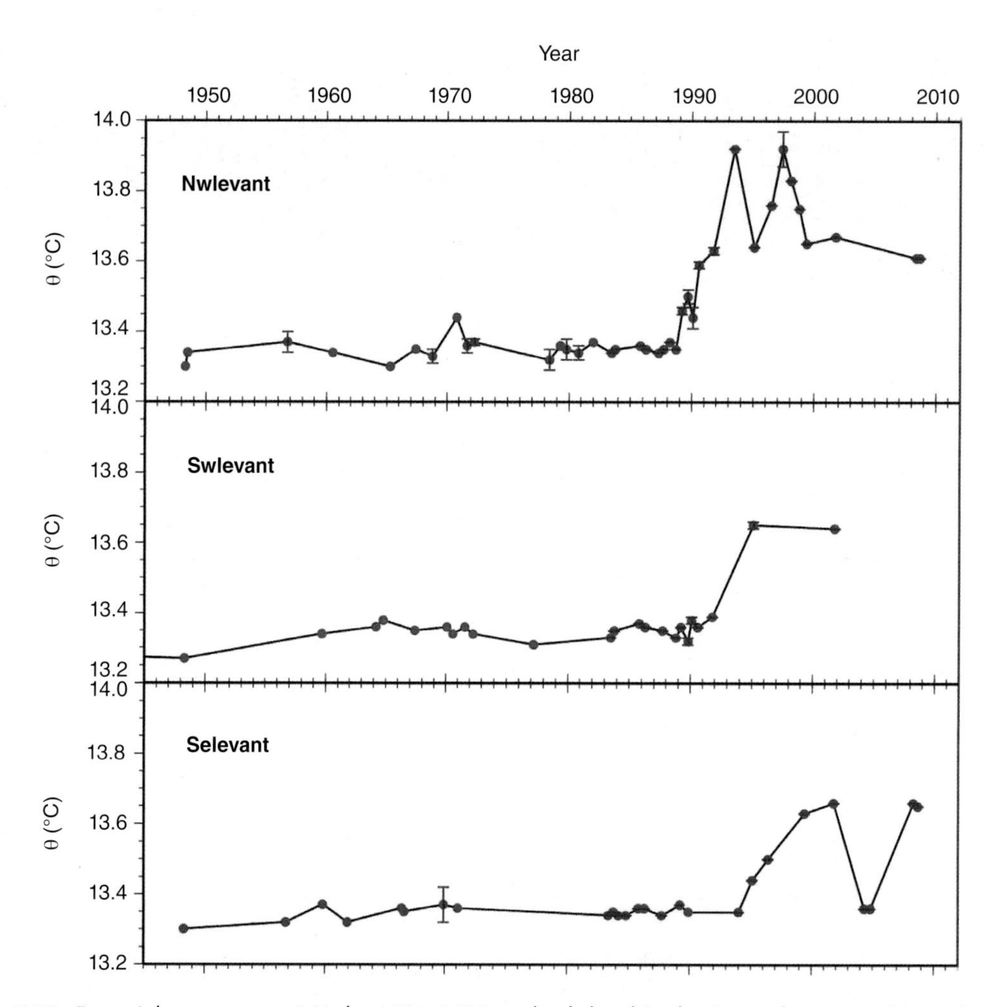

Figure 7.13. Potential temperature θ in the 1900–2100-m depth level in the (a) northwestern, (b) southwestern, and (c) southeastern Levantine subdomains.

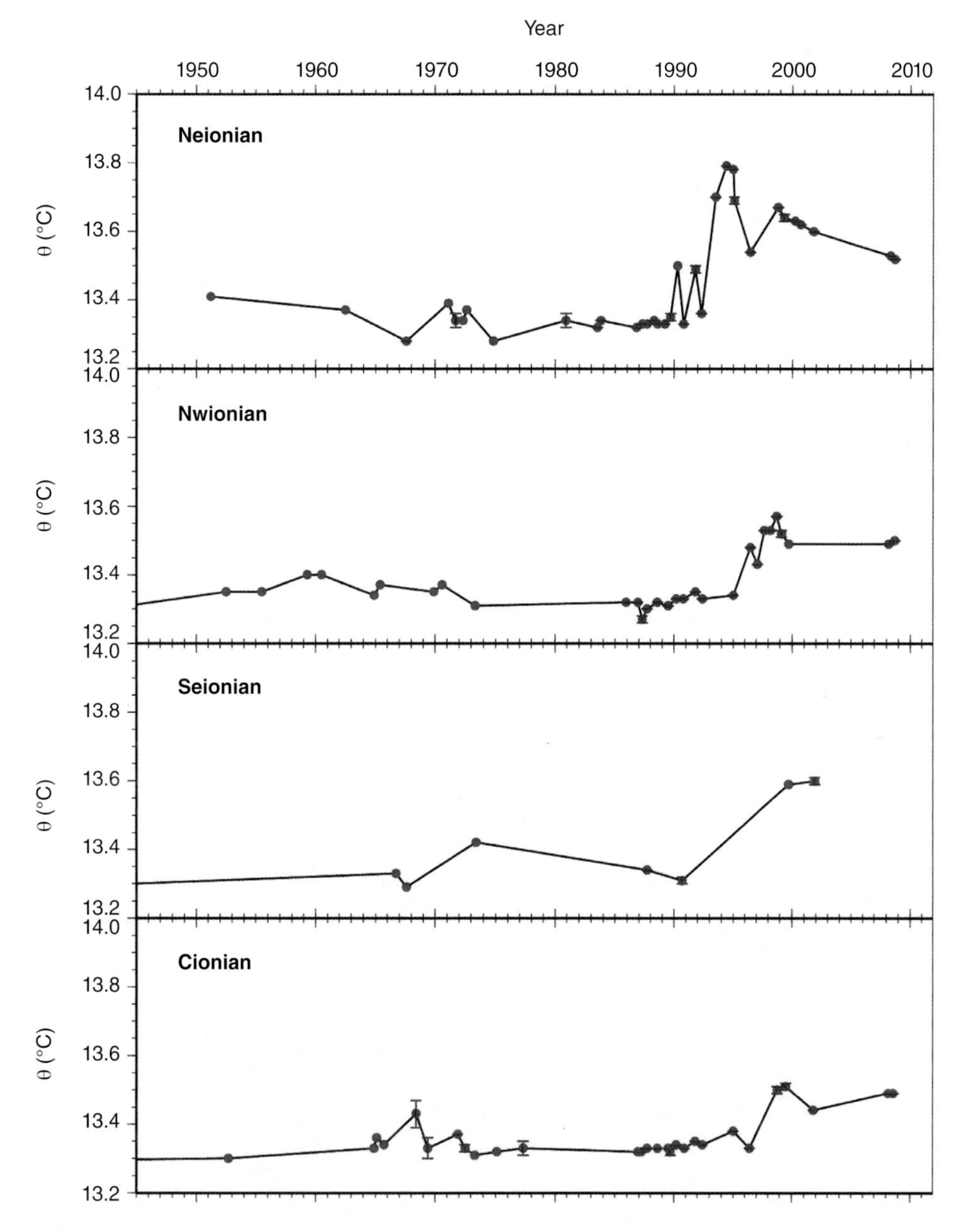

Figure 7.14. Potential temperature θ in the 1900–2100-m depth level in the (a) northeastern, (b) northwestern, (c) southeastern, and (d) central Ionian subdomains.

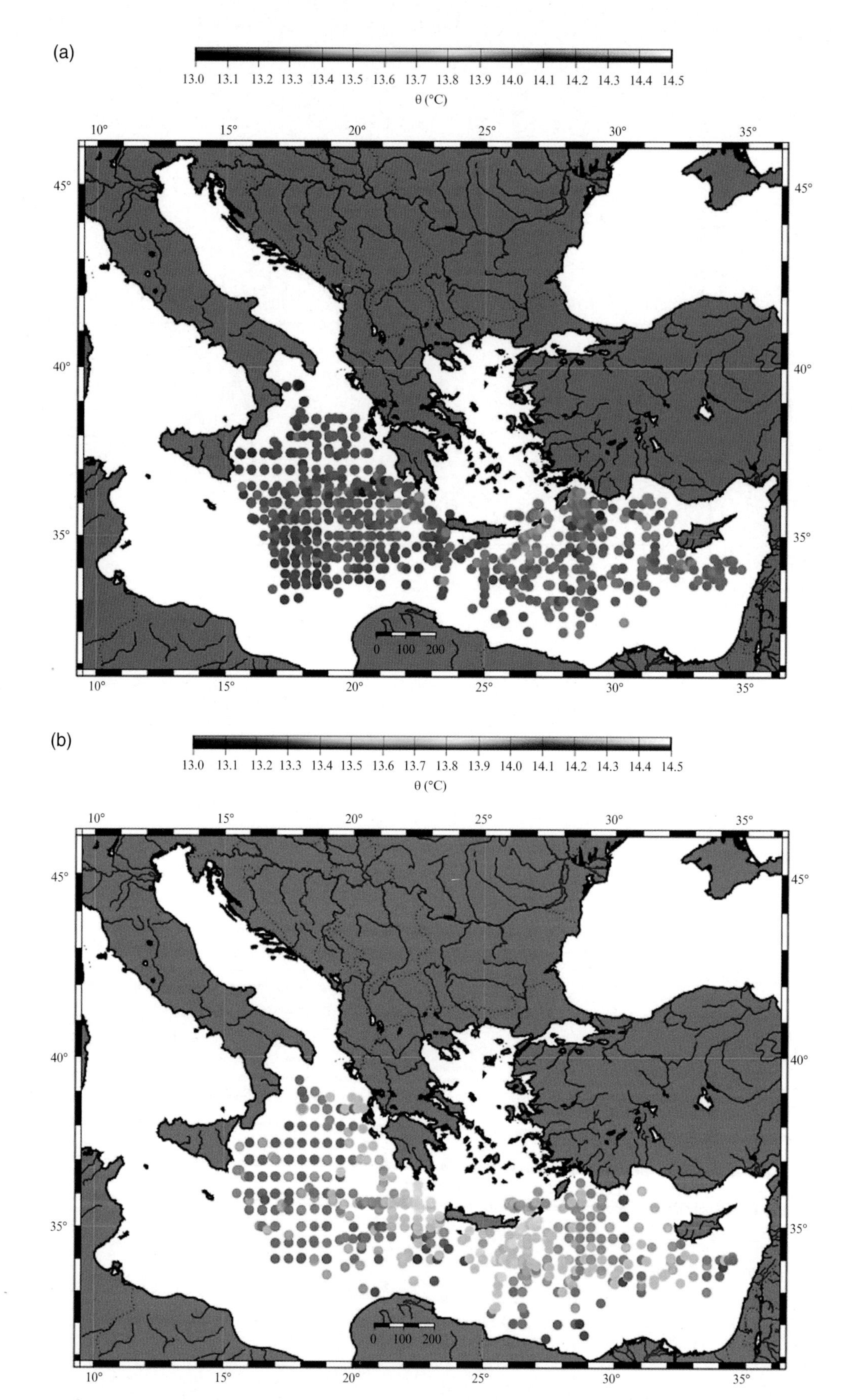

Figure 7.15. Temperature at 1900–2100-m depth interval binned and averaged at 0.1° x 0.1° latitude-longitude intervals during the (a) 1970–1990 and (b) 1990–2012 periods.

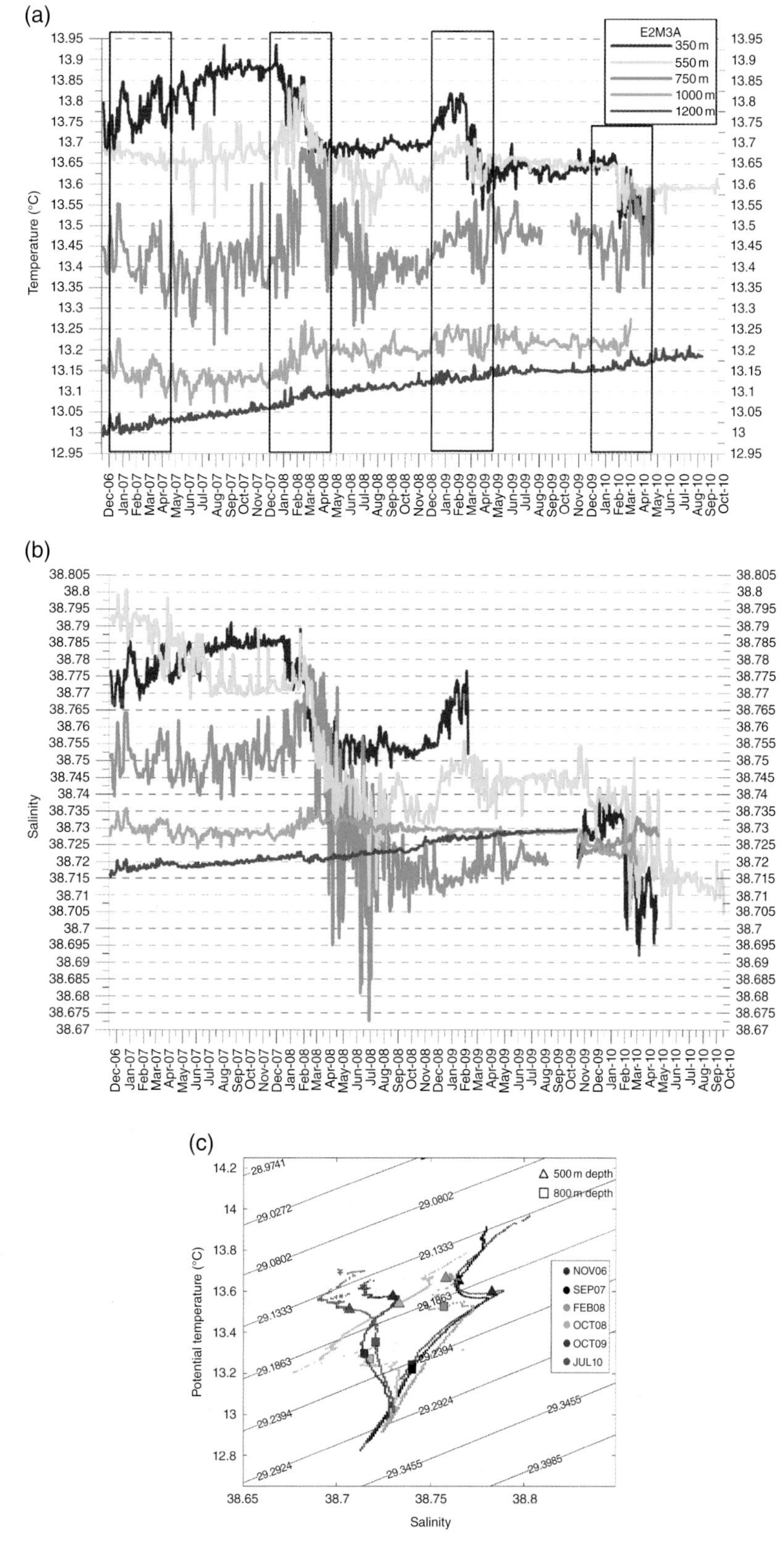

Figure 9.4. Time series of *in situ* (a) temperature and (b) salinity recorded by the CT and CTD sensors installed at E2M3A. Data were despiked and filtered with a 33-h Hamming filter. Winter periods are indicated in panel a. Potential temperature (θ) - Salinity (S) diagram of CTD profiles carried out during the oceanographic cruises in the proximity of the E2M3A (panel (c).

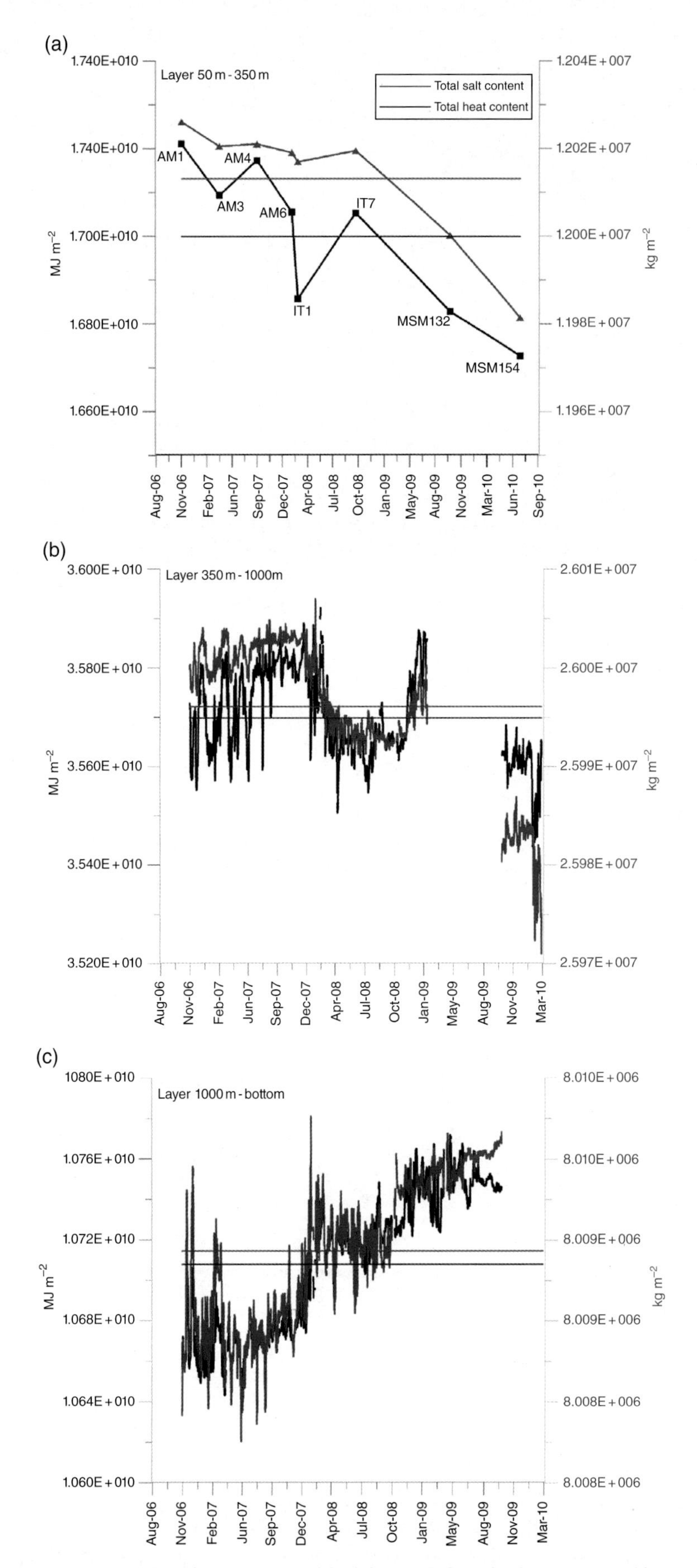

Figure 9.5. Temporal evolution of total heat content (black lines) and total salt content (red lines) at E2M3A. Panel (a) shows the results in the layer between 50 m and 350 m, as obtained from CTD casts in the proximity of E2M3A during its operational time; panels (b) and (c) show the results in the intermediate and deep layers, respectively.

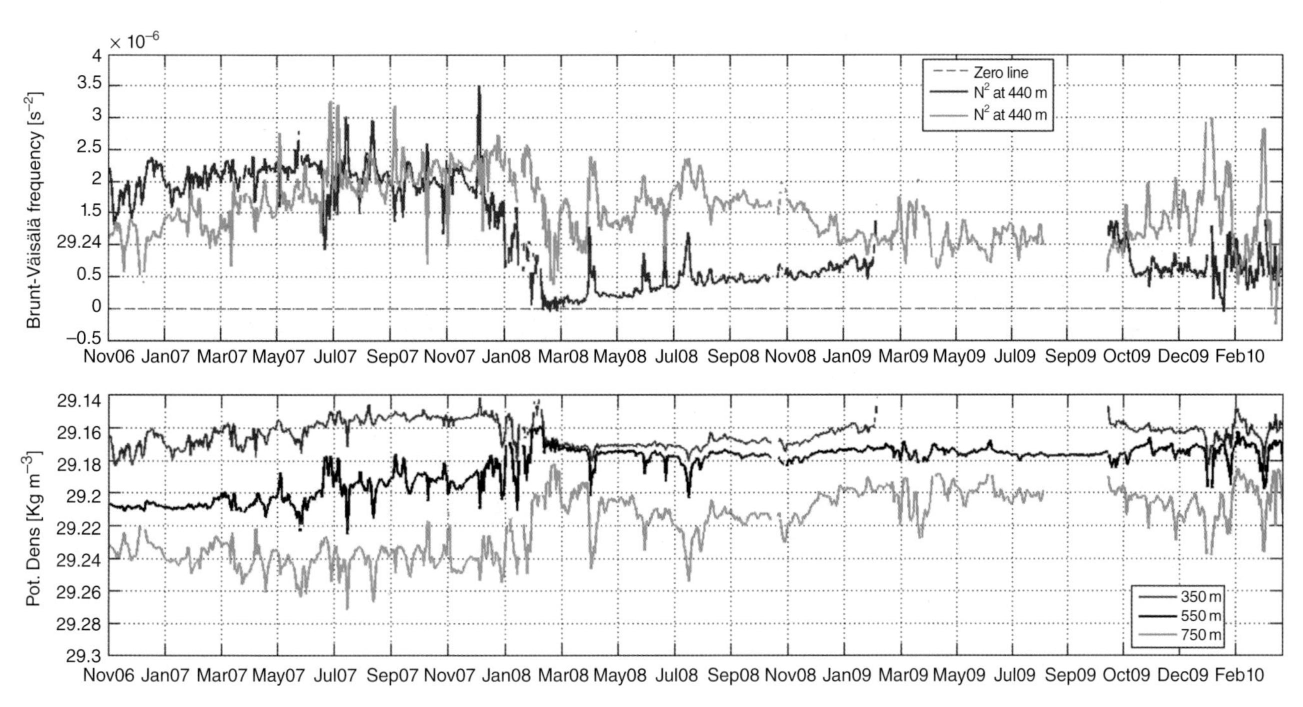

Figure 9.6. Stability of the water column calculated in the layers 350–550 m and 550–750 m (upper panel). The lower panel shows the time series of the potential density at 350 m, 550 m, and 750 m.

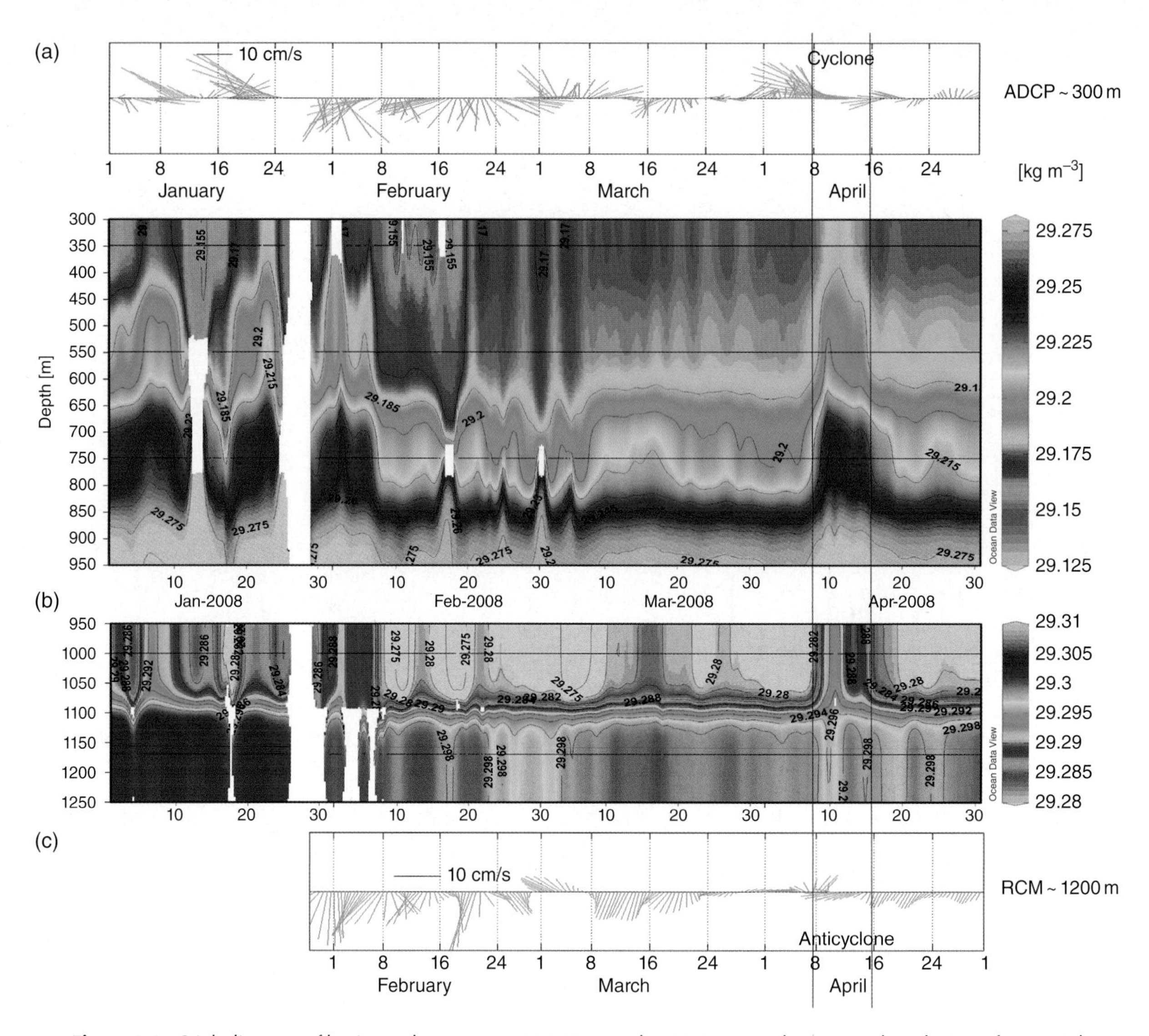

Figure 9.9. Stick diagram of horizontal currents at (a) 300 m and at (c) 1180 m; (b) temporal evolution of potential density throughout the water column. The passage of a cyclonic eddy in the intermediate layer and the concurrent passage of an anticyclonic eddy in the bottom layer are marked by the black lines. The data were recorded between 1 January 2008 and 30 April 2008 at E2M3A.

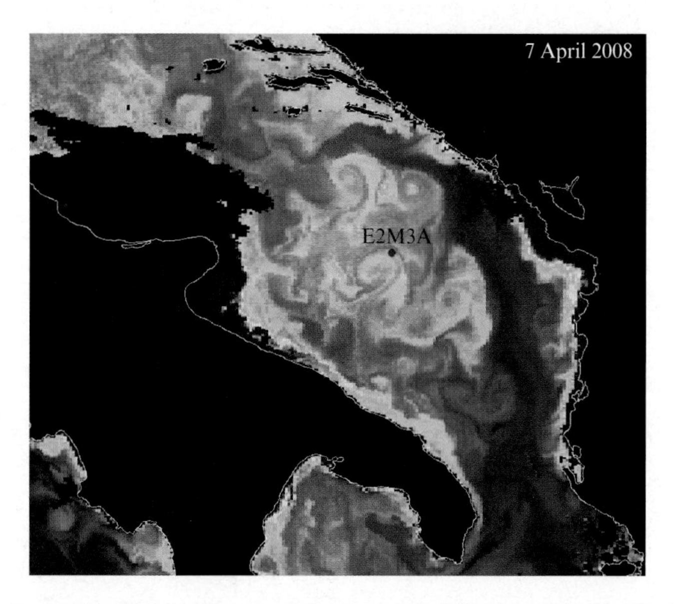

Figure 9.10. Moderate Resolution Imaging Spectroradiometer (MODIS) chlorophyll-*a* distribution in the Southern Adriatic Pit for 7 April 2008. The image has been processed at OGS with SeaWiFS Data Analysis System (Seadas). Data are courtesy of OceanColor database (freely available on http://oceancolor.gsfc.nasa.gov).

7

Deep-Water Variability and Interbasin Interactions in the Eastern Mediterranean Sea

Emin Özsoy[1], Sarantis Sofianos[2], Isaac Gertman[3], Anneta Mantziafou[2], Ali Aydoğdu[1], Sotiria Georgiou[2], Ersin Tutsak[1], Alex Lascaratos[2], Artur Hecht[3], and Mohammed Abdul Latif[1]

7.1. INTRODUCTION

Ocean–atmosphere-land interactions and consequent feedbacks between regional and global climate systems, could be disproportionately large in the Mediterranean as a result of contrasts between marine and continental climates and complex land-sea bottom topography [*Özsoy*, 1999]. The eastern basin of the Mediterranean is a remote area of the world ocean, the isolation of its various basins increased with distance from the Atlantic Ocean and with further constraints posed by the various straits.

The eastern Mediterranean and especially its easternmost basin, the Levantine Sea, were less well known in the first half of the last century, especially when compared with the western basin, though rapidly became the subject of advanced studies carried out since the Physical Oceanography of the Eastern Mediterranean [*POEM Group*, 1992] research program of 1985–1991. Despite continuing efforts under national and international programs, there appears to be a greater need for systematic observations in the whole of the eastern Mediterranean to understand its high level of climatic variability [*Özsoy*, 1999; *Lionello et al.*, 2006; *Hoepffner*, 2006; *CIESM*, 2008, 2011; *Malanotte-Rizzoli and the Pan-Med Group*, 2012].

The investigation of the deep-water characteristics and their variability on long timescales can help in the understanding of the possible mechanisms, sensitivity, localization, and frequency of water mass formation, and their links to atmospheric forcing. The need to document and understand deep-water variability is especially acute in the case of the eastern Mediterranean Sea, where unexpected recent changes have been observed in the thermohaline transport components connected with multiple sites of intermediate and deep convection.

The water masses in the ocean are usually identified with distinct water properties such as temperature and salinity pairs. The water masses are formed as a result of long-term circulation and mixing processes and in some way they are associated with the thermohaline circulation. For the Mediterranean basin, there are meridional and zonal vertical circulation belts. An open-ended, shallow zonal vertical circulation completes a circuit by the entry of Atlantic Water (AW) at Gibraltar, later transformed into Levantine Intermediate Water (LIW) in the eastern Mediterranean, which then returns to the Atlantic Ocean as a submerged flow. Superposed on this circulation, closed, deep meridional cells are created as a result of the deep-water mass formation in the northern parts of the Mediterranean basin (Gulf of Lions, Adriatic Sea, and Aegean Sea). The zonal cells have decadal timescales, while the meridional overturning cells have multidecadal timescales (50–80 years) [*Pinardi* and *Masetti*, 2000].

High salinity Levantine Intermediate Water (LIW) occupying the intermediate layer of the eastern Mediterranean basin (200–500 m) is formed in the permanent Rhodes Gyre [*Ovchinnikov and Plakhin*, 1984; *Lascaratos and Nittis*, 1998; *Pinardi and Masetti*, 2000; *LIWEX Group*, 2003], along the adjacent zones of the

[1] *Institute of Marine Sciences, METU, Erdemli, Mersin, Turkey*
[2] *University of Athens, Athens, Greece*
[3] *Israel Oceanographic and Limnological Research, Haifa, Israel*

The Mediterranean Sea: Temporal Variability and Spatial Patterns, Geophysical Monograph 202. First Edition.
Edited by Gian Luca Eusebi Borzelli, Miroslav Gačić, Piero Lionello, and Paola Malanotte-Rizzoli.

southern Aegean and northern Levantine, such as the Gulf of İskenderun and often simultaneously with deep waters [*Sur et al.*, 1993; *Özsoy et al.*, 1993].

The Eastern Mediterranean Deep Water (EMDW) is characteristic water mass of the deep eastern basin. Using limited amounts of data obtained in the late 1950s early 1960s, *Pollak* [1951], *Lacombe and Tchernia* [1960], *Wüst*[1961], *Plakhin* [1972], *Miller* [1974], and *El-Gindy and El-Din* [1986] have implied the Aegean Sea as a possible source contributing to the formation of the EMDW, although it has often been suspected that the quantity and density of the Aegean outflow would not be sufficient to contribute to the EMDW.

Based on historical data obtained since early last century [*Nielsen*, 1912], followed by others around mid-century [*Wüst*, 1961], and including those obtained by the extensive coverage during the POEM program, the Adriatic Sea was widely accepted as the main source of the EMDW. According to this dominant view of the past, the dense waters formed in winter in the relatively shallow Adriatic Sea, flowed to the bottom of the Ionian Sea, and spread farther to the Levantine Sea [*Wüst*, 1961; *Schlitzer et al.*, 1991; *Malanotte-Rizzoli and Hecht*, 1988]. The deep-water overturning time was estimated roughly to be about 100 years with an average formation rate of 0.3 Sv (1 Sverdrup = 10^6 m^3 s^{-1}) [*Roether et al.*, 1991].

Major changes in the deep-water formation and renewal occurred in the 1990s, when it became evident that the Aegean Sea acted as a new source of EMDW, instead of the widely held view on its Adriatic Sea origin. The first signs came during the *Meteor* cruise M25 of 1993, when an anomalously saline, warm-water mass was found in the deep Levantine Sea [*Heike et al.*, 1994]. The hydrographic surveys of 1994–1995 further revealed the Aegean Sea as a dominant source of deep water [*Roether et al.*, 1996, 2007; *Klein et al.*, 1999; *Lascaratos et al.*, 1999; *Malanotte-Rizzoli et al.*, 2003]. The average outflow rate from the Cretan Basin of the new deep water formed in the Aegean Sea was estimated to be about 1.2 Sv [*Roether et al.*, 1996], and later the estimate was revised to be about 3.0 Sv during the peak outflow period between 1992 and 1994 [*Roether et al.*, 2007], much greater than the former Adriatic outflow rate.

The mechanism creating the Cretan Dense Water (CDW) filling the Cretan Basin before its outflow to the eastern basin was not very well documented, although it was suspected that consecutive transformations between the shallow shelf area and the three deep basins of the Aegean Sea resulted in dense water flowing south from the northern reaches of the sea. Higher salinities and increased amounts of LIW entering the Aegean Sea from the Levantine Basin [*Malanotte-Rizzoli et al.*, 1999; Theocharis et al., 1999; *Zervakis and Georgopoulos*, 2002] aiding shelf mixing at the Samothraki and Lemnos plateaus of the north Aegean [*Ovchinnikov et al.*, 1990; *Theocharis and Georgopoulos*, 1993; *Lascaratos et al.*, 1999; *Zervakis et al.*, 2000a] and potentially at the Cyclades plateau of the central Aegean [*Theocharis et al.*, 1999; *Gertman et al.*, 2006] created the dense waters finally reaching the Cretan Basin. The dense water formed in the upper reaches of the Aegean Sea eventually fills the deep Cretan Sea Basin and results in outflows through the Cretan Sea Straits, triggering the EMT events. *Zervakis et al.* [2003] and *Androulidakis et al.* [2012] show that the air-sea interactions and lateral inputs of the low-density Black Sea Water (BSW) outflowing from the Dardanelles Strait can effectively modulate the dense-water production in the north Aegean Sea.

In about the same years that the main part of the POEM program was concluded, it had become clear through new observations that significant changes were taking place in the eastern Mediterranean. In the northeastern part of the sea, cold winters in 1985, 1987, 1989, and 1992–1993 created favorable conditions for Levantine Intermediate Water (LIW) formation on the periphery of the Rhodes Gyre, in the northern Levantine Basin, often simultaneously with deep-water formation at the center of the Rhodes cyclonic circulation [*Özsoy et al.*, 1989, 1991, 1993; *Sur et al.*, 1993; *Gertman et al.*, 2006]. During the same years, the last common field experiment of POEM in October 1991 indicated significant changes in the thermohaline circulation. Increased salinity of the Levantine surface water in this period [*Özsoy et al.*, 1993; *Hecht and Gertman*, 2001; *Gertman et al.*, 2006] coincident with the blocking of the Levantine circulation by large anticyclonic eddies, resulted in diversion toward the Aegean Sea of the saline water transported by the Asia Minor Current (AMC) along the Anatolian coast [*Malanotte-Rizzoli et al.*, 1999; *Theocharis et al.*, 1999]. Changes in the air-sea fluxes during the same period were associated with a series of observed changes in the circulation and hydrography [*Zervakis et al.*, 2000; *Josey*, 2003].

A number of studies used numerical modeling techniques to simulate the events in an effort to understand exact sequence of the events and the underlying physics. The role of changes in the Aegean Sea atmospheric forcing [*Samuel et al.*, 1999], the dense-water formation processes in the north Aegean Sea [*Androulidakis et al.*, 2012], the mixing effects of a series of cold winters during 1987–1995 on the Cretan Sea outflow dynamics [*Wu et al.*, 2000], and the details of LIW and EMDW production near the Rhodes Gyre [*Lascaratos et al.*, 1999; *Nittis et al.*, 2003] were investigated by numerical simulations. Other numerical experiments [*Stratford and Haines*, 2002; *Beuvier et al.*, 2010] showed sensitivity to successive cold winters and changes in atmospheric forcing, resulting in dense-water outflow from the Aegean Sea.

A box-model with interbasin coupling [*Ashkenazy et al.*, 2011] showed multiple states of the nonlinear system, implying underlying instabilities.

Despite all experimental and numerical evidence, it is not yet clear how frequent is the switching between deep-water sources, or how persistent are the described thermohaline circulation cells of the Mediterranean Sea. It is also not clear how the events at near-surface or deep levels are connected to the surface climatological forcing by the atmosphere. In this study, however, we are presently not much concerned with the connection to the atmosphere, as we will only be analyzing the deep-water characteristics independent of the changes in the upper-water column.

In the following, we provide the data sources and methodology (section 2), followed by an analysis of the changes in basinwide properties linked with the interactions between individual basins of the eastern Mediterranean (section 3), and the intrabasin variability (section 4). We provide a general discussion in section 5.

7.2. METHODOLOGY

In this work, we tried to collect the most complete temperature and salinity profile dataset in the eastern Mediterranean (1912–2010). A comprehensive dataset was made available by ISRAMAR, the Israel Marine Data Center, National Institute of Oceanography of the IOLR, Israel. This dataset is based on MEDAR/MEDATLAS II collection (http://www.ifremer.fr/medar) and the MATER project collection (http://www.ifremer.fr/sismer/program/mater). Additional POEM basinwide cruises (1985–1991) obtained in several multinational cruises in the eastern Mediterranean as well as Soviet cruises (1987–1990) in the eastern Mediterranean [*Hecht and Gertman*, 2001] were included. The dataset was extended significantly by POEM and Soviet cruises data, and also in the framework of SESAME EU project, including recent cruise and ARGO floats data collected from publicly available oceanographic databases ("WOD05"— http://www.nodc.noaa.gov/OC5/WOD05/pr_wod05.html; "Coriolis" —http://www.coriolis.eu.org; "ICES"— http://ocean.ices.dk), as well as cruise data collected in the framework of the SESAME project. The data are inclusive of the SeaDataNet CTD and bottle database, except that the ISRAMAR database has the above additions with a larger number of casts and improved quality control of the data after 2000. Data obtained by bottles, CTD, and Argo floats were accepted for analysis. XBT data were rejected due to their relatively low accuracy, which is comparable with the variability in deep layers. The vertical resolution of the analyzed data varies as a function of the dates of data collection, instrument type, etc., but the CTD data available after the 1980s typically have a resolution of about 1 m.

The Mediterranean and especially the eastern Mediterranean Sea is trapped among three continents and divided into several basins by the geometry of the main landmasses and islands. Because each region in the Mediterranean has specific climate and each part of the sea responds differently to the forcing, it is required to discover these differences and to investigate whether particular changes are triggered in some regions and if later the effects are transmitted to other regions. We therefore analyze the data with respect to the subdomains designed in Figure 7.1. We base our analysis on all deep stations (deeper than 450 m) in the database, displayed in Figure 7.1a.

Although the stations shown in Figure 7.1a have been filtered for stations deeper than 450 m, few stations appear in shallower areas, apparently as a result of errors in entering coordinates in the database. In addition, it should be obvious that not all of these stations are used in the analysis because of the quality and depth intervals filtering described in the following.

To observe long-term climatic influences, we first look at the deep data. We have grouped the data according to subdomains shown in the map, and have then selected the data in depth intervals that were arbitrarily defined but at the same levels in all basins. We selected data below 450 m, that is, depths increasingly isolated from seasonal surface processes, specifically to study climatic influences. The depth intervals compared were selected to be 450–550 m, 900–1100 m, 1900–2100 m and 2900–3100 m, centered respectively at depths of 0.5, 1, 2, and 3 km. In shallower basins such as the Aegean and Adriatic seas, not all of these depth intervals had adequate amount of data, producing results only at depths centered at 0.5 and 1 km in the Adriatic Sea and at 0.5, 1, and 2 km in the Aegean Sea.

The data quality flags for the depth, temperature, and salinity values were checked and used to filter out data with unsatisfactory individual (flag value ≥ 2) or overall data quality assignments. The computed potential temperature (referenced to the surface), salinity, and potential density (referenced to the surface) data were collected in half-year time bins and the average, standard deviation, and confidence limits were calculated in each bin. The grouping of data in half-year bins allows sufficient time resolution to better detect dense-water spreading events expected at the cold part of the year, but relatively free of the seasonal signal diminished at depths below 500 m.

Average properties are represented by the data points in the time-series plots of Figure 7.2 an later, while confidence intervals of the mean estimates are shown by vertical bars. The calculation of the double-sided 95% confidence intervals of the estimated mean properties in

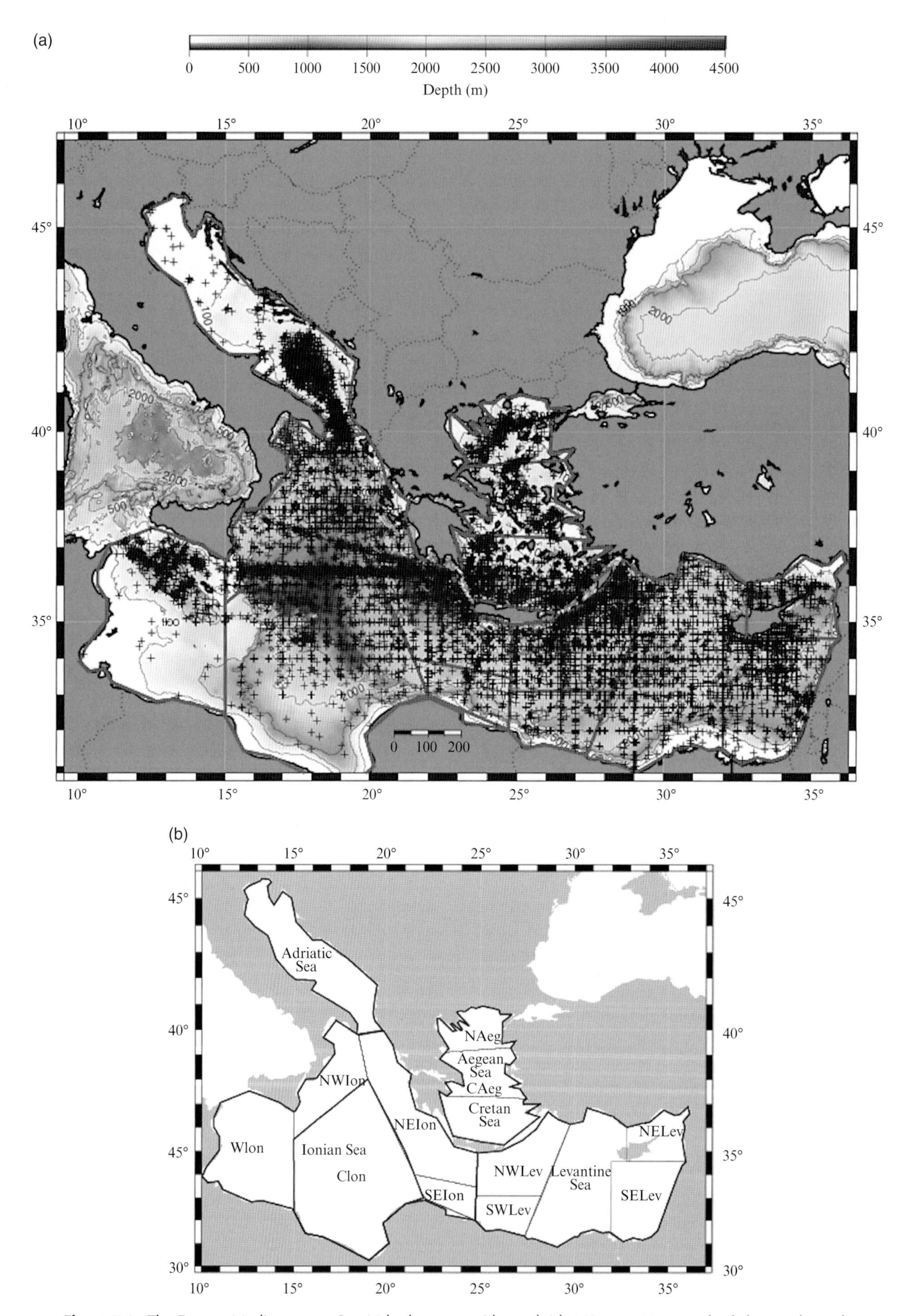

Figure 7.1. The Eastern Mediterranean Sea (a) bathymetry with overlaid station positions and subdomain boundaries, (b) subdomains of the study area. For color detail, please see color plate section.

the bin intervals were based on Student's t-test [*Emery and Thompson*, 2001], making use of the Numerical Recipes [*Press et al.*, 2007] library functions INVBETAI, BETAI, BETACF, and GAMMLN. The confidence interval, multiplying the standard deviation by a strong function of the number of samples, measures the reliability range of the estimated mean value, and therefore it is much reduced when there are a large number of observations proportional with the sample variance. The longer error bars in the plots appear when the number of observations or sample variance is smaller. In most cases, when there is a sufficiently large number of observations, the error bars are diminished and not clearly seen in the plots. However, we should also note that only a circular symbol for the data point without the error bar is displayed for the trivial case of a single observation in a bin, which is noted by not having a bar in the bar graphics showing the logarithm of the number of observations.

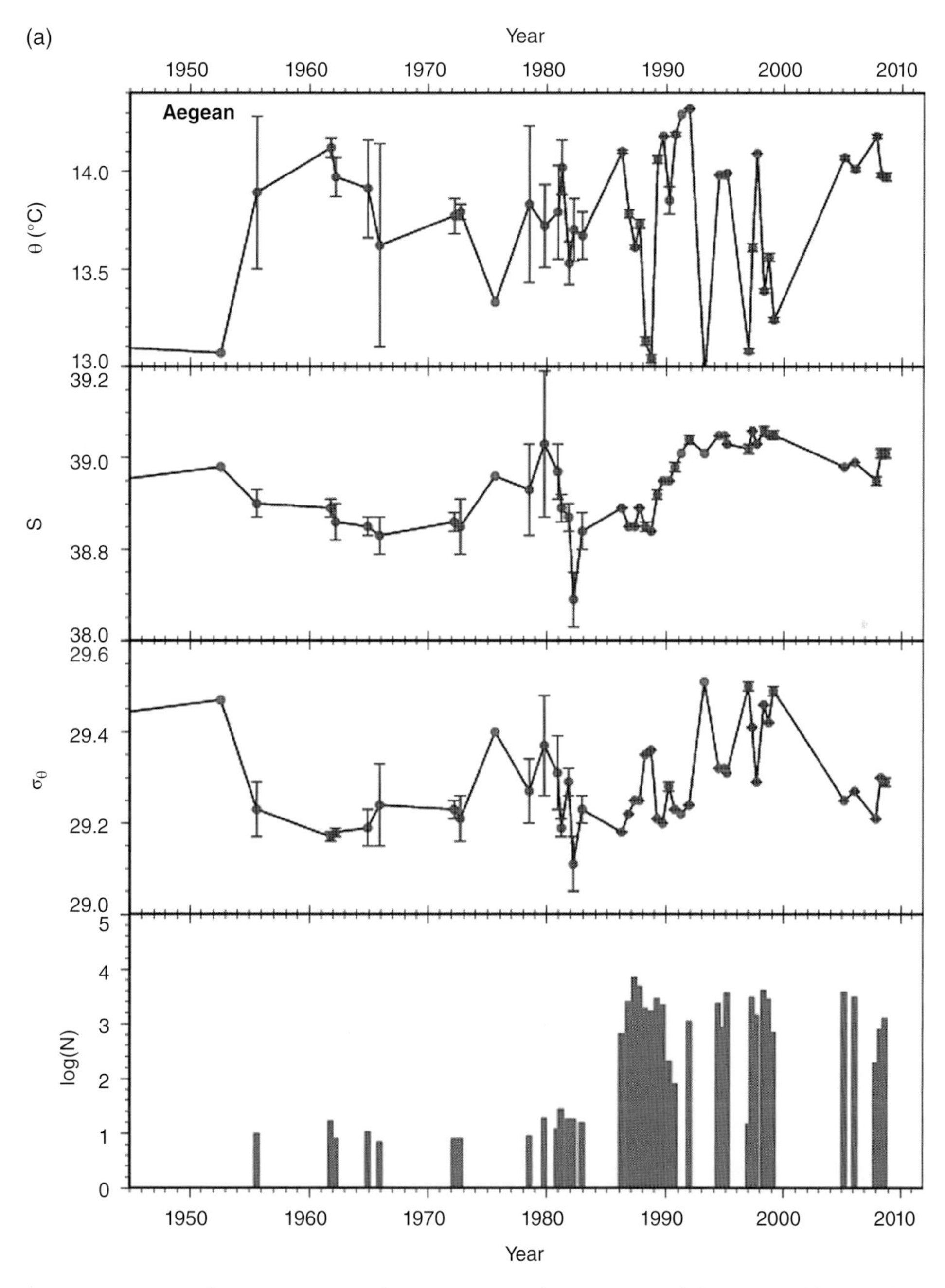

Figure 7.2. (a) Time series of average potential temperature, salinity, potential density, and log(N), the logarithm of the number of data points N in the 900–1100-m depth level, binned at half-year intervals in the Aegean Sea (vertical bars with end caps denote 95% confidence limits for data in each bin);

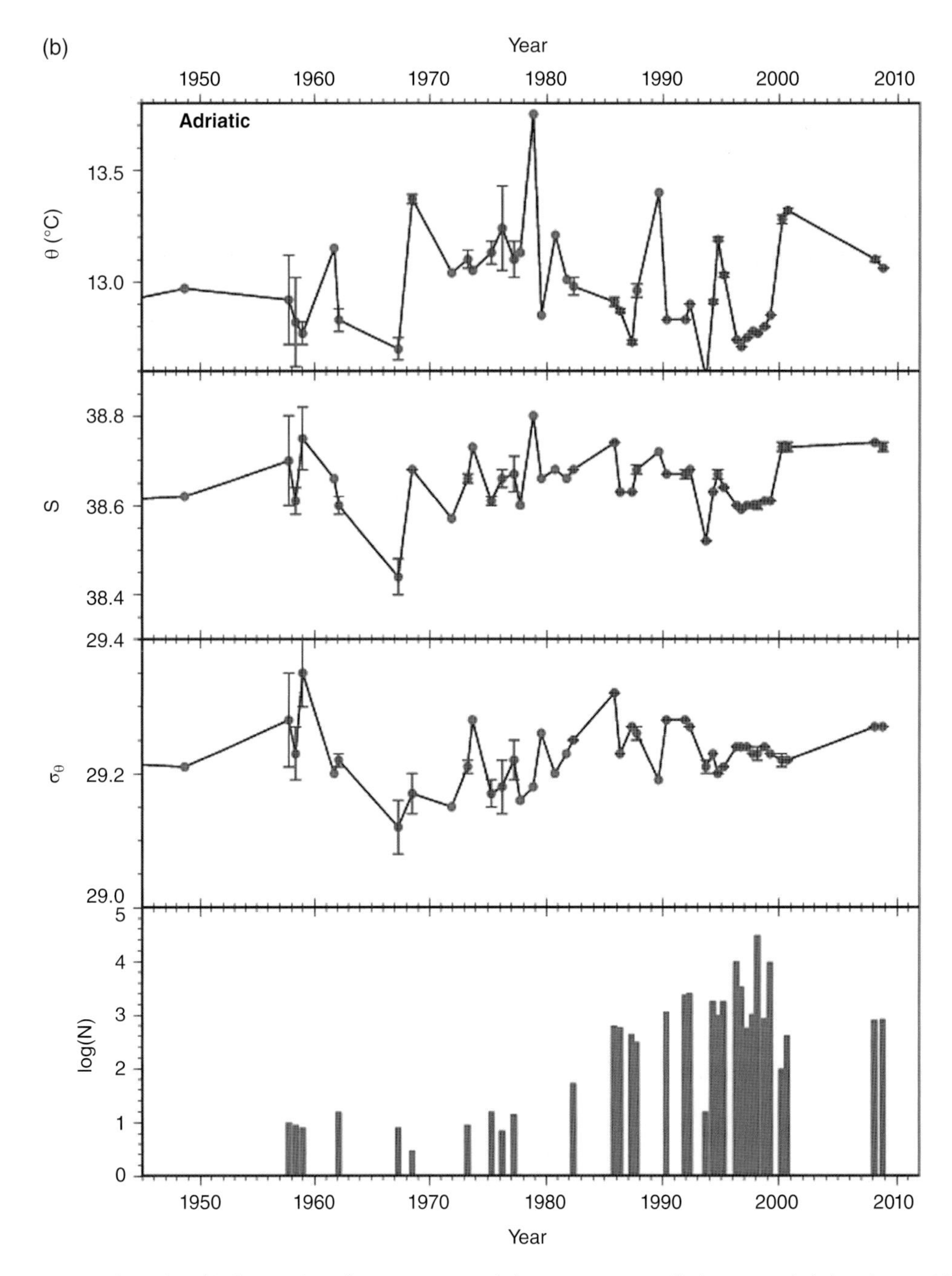

Figure 7.2. (continued) (b) time series of average potential temperature, salinity, potential density, and log(N), the logarithm of the number of data points N in the 900–1100-m depth level, binned at half-year intervals in the Adriatic Sea (vertical bars with end caps denote 95% confidence limits for data in each bin). For color detail, please see color plate section.

7.3. BASINWIDE AND INTERBASIN VARIABILITY

We first compare the general deep-water characteristics of the main subbasins of the eastern Mediterranean, namely, the Aegean and Adriatic shelf seas and the deep Levantine and Ionian basins, through a discussion of the water properties at the most relevant depth for each basin, 1 km for the Adriatic and the Aegean and 1–2 km for the Levantine and Ionian basins.

At depths centered at 1 km (Figure 7.2), the Aegean and the Adriatic seas have large oscillations of temperature, respectively with amplitudes of 1.5°C in the Aegean and 0.5°C in the Adriatic Seas. The cooling periods and patterns do not coincide in all cases. Recurrent decadal to multidecadal cooling periods with superposed shorter term events in 1950, 1970–1980, and 1990s are evident in both regions, with some mismatch between the two basins (see also Figures 7.6–7.9). In comparison, the Levantine and Ionian seas temperature centered at 1-km depth (not shown) displays much smaller interannual oscillations with amplitudes of 0.1–0.2°C. The higher amplitude response of the Aegean and Adriatic to surface effects is mostly related to the size of the basins, the associated sensitivity to atmospheric forcing, and the renewal timescales of the deep waters, which is much greater in the Levantine and Ionian seas.

In the Aegean Sea (Figure 7.2a), the most prominent feature in the time series is the steady rise of salinity continuing from the mid-1980s until the mid-1990s, which has been identified to be the result of a change in the Levantine circulation, the blocked circulation diverting saline water into the Aegean Sea [*Malanotte-Rizzoli et al.*, 1999]. The temperature is increasing along with salinity in the initial phase of this rise, but then the rapid cooling events in the early 1990s are ideal conditions for the massive formation of dense water in the Aegean Sea, as both the temperature and the salinity contribute to the abrupt density increase that has led to the EMT [*Özsoy and Latif*, 1996; *Theocharis et al.*, 1999].

On the other hand, at 2-km depth in the Levantine and Ionian basins (Figure 7.3), smaller oscillations of less than 0.1°C amplitude are detected until the early 1990s when abrupt, dramatic changes of more than 0.5°C in the Levantine and of 0.4°C in the Ionian basins occur in the form of interannual oscillations superposed on a stepwise change influencing the basin for the next two decades. The average temperature in the Levantine basin (Figure 7.3a), in fact, started to rise from 1988 onward, to reach a peak in 1993, followed by a secondary peak in 1997 and other oscillations in the following years.

The average temperature in the Ionian basin (Figure 7.3b) rose relatively more abruptly in 1992–1994, followed by a secondary peak in 1998–2001 in transient oscillations that seem to settle at about half the initial temperature rise. It appears that the outflow was felt immediately in both the Levantine and Ionian basins, but has led to greater changes in the Levantine Basin. Similar variations are detected in salinity, with an initial overshoot in 1993 that is larger than the stepwise change that follows. The overshoot signal in 1993 is present both in salinity and density, induced by the water sinking to these depths with the gravity current, although very little change persists in density in the later years, as the spreading of the anomalous waters appears to reach near equilibrium with the local density in the deep basin. We identify these changes to be tied to the cascading and spreading of the new dense water formed in the Aegean, verifying the long-term variations known as the Eastern Mediterranean Transient (EMT).

In addition to the temperature and salinity changes accompanying the EMT in the last two decades, there appear other significant peaks of salinity that influence the density in about 1980 in the Levantine Sea (Figure 7.3a), however without any significant changes in temperature. Similar small peaks of salinity without a signature in temperature occur in 1986 in the Ionian Sea (Figure 7.3b). These changes could be related to the intrusion of a water mass with salinity anomaly alone, or it is likely that some measurement errors could be involved in these cases.

Because salinity measurements have larger uncertainties and drifts due to instrumentation and sampling methods, we tend to rely more strictly on the temperature signals (see further discussion below), while evaluating salinity and density variations in parallel. In fact, in the case of the 1980 spike in the Levantine Sea (Figure 7.3a), the number of data points averaged is only about 10, reflected by the larger error bars (confidence intervals) in salinity and density in this period. In contrast, the 1993 salinity spike in the same figure provides much more confidence with several thousands of data points averaged, also coincident with the beginning signal of the EMT surviving later in the temperature anomaly. Other much smaller signals of simultaneous temperature and salinity variations affecting the density occur in the late 1960s and early 1970s both in the Levantine Sea and in the Ionian Sea.

In the deeper basins, the available temperature data centered at 3-km depth (Figure 7.4) shows a rapid rise of about 0.5°C in the Levantine Basin in 1992–1994, while a rise of about 0.2°C is indicated in the Ionian Basin in 1992–1995, followed by a secondary peak in 1998. Smaller peaks occur in the late 1960s and early 1970s in the Ionian Sea, but their significance is possibly limited, compared to the signals in the 1990s characterizing the EMT event influencing both deep basins of the eastern Mediterranean.

After the above description of clear signals in the deep temperature data, we turn our attention to the coevolution of water properties in the main basins. The potential temperature, salinity, and $\sigma\theta$ density at 0.5-, 1-, and 2-km depth layers, and the logarithm of the number of data points averaged for each bin at the respective depths in the Aegean Sea are displayed in Figure 7.5. For reasons of better visualization in a series of plots comparing properties at different depth intervals of each region, an offset factor proportional to n (where n = 0, 1, 2, 3, corresponds to the number assigned to each depth interval) is subtracted from the parameters' values. In the bottom panel of each plot, the number of data points entering the half-year time bins is shown in logarithmic scale (e.g., log(N) = 3 corresponds to data averaged from 1000 measurement points).

Temperature in the Aegean Sea, at the first two levels of Figure 7.5, shows interannual and decadal oscillations resulting in cooling by about 1.5°C for each of the events in the 1987–1989, 1993, 1997–1999 periods, while the third level at 2-km depth is relatively constant with a higher value between 1986–1994. This is almost the same period when salinity is on the rise at all displayed depth levels, that is, the period of saline water entry into the Aegean Sea noted in the literature [*Theocharis et al.*, 1999; *Malanotte-Rizzoli et al.*, 1999]. We should note, however, that the sawtooth pattern in temperature of the first two depth levels after the 1990s is a result of the uneven distribution of data in the whole of the Aegean Sea, with the colder observation points located in the northern and central Aegean subdomains (compare with Figure 7.10).

The uniform increase of salinity at all deep layers of the Aegean Sea during the 1985–1995 period is very interesting. If the increase were to be attributed to the LIW import from the Levantine Sea alone, it would be hard for the rather shallow LIW to influence the deeper layers. This observation, that needs further investigation, implies that deep-water formation processes in the

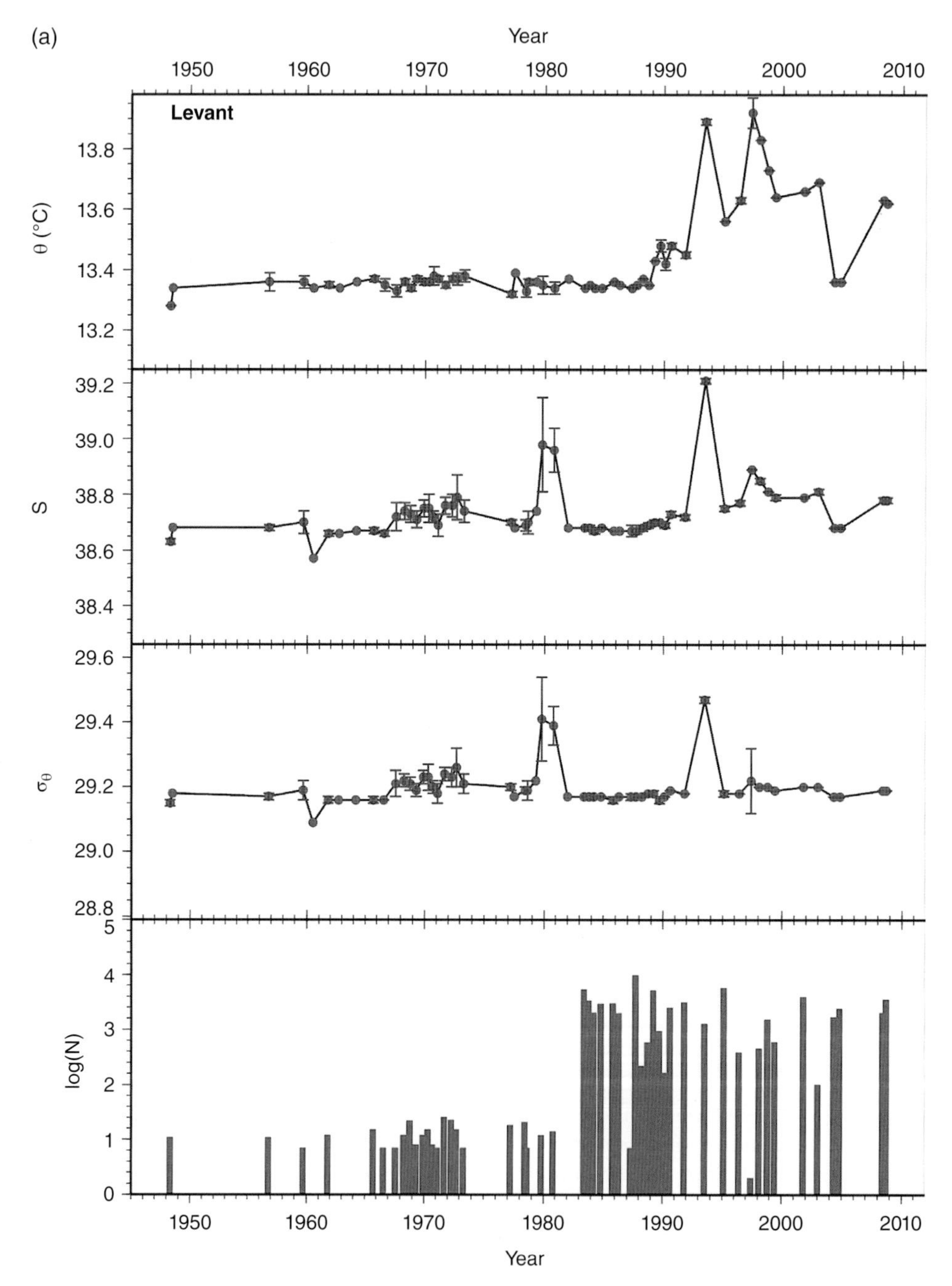

Figure 7.3. (a) Time series of average potential temperature, salinity, potential density, and log(N), the logarithm of the number of data points N, in the 900–1100-m depth level, binned at half-year intervals in the Levantine Sea (vertical bars with end caps denote 95% confidence limits for data in each bin);

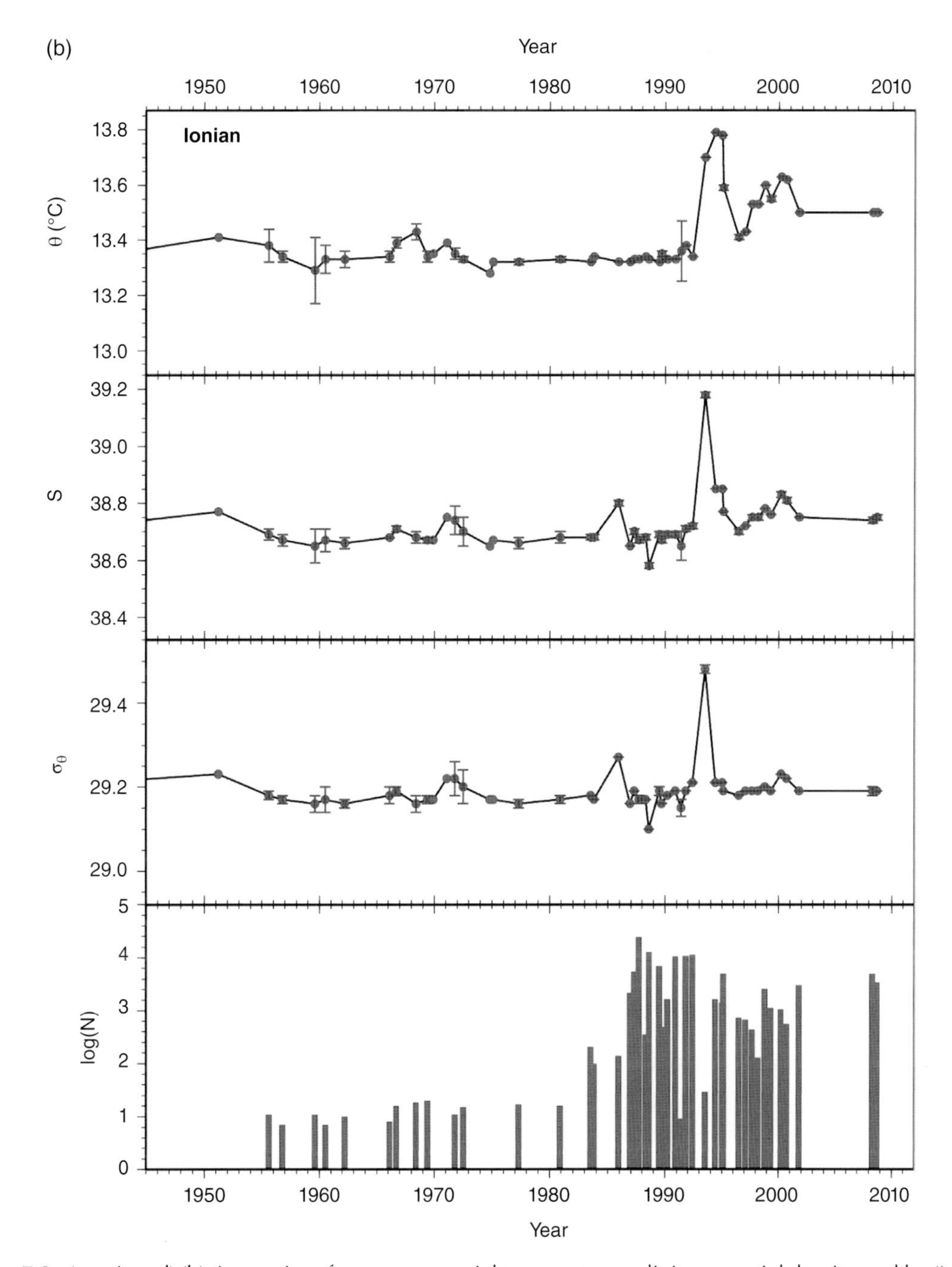

Figure 7.3. (continued) (b) time series of average potential temperature, salinity, potential density, and log(N), the logarithm of the number of data points N in the 900–1100-m depth level, binned at half-year intervals in the Ionian Sea (vertical bars with end caps denote 95% confidence limits for data in each bin). For color detail, please see color plate section.

Aegean Sea were affecting the whole water column. The salinity at the first two levels has marked interannual to decadal oscillations before and after this period, but the last level of 2-km depth reaches almost an asymptotic value after a salinity rise of about 0.2 in the period 1985–1995. The density increase to peak values in the late 1990s is recovered in later years at the first two levels but not in the deepest layer. The waters in the deeper parts of the basin (that are isolated by abrupt topography) are not yet affected by the post-EMT temperature and salinity variations in the shallower parts of the basin. This was also recorded in two cruises (2005–2006) with observations at the deepest parts of the Aegean Sea [*Vervatis et al.*, 2011].

The Adriatic Sea in Figure 7.6 displays interannual to decadal oscillations, but much smaller than those

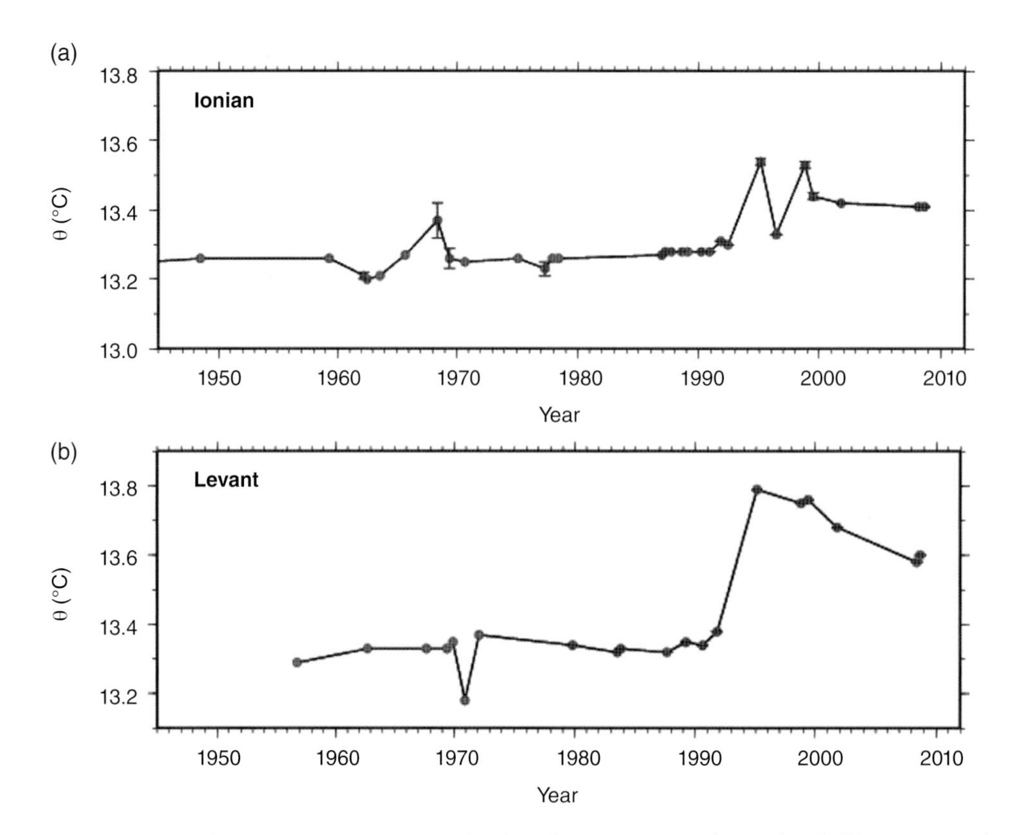

Figure 7.4. Time series of average temperature at depths of 2900–3100 m binned at half-year intervals in the (a) Ionian and (b) Levantine Seas. For color detail, please see color plate section.

observed in the Aegean Sea. A stepwise rise of salinity or density resembling the one in the Aegean Sea is not discernible during the observation period. If any dense-water production event should be responsible for an Adriatic contribution to the EMDW, it would have to be linked to these oscillations. We note temperature and salinity decreases in the 1960s, in 1987, and in the 1989–1994 periods, especially in the 0.5-km depth layer. The latter period of 1987–1994 coincides with the EMT period, and overlaps with the 1988–1997 period of anticyclonic circulation in the Ionian Sea, when a decrease has been observed in the Adriatic salinity [*Gačić et al.*, 2010]. On the other hand, temperature and salinity increases are found in the 1973–1983 and 1998–2001 periods, partially overlapping with the 1998–2006 period of cyclonic circulation in the Ionian Sea when an increase is observed in the Adriatic salinity [*Gačić et al.*, 2010]. The density in the 1.0-km layer appears to increase in the 1970s, early 1980s, and early 1990s periods.

Because the EMT appears dominating in the time history of the deep Levantine and Ionian seas properties, we focus our attention on the immediate neighborhood of the Cretan Sea source, *that is*, the northwestern Levantine and the northeastern Ionian seas regions where most of the variability is encountered, rather than providing basinwide changes in this section. The basinwide changes in properties at 2-km depth have already been reviewed in Figure 7.3.

In the northwestern Levantine Sea (Figure 7.7), a cooling event is detected at the first level of 0.5-km depth in the years 1993–1994, coinciding with other changes in the deeper layers in the ensuing EMT period. Stacked plots of properties indicate stepwise increases of temperature and salinity at depths centered at 2- and 3-km depths in the early 1990s following the EMT. Salinity spikes at the depth layers 0.5, 1, and 2 km are coincident with beginning of EMT anomalies at the same depths, while only a stepwise increase of salinity is detected in the 3-km depth layer. This is because the introduction of the new water masses into the Levantine basin during the EMT event occurs in the form of a deep overflow from the Aegean straits, which then spreads in the entire Levantine Basin and continues to influence the basin throughout the next two decades.

It is also interesting that there is no discernible net increase in density after the EMT event except the transient anomalies and the slight increase in the 3-km depth layer, suggesting that the overflow from the Aegean Straits largely interleaves into the existing stratification without actually changing the density significantly, because of the compensating nature of the warm temperature against the higher salinity of the outflow. *Roether*

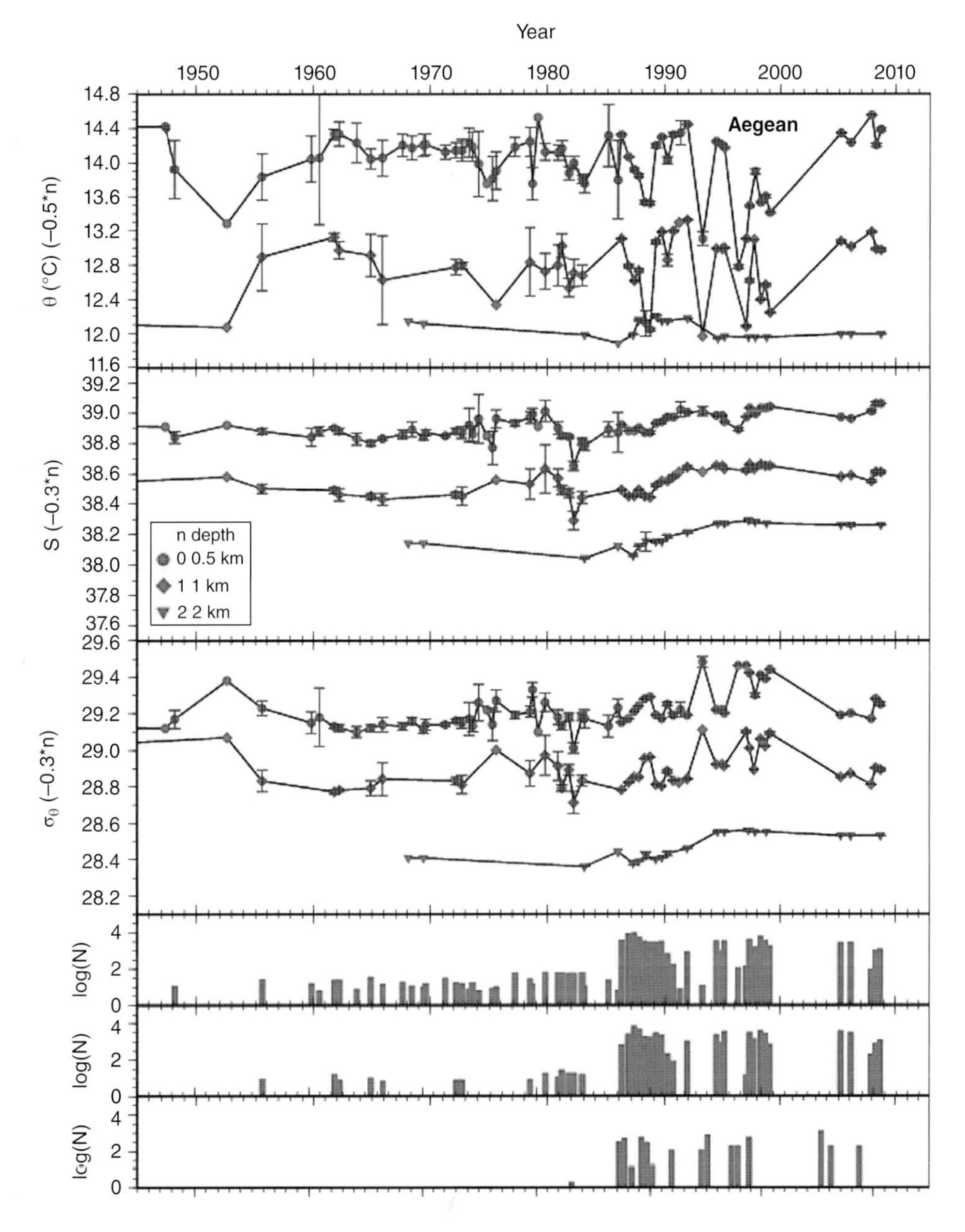

Figure 7.5. Aegean Sea potential temperature θ, salinity, and σ_θ density at depth layers of 450–550 m (n = 0), 900–1100 m (n = 1), and 1900–2100 m (n = 2) (vertically shifted down by a constant times n) and log(N), the logarithm of the number of data points N averaged for each half-year bin interval. (Vertical bars with end caps denote 95% confidence limits for data in each bin.) For color detail, please see color plate section.

et al. [2007] have argued that the dominance of the Aegean source was primarily a result of rates and much less of density. They find that after 1994, the near bottom flow was driven by extremely low lateral density gradients that were also nearly invariant in time.

While the EMT associated temperature increase in the deepest layers seems to have occurred only after the 1990s, and an absolutely stable temperature record is found before this period without a notable trace of change up until the late 1980s, the same cannot be said for salinity and density. It seems that short periods of increased salinity occurred around 1960, 1970, and the 1980s, but without a trace in temperature, which means that peaks in density follow the influence of salinity. The salinity peaks producing effects on density without a compensating temperature change in around 1980 have higher error bars based on few stations available, as discussed earlier in relation to Figure 7.3.

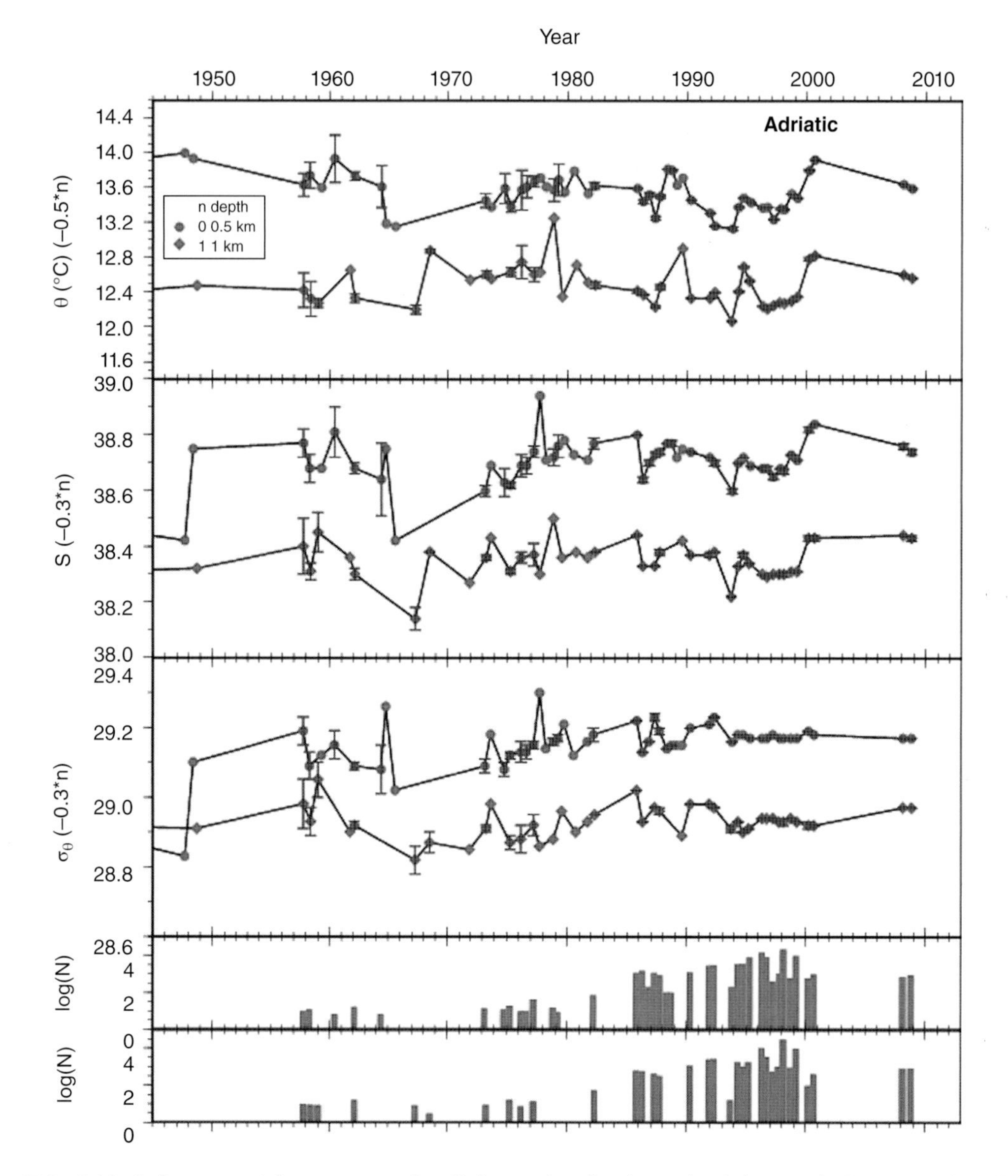

Figure 7.6. Adriatic Sea potential temperature θ, salinity, and σ_θ density at depth layers of 450–550 m (n = 0) and 900–1100 m (n = 1) (vertically shifted down by a constant times n), and log(N), the logarithm of the number of data points N averaged for each half-year bin interval. For color detail, please see color plate section.

We next present results for the northeastern Ionian Sea (Figure 7.8), where salinity peaks occur at all displayed depth levels, coinciding with the largest signal in temperature at the 2- and 3-km depth levels, marking the outflow of dense water from the Aegean into the Ionian Sea during the EMT. The first level at 0.5-km depth interval shows a continuous cooling trend from the 1980s onward, until a stronger cooling event is observed in the first half of the 1990s, coinciding with a peak in salinity at this and lower layers.

The positive temperature and salinity anomalies observed at the deep layers during the EMT events in some time-series plots may appear unnatural when cooling is expected from an event of convective origin. However, this is related to the production of a new deep-water mass during the EMT, when waters of greater salinity in the Aegean Sea did not have to undergo very extensive cooling in order to reach very large densities and sink and spread in almost all the eastern Mediterranean Sea. Turbulent entrainment processes also influence the final characteristics of the spreading water mass, as indicated by the higher salinity and temperature of the anomalous waters arriving at the deeper levels. The positive anomalies observed at depths greater than 0.5 km in the Levantine and Ionian seas illustrate the fact that the source of water is external to these basins, as it clearly originates from the neighboring Aegean Sea.

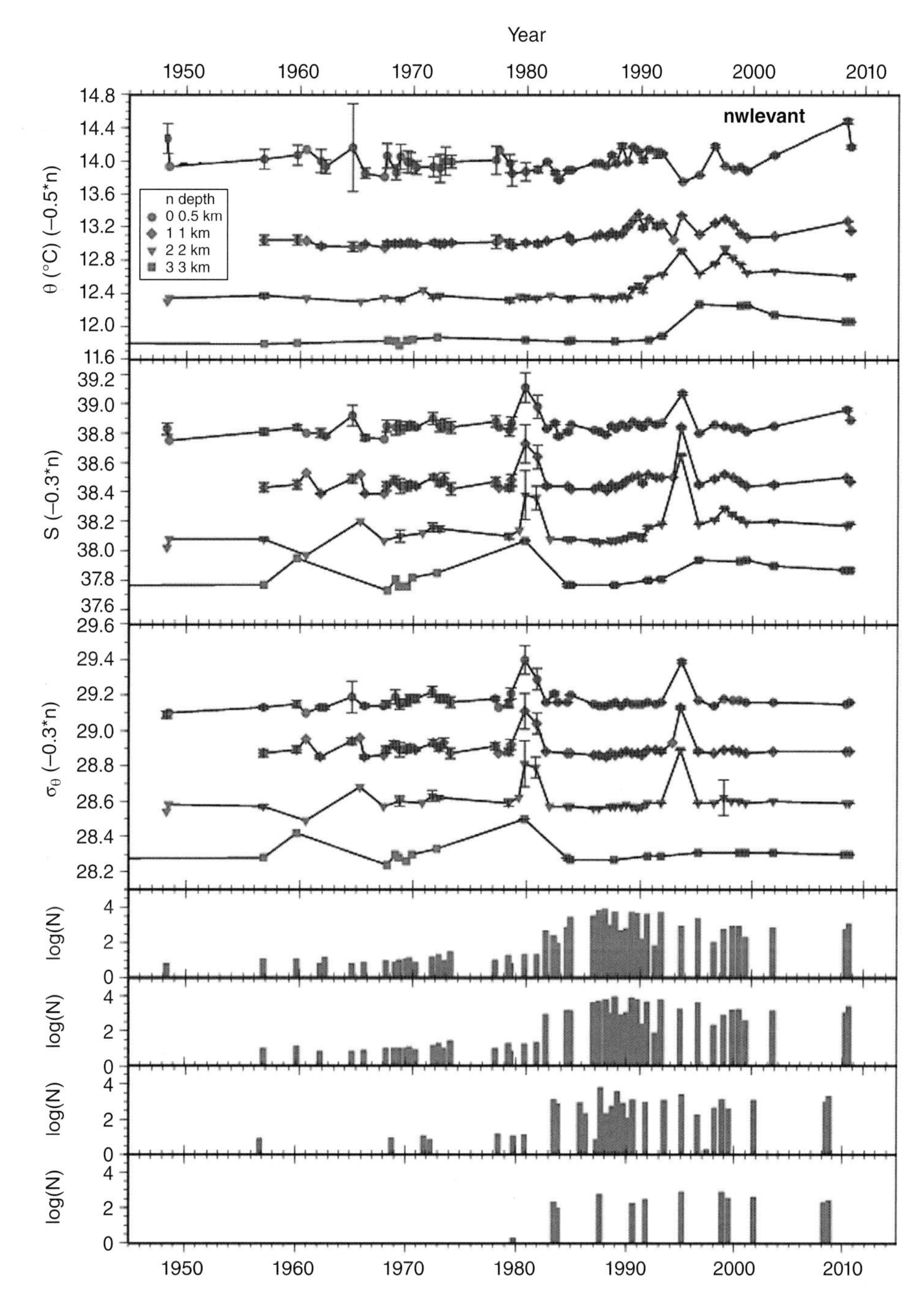

Figure 7.7. Northwestern Levantine Sea potential temperature θ, salinity, and σ_0 density at depth layers of 450–550 m (n = 0), 900–1100 m (n = 1), 1900–2100 m (n = 2), and 2900–3100 m (n = 3) (vertically shifted down by a constant times n) and log(N), the logarithm of the number of data points N averaged for each half-year bin interval. For color detail, please see color plate section.

In Figure 7.9 we provide the density variations at 4-km depth in the deepest troughs of the Levantine and Ionian seas, respectively in the northwestern Levantine and central Ionian subdomains. A decrease in bottom layer density is detected in the northwestern Levantine (Rhodes) depression after the EMT period, while an increase is detected in the central Ionian depression, the density being equalized at a value of about 29.2 in both basins after the 1990s. We should note, however, that the number of observations prior to the 1990s is limited to a single one for each time bin, while it is increased to about 100 profiles per bin in the latter period, increasing the reliability of the observations in the last two decades.

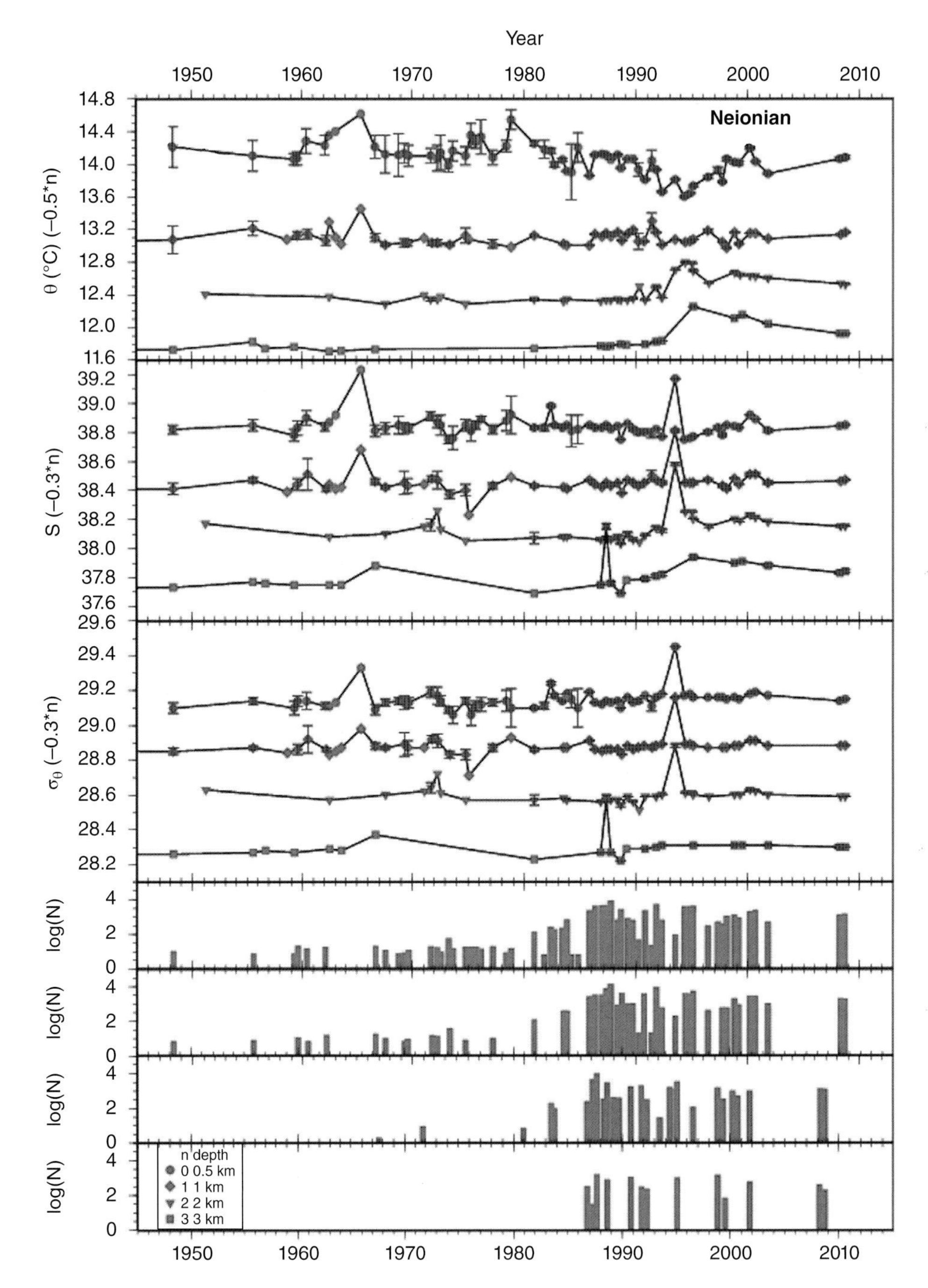

Figure 7.8. Northeastern Ionian Sea potential temperature θ, salinity, and σθ density at depth layers of 450–550 m (n = 0), 900–1100 m (n = 1), 1900–2100 m (n = 2), and 2900–3100 m (n = 3) (vertically shifted down by a constant times n), and log(N), the logarithm of the number of data points N averaged for each half-year bin interval. For color detail, please see color plate section.

7.4. INTRABASIN VARIABILITY

We next review the behavior of the subdomains within the Aegean Sea in Figure 7.10. The Aegean Sea played a key role in the largest event of deep-water variability in the eastern Mediterranean Sea. Furthermore, its geography and topography is the most complicated in the region, with thousands of islands and islets and a large number of depressions and sills. The Aegean Sea is mainly a shelf sea, with most of the basin having shallow depths except the three deep basins of the north Aegean (the North Aegean trough consisting of the Sporades, Athos, and Lemnos basins), the central Aegean (Skiros and Chios basins), and the southern Aegean Sea (Cretan Basin).

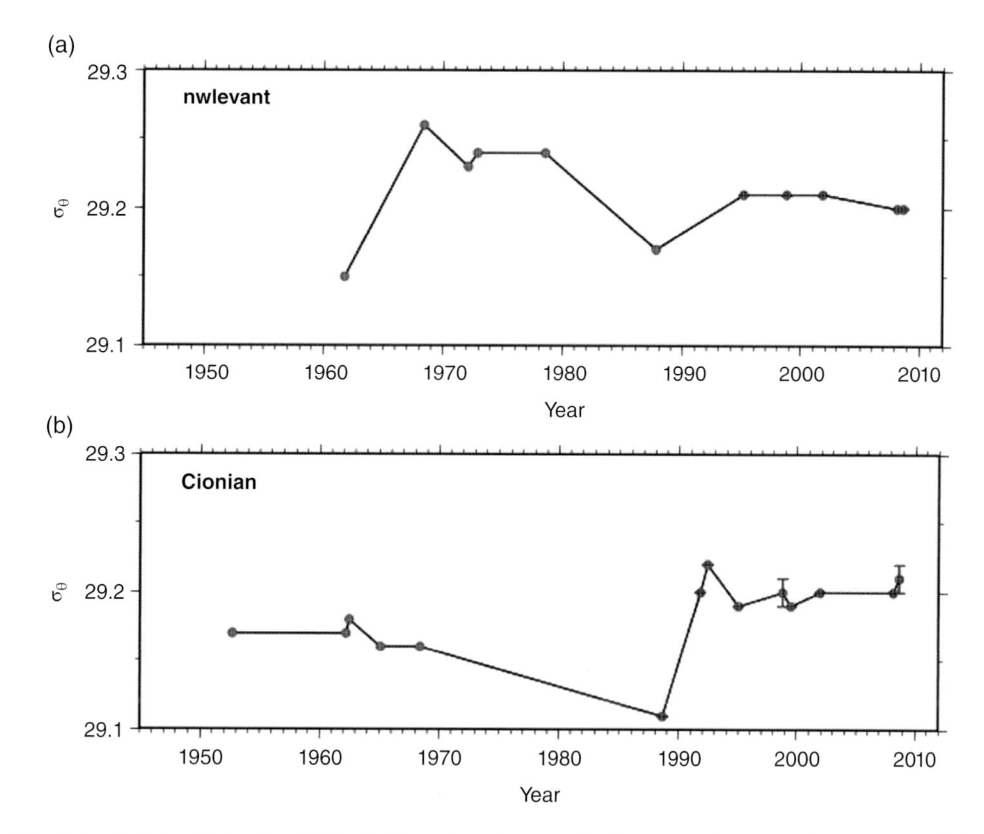

Figure 7.9. Potential density in the 3900–4100-m depth interval in the (a) northwestern Levantine and (b) central Ionian subdomains representing the deepest areas of the Levantine and Ionian seas. For color detail, please see color plate section.

In the north Aegean Sea (Figure 7.10a) at depth layers centered at 0.5- and 1-km depth, a very strong cooling event occurs in the late 1980s followed by a steady increase in salinity, leading to the highest densities in the late 1990s. In fact, the mean potential density values of up to 29.53, reached in the bottom waters of the north Aegean Sea in the mid-1990s, are the highest values observed anywhere in the entire Mediterranean Sea.

A rather stronger cooling event occurred at the earlier period of 1987–1989 in the northern and central basins [*Gertman et al.*, 2006], but it is absent in the Cretan Sea, which together with the increasing trend of salinity in the entire Aegean Sea, may have served in the preconditioning of properties. In the north Aegean (Figure 7.10a), data are insufficient to show a strong cooling event in the years 1992–1993, but there are enough data to show this in the central Aegean basin (Figure 7.10b).

A stagnant period in the years 1994–2000 followed the density maximum in the mid-1990s and survived till many years later [*Zervakis et al.*, 2003; *Androulidakis et al.*, 2012], as a result of the increased stability and the combined influences on surface buoyancy created by the net water fluxes at the sea surface and the changes in the flux of Black Sea Water (BSW) outflowing from the Dardanelles Strait.

In the Cretan basin layers of 1 and 2 km (Figure 7.10c), interdecadal oscillations of 10–15-year periods are observed both in temperature and in salinity and density. These oscillations reach a plateau after the early 1990s EMT event, following a period of steady increase in salinity in the 1980s and the cooling event in 1992–1993. The high-density deep waters filling the deeper part of the Cretan Basin seem to have reached a rather stable situation after the 1990s, at least until the present. According to *Theocharis et al.* [1999], the deep basin of the Cretan Sea had been filled by dense Cretan Deep Water (CDW) starting in 1987 and lasting until 1992–1993, when the first dense-water outflow from the Cretan Strait was initiated. Continued observations reported by *Theocharis et al.* [2002] seemed to indicate a return to pre-EMT conditions after the mid-1990s, with continued but smaller overflows contributing only to the midlevels of 1.5–2-km depth in the exterior region.

Our analysis in Figure 7.10c shows that the increase in salinity started after the strong cooling period in the early 1980s and lasted until mid-1990s. The salinity and density in the deeper basin at 2 km stayed stationary, but decreased slightly after the mid-1990s maxima. Further review of Figures 7.11 and 7.13 will emphasize the fact that a net stepwise rise in the temperature signal has been preserved at both 1- and 2-km depth until the last decade, although large fluctuations are superposed on this rise.

On the other hand, stronger cooling events seem to have taken place during the 1970s and 1980s, but in these cases the cooling periods were also associated with lower salinity, and therefore without any increase in the density. The difference in the early 1990s case was that an increase in salinity occurred throughout the entire later part of the 1980s until 1991 (preconditioning phase), and with the cooling imposed on this situation it probably created sufficiently high density to initiate the outflow from the Cretan Sea into the adjacent seas.

It may also be observed in Figure 7.10 that the density is always higher in the northern basin of the Aegean as compared to the central and southern basins, but in the case of the 1990s EMT period, density in the deeper northern and central basins approach each other almost to be equalized, though remaining much higher than the

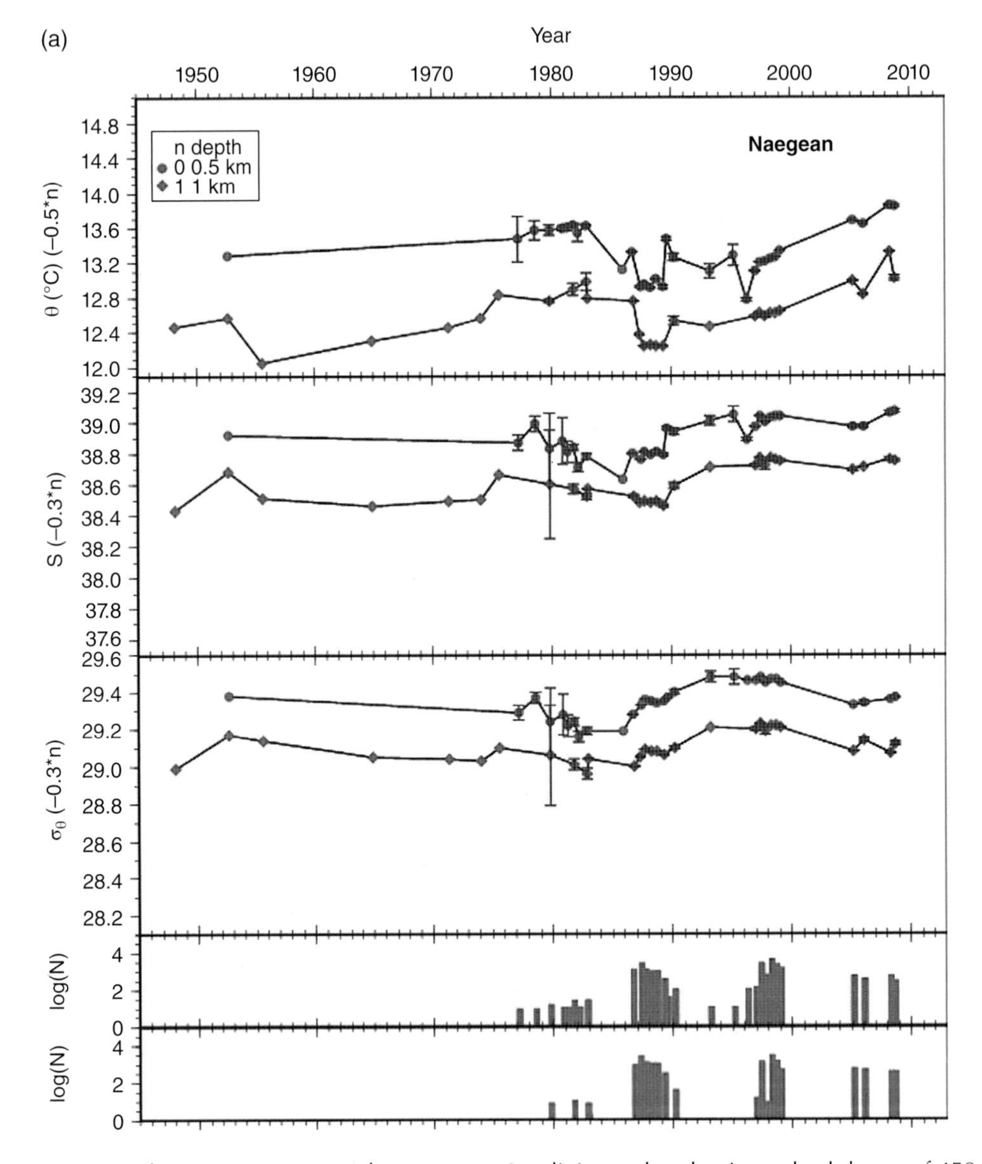

Figure 7.10. (a) North Aegean Sea potential temperature θ, salinity, and σ_θ density at depth layers of 450–550 m (n = 0), 900–1100 m (n = 1), and 1900–2100 m (n = 2) (vertically shifted down by a constant times n) and log(N), the logarithm of the number of data points N averaged for each half-year bin interval; (b) Central Aegean Sea potential temperature θ, salinity, and σ_θ density at depth layers of 450–550 m (n = 0), 900–1100 m (n = 1), and 1900–2100 m (n = 2) (vertically shifted down by a constant times n) and log(N), the logarithm of the number of data points N averaged for each half-year bin interval; (c) Southern Aegean Sea (Cretan Sea) potential temperature θ, salinity, and σ_θ density at depth layers of 450–550 m (n = 0), 900–1100 m (n = 1), and 1900–2100 m (n = 2) (vertically shifted down by a constant times n), and log(N), the logarithm of the number of data points N averaged for each half-year bin interval. For color detail, please see color plate section.

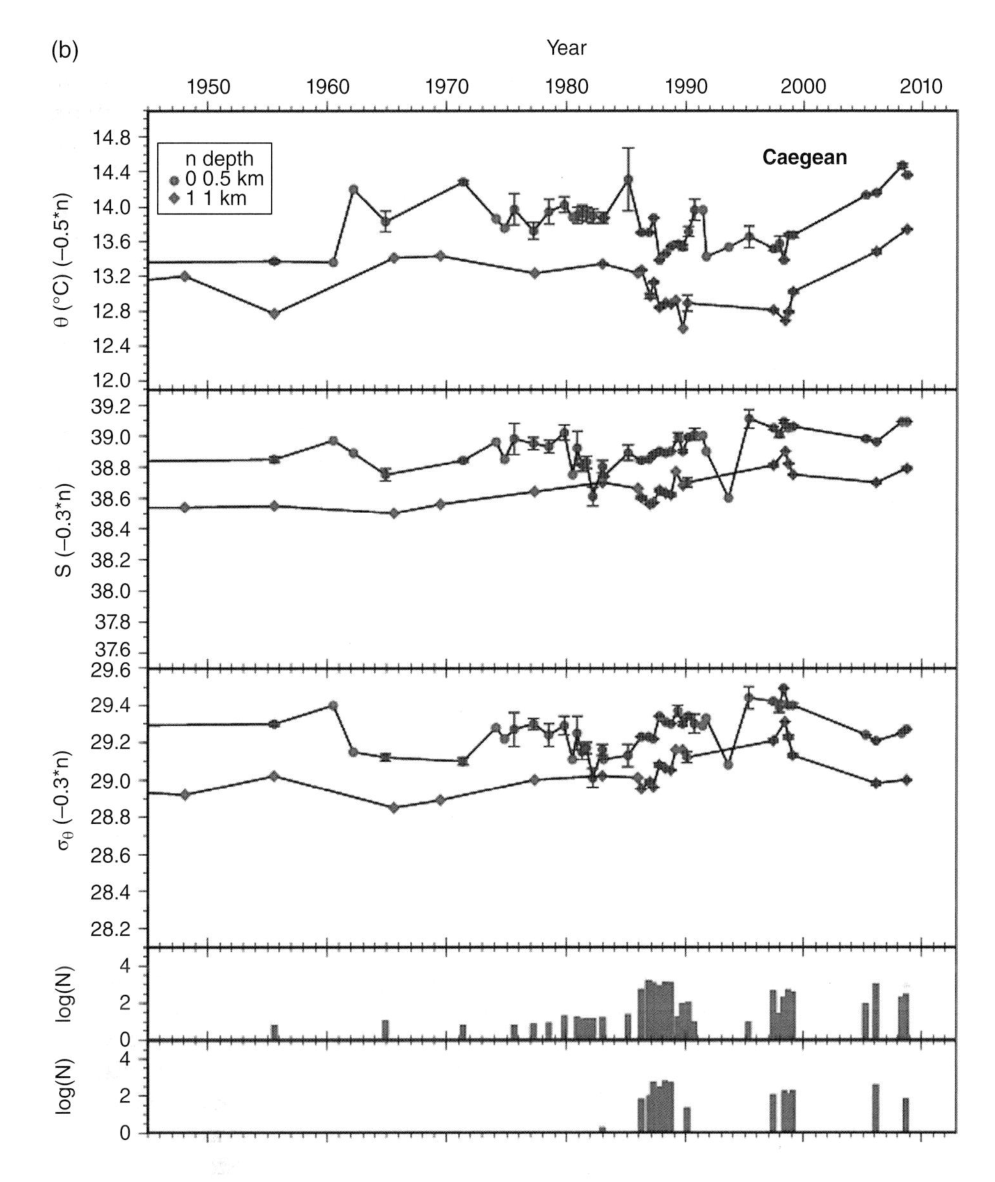

Figure 7.10. (continued)

density at the same depths in the Cretan Sea. It is also discernible in Figure 7.10 that the stratification in the Cretan basin increases in parallel with the period of steep increases in salinity and density during the 1983–1993 period. The same can be said for the central Aegean basin only in the 1980–1990 period, though the stratification appears weaker during the rest of the observed period.

Sayın and Beşiktepe [2010] confirm some of these results based on measurements made in the north Aegean on the eastern side of the north Aegean trough (Saros deep and Limnos north), north of the central Aegean (Limnos south), and on the eastern side of the Cretan Sea (south Aegean). Through analyses of these partial data, they have shown the mixing to a depth of 200 m producing water of density greater than 29.5 south of Limnos Island in the central Aegean region, but they have missed completely the same levels of density reached in the north Aegean. *Sayın and Beşiktepe* [2010] also note the cyclonic circulation of the central Aegean aiding northerly transport of saline water on the eastern flank and the transport of dense water south by the same circulation. This could be an additional factor for transporting dense water to the Cretan Sea while withdrawing more saline water from the south, eventually contributing to the dense-water formation in the north.

The role of the Central Aegean Sea in the regional thermohaline circulation and deep-water formation was also investigated during the 2005–2006 cruises [*Vervatis et al.*, 2011]. They recorded dense-water formation processes and found that the central part of the basin plays a very important role for the whole basin. Due to the presence of high salinity waters of Levantine origin,

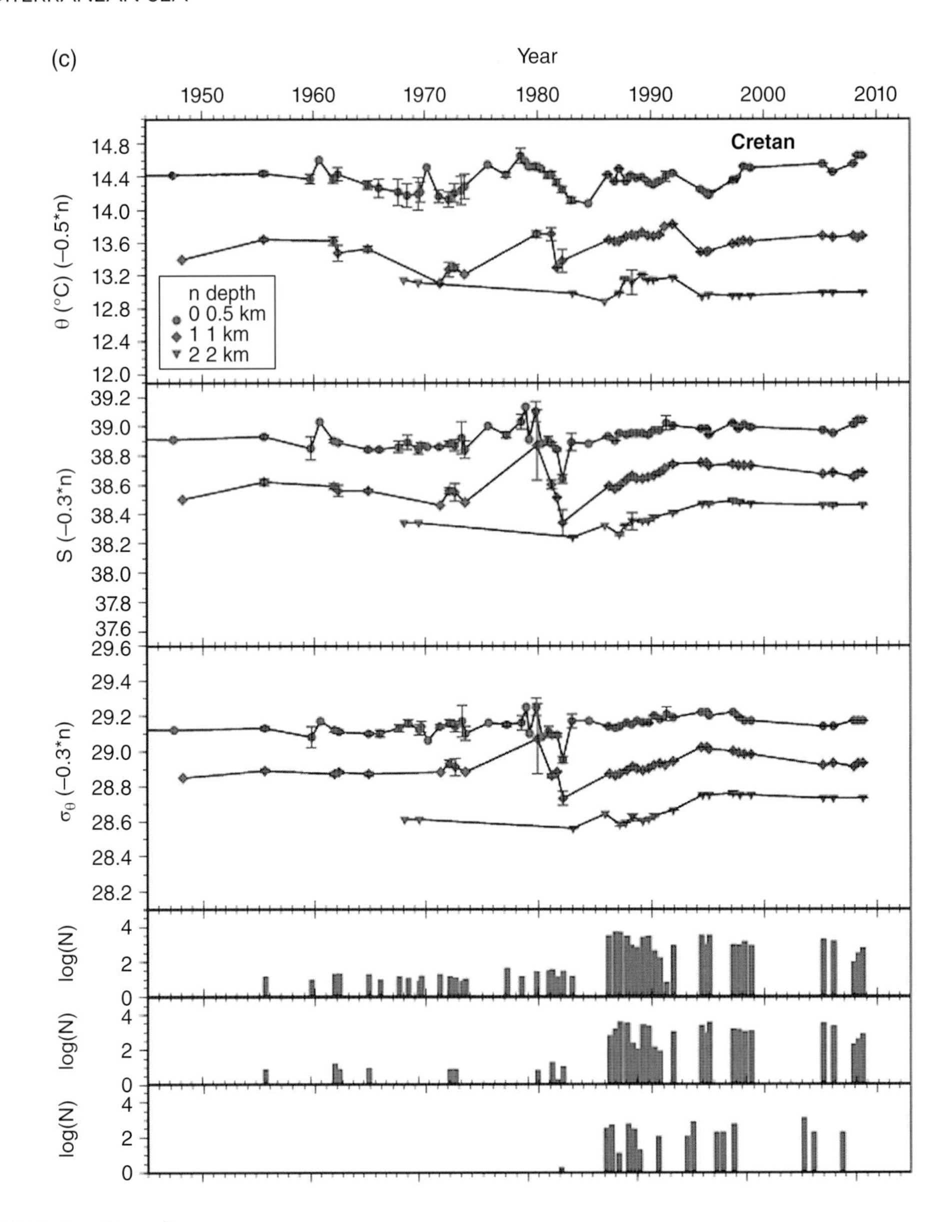

Figure 7.10. (continued)

the strong atmospheric forcing and its shape, it is the most prominent area for dense-water formation and can act as a reservoir of dense waters for the adjacent Aegean subbasins. This contribution of the central Aegean Sea to the intermediate and deep layers of the Aegean Sea, also identified by two cruises in the late 1980s and early 1990s [*Gertman et al.*, 2006], can also explain the larger salinity increase signal in the central and northern parts of the Aegean Basin and the amplified signal in the deeper levels of the Cretan Sea.

We believe the salinity increase that led to the density increase in the deep Northern Aegean Trough resulted from the LIW entering into the Aegean Basin and penetrating to the central and north Aegean during this period, as a result of the blocked circulation in the Levantine Sea as observed in October 1991 [*Theocharis et al.*, 1999; *Malanotte-Rizzoli et al.*, 1999]. Contributions of increased E-P [*Josey et al.*, 2011] and reduced BSW inflow in the Aegean Sea [*Zervakis et al.*, 2000] were also proposed as mechanisms for the density increase observed in the basin. Once the wheel starts to turn, that is, when the outflow of dense water out of the Aegean starts, there would be more saline water entering in replacement as *Sayın and Beşiktepe* [2010] proposed. However in Figure 7.9, we see that the increase in salinity starts much earlier in the late 1980s, possibly after the extensive cooling in the north Aegean, which may have initiated at least one of the triggers.

The analysis of spatial variations across the deep Levantine and Ionian basins is carried out on the basis of

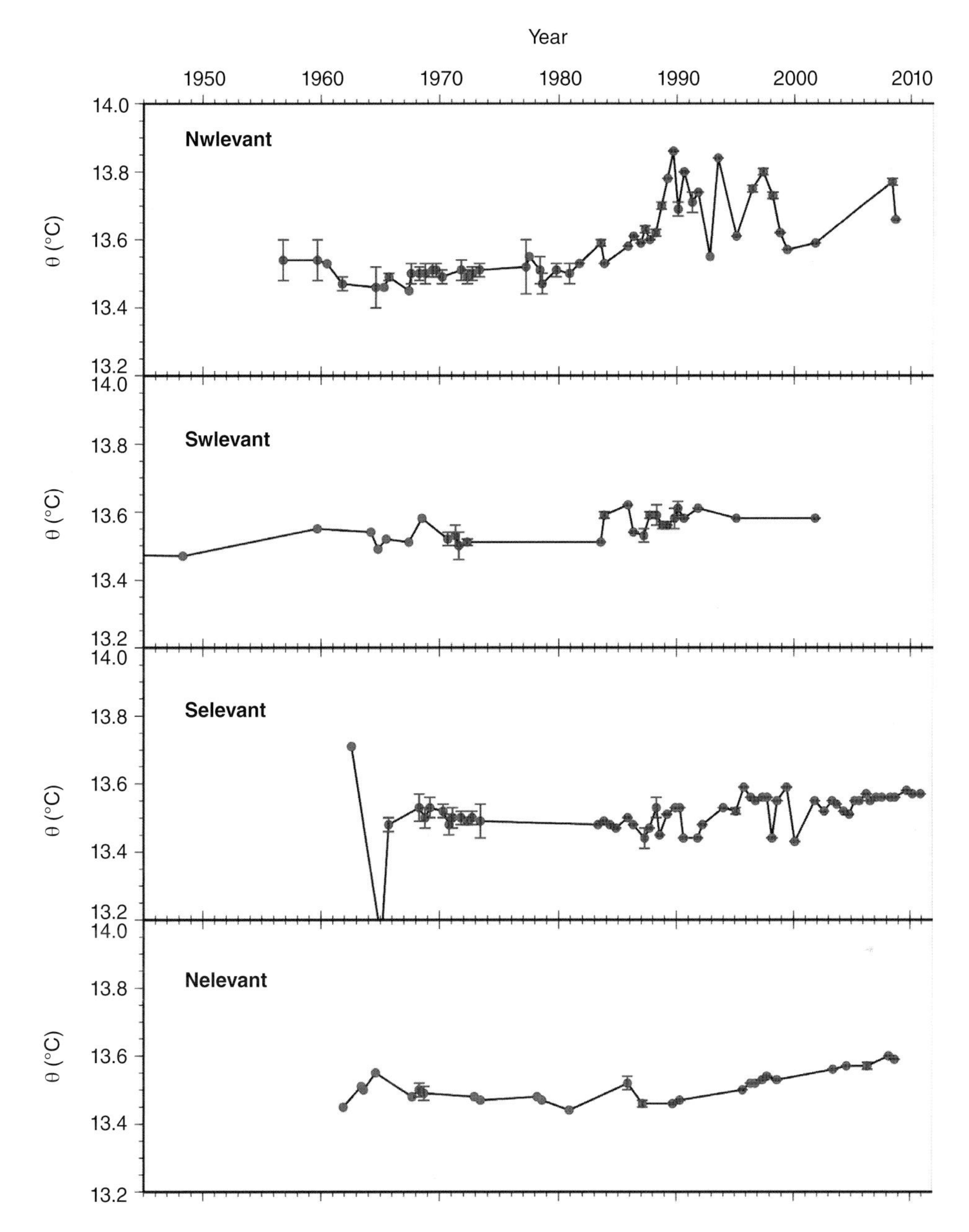

Figure 7.11. Potential temperature θ in the 900–1100-m depth level in the (a) northwestern, (b) southwestern, c) southeastern, and d) northeastern Levantine subdomains. For color detail, please see color plate section.

temperature data, built upon the confidence we have on the measurement of temperature. Although salinity is usually considered as a conservative tracer in the ocean, in this case we prefer to use temperature as a more reliable tracer for better detection of deep-water changes. We trace these changes based on the comparison of time-series between subdomains of the Levantine and Ionian seas at the 1- and 2-km depth levels in Figures 7.11–7.14, respectively.

The distinct pattern of the temperature time-series at 1-km depth in the northwestern Levantine region (Figure 7.11a) is clearly differentiated from the other time-series in the various Levantine Basin subdomains at the same depth (Figure 7.11). In this region, the temperature starts rising continuously from the early 1980s onward, reaching a peak in 1989 and oscillating afterward, with other peaks in 1992, 1997, and 2008, by a total temperature rise of about 0.4°C. Combined with the interpretation of Figure 7.7 showing parallel changes in temperature and salinity at all depths, the enlarged plot in Figure 7.11a shows that the preconditioning evolution that leads to the EMT starts in the second half of the 1980s, much earlier than 1992. It can also be understood that the overflow of anomalous waters from the Cretan

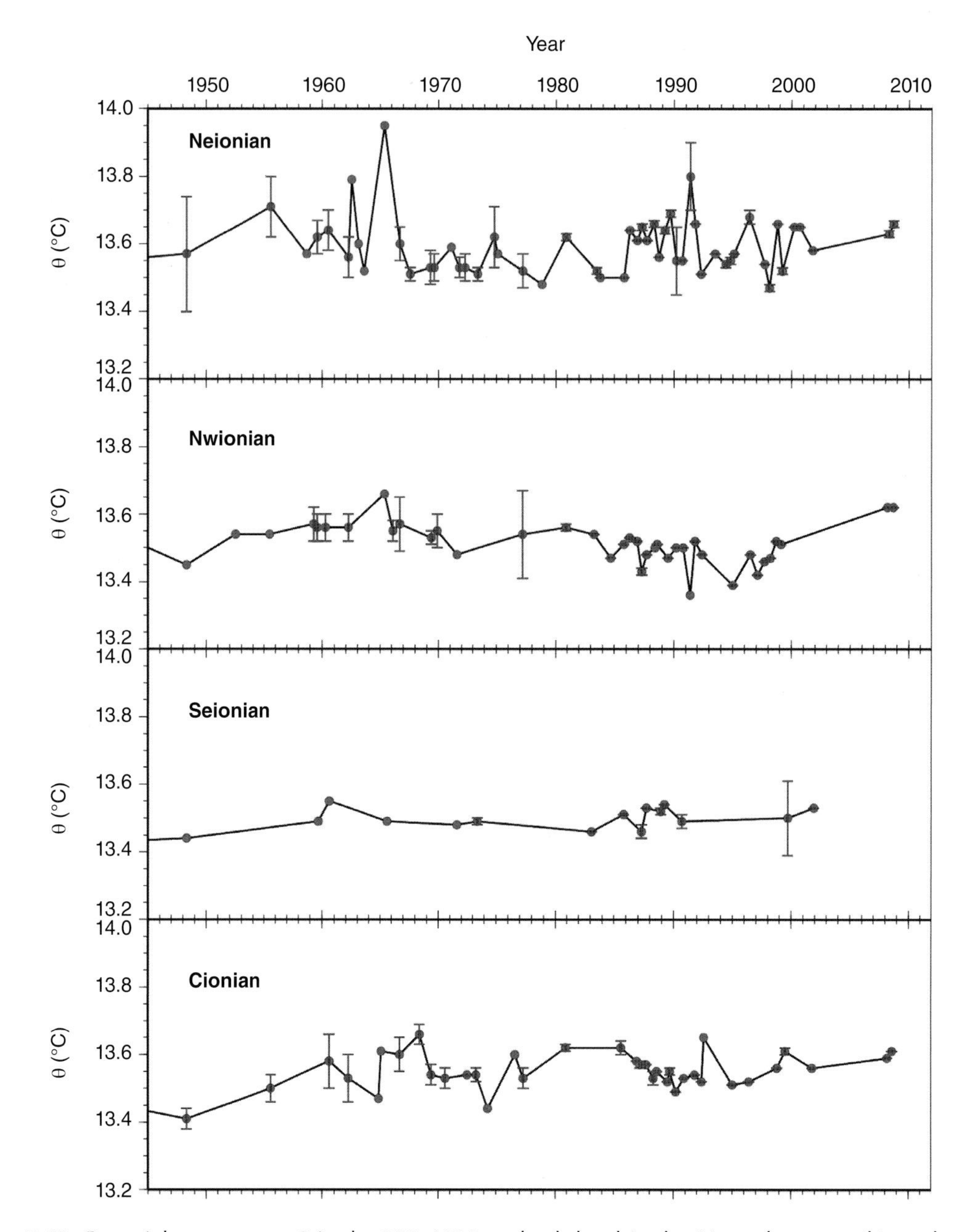

Figure 7.12. Potential temperature θ in the 900–1100-m depth level in the (a) northeastern, (b) northwestern, (c) southeastern, and (d) central Ionian subdomains. For color detail, please see color plate section.

Sea has started much earlier than the peak of the EMT signal in 1992–1993.

Barely noticable increases of 0.1–0.2°C are indicated in the southwest and southeast regions (Figures 7.11b,c) almost synchronous with the northwest after the 1980s, but not followed by the later sequence of peaks and the clear stepwise increase observed there. The northeast box indicates only a continuous trend of increasing temperature during the same period (Figure 7.11d). The Ionian Basin northeast, northwest, southeast, and central subdomains (Figure 7.12a–d) all have different patterns of similar magnitude, none of them indicating a pattern that comes close to the signature in the nortwestern Levantine box from starting in the second half of the 1980s.

The above discussion shows that the largest anomalies occur only in the northwest Levantine area in the 1-km depth level, reflecting the influence of the new dense water from the Cretan Sea into the Levantine Sea in its neighborhood. Studies carried out in the pre-EMT and post-EMT periods show that the stratification of the Aegean Sea and its adjacent basins are influenced by the exchange fluxes in the Cretan Arc straits system [*Zodiatis*, 1992, 1993; *Kontoyiannis et al.*, 2005]. It is probably through the Kasos Strait, with a sill depth of 1000 m at its deepest point east

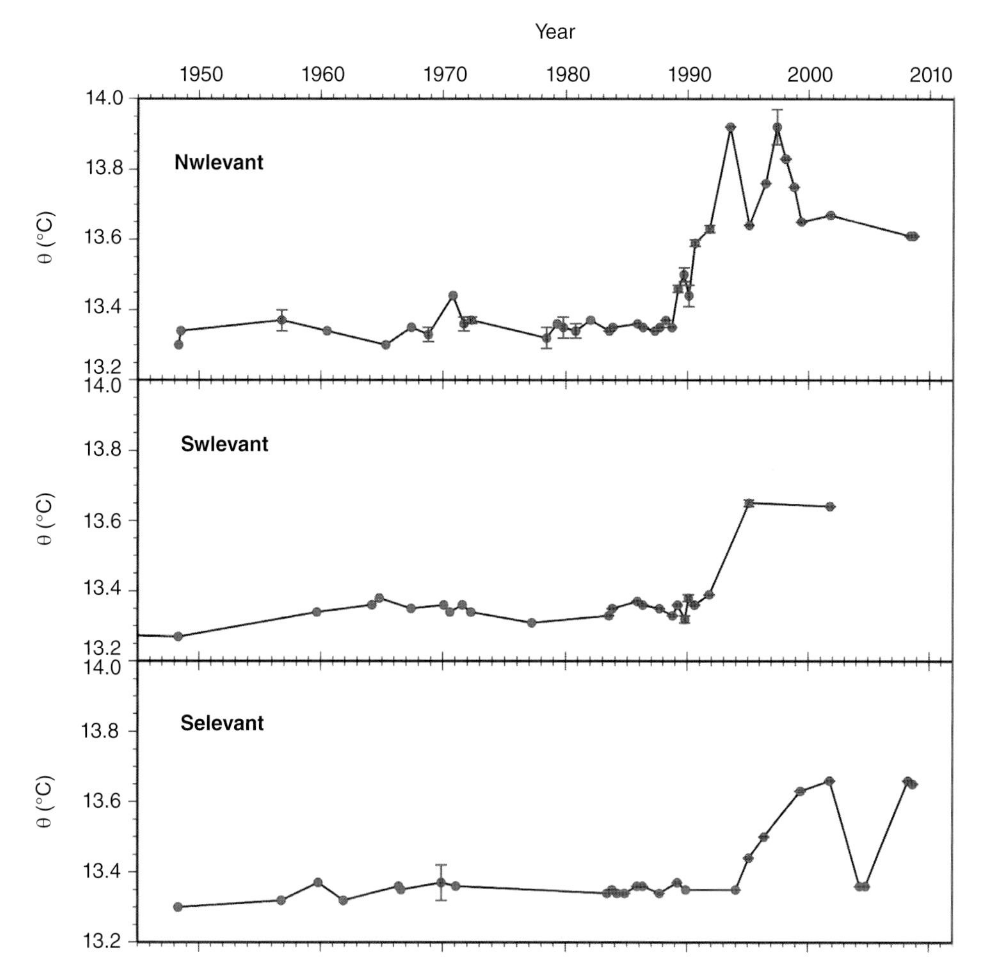

Figure 7.13. Potential temperature θ in the 1900–2100-m depth level in the (a) northwestern, (b) southwestern, and c) southeastern Levantine subdomains. For color detail, please see color plate section.

of Crete, that the stronger EMT signal is conveyed to the Levantine Basin. Since the other straits connecting the Aegean Sea to the eastern Mediterranean are shallower, notably the Antikithira Strait with 560 m sill depth, it is confirmed by the above review of properties in Figures 7.11 and 7.12 that the main outflow route of dense water out of the Cretan Sea is the Kasos Strait, confirming results from the earlier studies [*Roether at al.*, 2007].

The behavior at 2-km depth is different. In the northwest Levantine area adjoining the Kasos Strait (Figure 7.13a), a large temperature signal of 0.6°C is again the largest rise observed among the different subdomains at this depth. A stepwise increase of temperature of about 0.35°C marking the arrival of the EMT signal in the southwest Levantine region (Figure 7.13b) follows closely the northwest region, while in the southeast Levantine, the peak values are reached much later at the end of the 1990s (Figure 7.13c).

Now comparing the Ionian Basin subdomains at 2-km depth, the northeastern Ionian box stands out (Figure 7.14a) indicating a stepwise response with superposed oscillations reaching an amplitude of about 0.5°C in 1993. Smaller amplitude temperature steps of 0.2°C are found at the other subdomains, but with later arrival times of the peak change in about 1995 in the northwest Ionian (Figure 7.14b) and southeast Ionian (Figure 7.14c), and in1998 in the central Ionian Sea (Figure 7.14d).

The above observations suggest direct influence of the EMT on the northwestern Levantine Basin area, from where the anomaly spreads to the southern Levantine Basin and the Ionian Sea. The presence of a large anomaly only in the northwest Levantine area adjacent to the Cretan Sea at 1-km depth (Figures 7.11 and 7.12) confirms the outflow from Kasos Strait at this depth. At the deeper level of 2 km (Figures 7.13 and 7.14), which is more representative of the core of the CDW outflow in the Levantine Basin, the source driven anomaly is present at all subdomains with reduced amplitude and with delays. Due to the basin's size and topographic constraints (obstacles imposed by sea mounts), the EMT

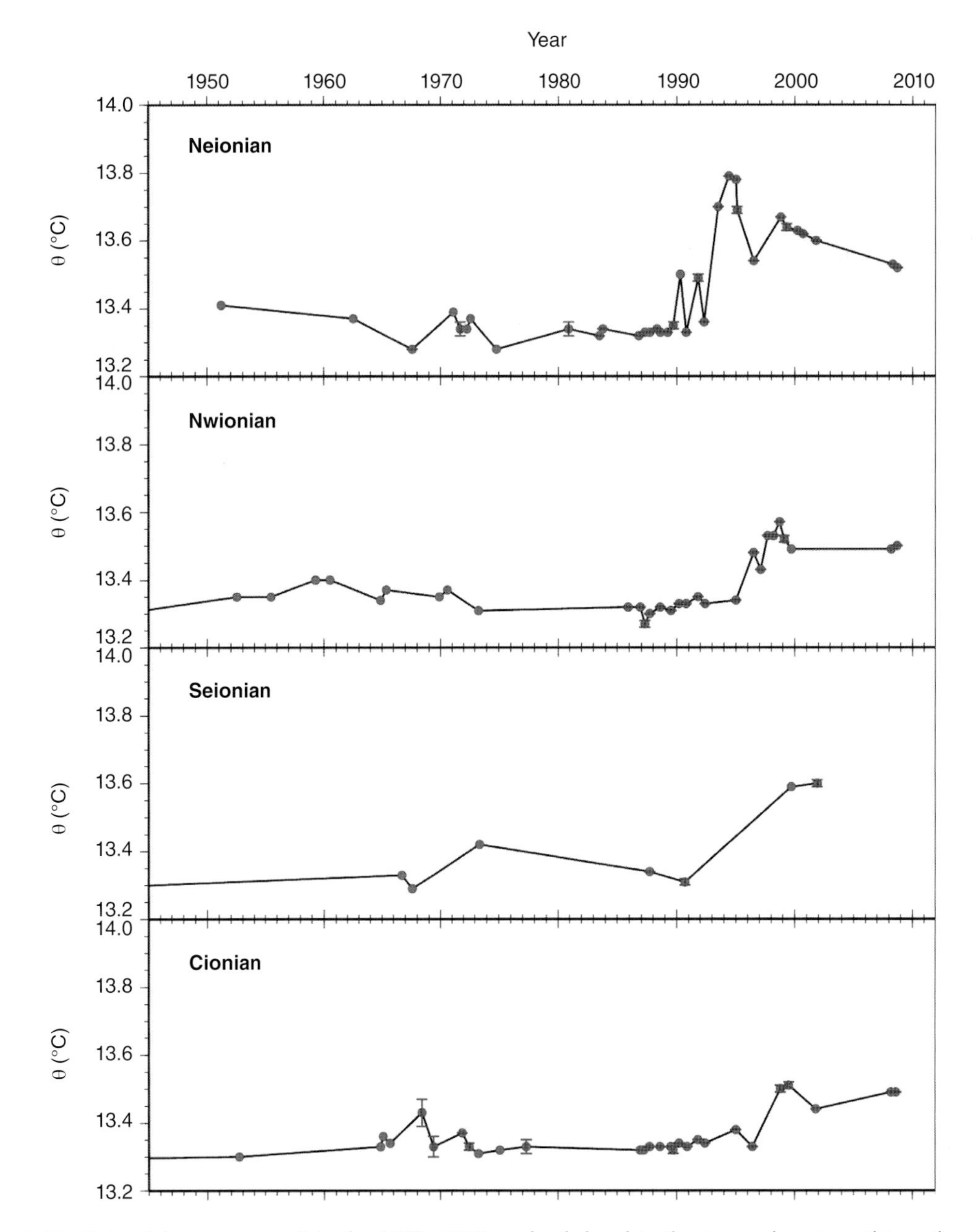

Figure 7.14. Potential temperature θ in the 1900–2100-m depth level in the (a) northeastern, (b) northwestern, (c) southeastern, and (d) central Ionian subdomains. For color detail, please see color plate section.

signal arrives at the southeast Levantine area with a delay of a few years. The origin of outflow in the northwest Levantine area near Kasos Strait at 1 km and the existence of an anomaly of similar amplitude only at 2-km depth in the northeastern Ionian area near Antikithira Strait, with later arrivals elsewhere, confirm the outflow scheme proposed by *Roether et al.* [2007] in his Figure 14: the dense water outflows from Kasos Strait and encircling the deep Hellenic Trench south of Crete arrives in the Ionian Sea, while its southward extension recirculates to both basins along the southern coast after overflowing the eastern Mediterranean Ridge.

Analyses based on the horizontal distributions of properties at any particular time are probably not justified for a historical profiles database of uneven spatial and temporal distribution such as ours. However, by averaging the data roughly for the pre and post-EMT periods, 1970–1990 and 1990–2012, respectively, in Figures 7.15a and 7.15b, we attempt to show the main changes between these two periods, again using deep potential temperature at the 2-km depth layer as tracer.

While the pre-EMT period shows lower temperatures in the Ionian and Levantine basins clearly differentiated from those in the Cretan Sea in Figure 7.15a, the spreading of the warmer and more saline water from the Cretan Sea out into the adjacent basins in the post-EMT period (Figure 7.15b) modifies the deep-water properties throughout the entire area of the Levantine and Ionian basins, with the influence spreading from Kasos Strait through the paths identified above.

7.5. SIMPLE STATISTICS

The number of stations and observation points (scans) in a depth range, and the computed mean and standard deviation and trends computed for the full length of the observation period in each of the maxin subbasins of the eastern Mediterranean are given in Table 7.1.

It is interesting to note that the highest deep salinities and densities occur in the Aegean Sea in the observed period, although partially compensated by potential temperature values slightly higher than the other regions. The coldest deep waters at a given depth interval occur in the Adriatic Sea, with density at 1-km depth remaining higher than the deep Ionian and Levantine sea waters at 4-km depth. Still the highest density occurs in the Aegean Sea, exceeding that of the Adriatic Sea.

Table 7.1 also provides trends in the observed data, calculated by fitting regression lines to the data. Long-term trends appear in the observation periods, including transients and stepwise increases imposed by the EMT in the last two decades. In all the basins, there is a positive trend in all variables at the deeper levels, which is a result of the increases in salinity partially compensated by increased potential temperature. The highest temperature trend partly compensating the salinity increase appears in the Levantine Basin deep waters where the density trend is the lowest. A relatively higher trend in density occurs in

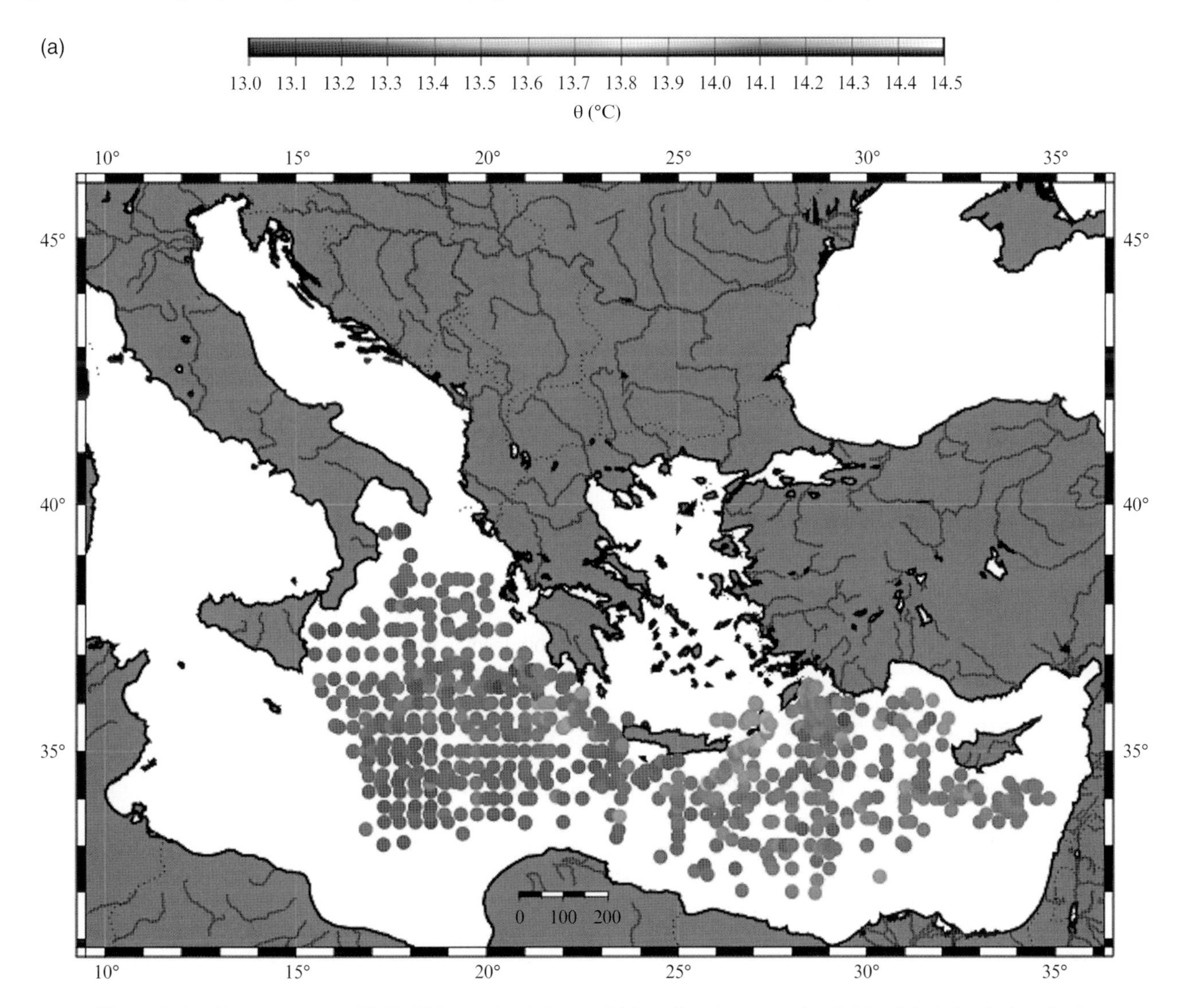

Figure 7.15. Temperature at 1900–2100-m depth interval binned and averaged at 0.1° x 0.1° latitude-longitude intervals during the a) 1970–1990 and

(b)

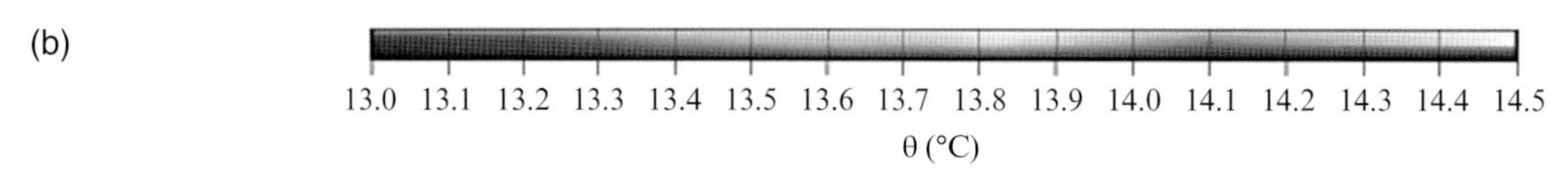

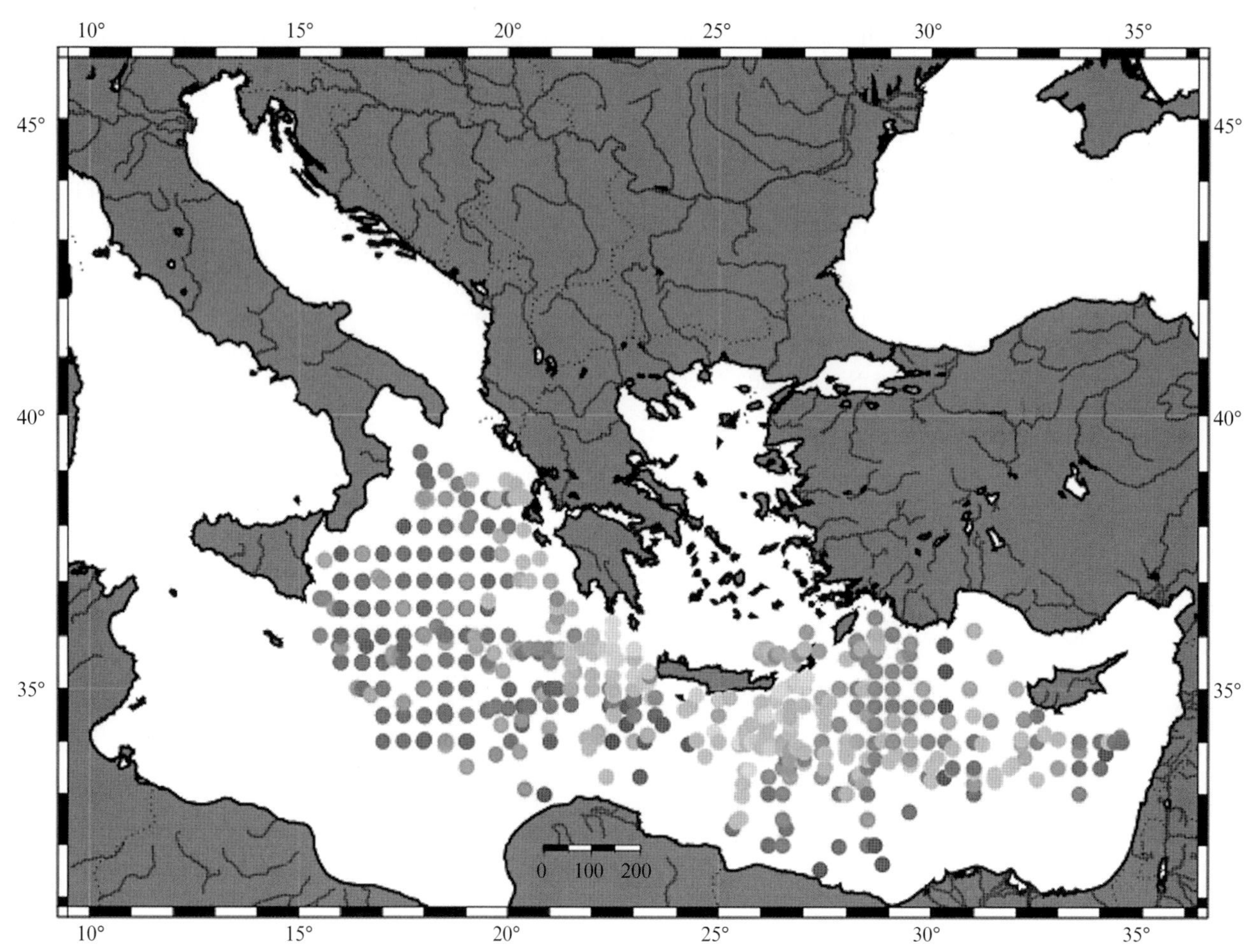

Figure 7.15. (continued) b) 1990–2012 periods. For color detail, please see color plate section.

the Ionian Sea. The highest increase in density occurs in the deep waters of the Aegean and Adriatic seas, where it is associated with cooling effects, with a higher rate in the Aegean Sea.

7.6. DISCUSSION AND CONCLUSIONS

Apart from the very long-term changes, decadal variability is observed to dominate the temporal evolution of the deep waters. It is also observed that the Eastern Mediterranean Transient (EMT) stands as the most extraordinary deep-water episode, while other interdecadal changes of varying strength are predominantly detected in certain areas of the eastern Mediterranean. The largest interdecadal variability is observed in the Aegean and Adriatic basins, where changes are continuously observed during the second half of the last century. An abrupt increase of salinity and density with an associated drop in temperature in the Aegean Basin marks the EMT starting in the early 1990s. In the deep Levantine Basin, the EMT event starts at the same time, particularly after the 1992–1993 cooling event, but lasts throughout the 2000s for about two decades, with stepwise increases in temperature, salinity, and density, attributed to cascading out of the southern Aegean Sea into the adjacent area. In the deep Levantine and Ionian basins, salinity changes dominate the interdecadal variability most of the time. Yet potential temperature proves to be a useful tool to trace these variations, allowing the construction of a basic description of the changes, especially in relation to the EMT. The evolution of the eastern Mediterranean deep-water characteristics suggests a strong interaction between the various subbasins.

The evolution history some years past the EMT can be found in the later literature, showing that the density of the

Table 7.1. Trends in water properties at selected depth intervals in the selected subdomains of the eastern Mediterranean Sea (1948–2010).

	Number of averaged		Potential temperature		Salinity		Potential density	
Depth Interval (m)–Region	stations	scans	mean ± std dev (°C)	trend (10^{-3} °C /yr)	mean ± std dev	trend (10^{-3} /yr)	mean ± std dev (kg/m^3)	trend (10^{-3} kg /m^3/yr)
Adriatic Sea								
0450– 0550	1627	84450	13.41±0.25	1.26	38.72±1.28	1.67	29.18±0.22	1.07
0900– 1100	433	14835	13.04±0.18	−2.33	38.67±0.42	1.53	29.23±0.14	1.66
Aegean Sea								
0450– 0550	2699	70225	14.01±0.51	−0.35	38.96±0.65	2.41	29.25±0.25	1.97
0900– 1100	926	49184	13.89±0.39	4.75	38.98±0.62	2.40	29.29±0.19	0.80
1900– 2100	91	5570	14.03±0.09	−3.98	39.02±0.25	5.16	29.29±0.17	4.83
Levantine Sea								
0450– 0550	7575	290072	13.99±0.32	−2.29	38.92±2.99	−0.22	29.09±1.53	0.29
0900– 1100	5978	332038	13.56±0.32	1.97	38.79±1.12	0.43	29.09±1.93	−0.05
1900– 2100	1748	65806	13.46±0.07	6.33	38.72±0.53	2.30	29.18±0.84	0.40
2900– 3100	112	6125	13.59±0.20	8.75	38.78±0.21	2.77	29.20±0.24	0.40
3900–4100	15	968	13.62±0.12	8.23	38.82±0.15	1.94	29.21±0.10	0.30
Ionian Sea (10273 stations)								
0450– 0550	6014	156955	13.90±0.23	−0.62	38.81±1.08	0.86	29.14±0.30	0.78
0900– 1100	3993	204853	13.57±0.28	0.44	38.67±2.57	0.58	29.15±0.53	0.36
1900– 2100	1982	112464	13.40±0.38	4.20	38.71±0.59	1.82	29.18±0.61	0.45
2900– 3100	552	46314	13.33±0.18	3.23	38.70±0.88	1.70	29.20±1.09	0.62
3900–4100	47	3270	13.40±0.17	2.32	38.73±0.24	1.72	29.20±0.18	0.79

deep water in the Cretan Sea has been decreased back to normal values after the trigger event of the EMT, but the deep water characteristics in the northwestern Levantine Sea region have been shown to survive long after the event [e.g., *Theocharis et al.*, 1999; *Kovačević et al.*, 2012].

Although a description of the climate variability aspects would add significantly to a more basic understanding of the observed deep-sea variability of the eastern Mediterranean, we have excluded the related discussion in this paper, keeping such analyses for future studies. There is already a number of studies on the climatic effects, not fully cited here. The Mediterranean response to climate forcing is rather complex. A synthetic discussion of the climate variability effects and a description of the EMT have been attempted by *Lionello et al.* [2006], among others. Yet, it appears still quite complicated to define how the surface effects are transmitted to deep water by either convective events or lateral fluxes and cascading.

A significant degree of synchronism, possibly imposed by large-scale controls, is often found between the Levantine, Black ,and Caspian seas [*Özsoy*, 1999], implying relationships with the Southern Oscillation (SO) in some years (e.g., 1982–83, 1986–87, the 1990s) or the North Atlantic Oscillation (NAO) in others (1983, 1986–87, 1989–90, 1992–93). Later, principal components analyses performed by *Gündüz and Özsoy* [2005] showed a greater influence of the North Sea Caspian Pattern (NCP) in the Levantine-Aegean-Black seas region compared to the NAO playing a greater role in the western Mediterranean. Especially in the Aegean Sea region, air temperature and surface heat fluxes were very strongly correlated with the NCP dipole pattern during the years 1964, 1975–1976, and 1992–1993. Some of these signals are evident in the analyses performed here. But again, the details and cause-and-effect relationships would probably be better exposed by continuing analyses and modeling efforts.

REFERENCES

Androulidakis, Y. S., V. H. Kourafalou, Y. N. Krestenitis, and V. Zervakis (2012), Variability of deep water mass characteristics in the North Aegean Sea: The role of lateral inputs and atmospheric conditions, *Deep-Sea Res., I 67*, 55–72.

Ashkenazy, Y., P. H. Stone, and P. Malanotte-Rizzoli (2011), Box modeling of the eastern Mediterranean Sea, *Physics A*, *391*, 1519–1531.

Beuvier, J., F. Sevault, M. Herrmann, H. Kontoyiannis, W. Ludwig, M. Rixen, E. Stanev, K. Béranger, and S. Somot (2010), Modeling the Mediterranean Sea interannual variability during 1961– 2000: Focus on the eastern Mediterranean Transient, *J. Geophys. Res.—Oceans*, *115*, C08017, doi:10.1029/2009JC005950.

CIESM (2008), *Toward an integrated system of Mediterranean marine observatories*, No. 34 in CIESM Workshop Monographs, ed. F. Briand, Monaco.

CIESM (2011), *Designing Med-SHIP: A program for repeated oceanographic surveys*, No. 43 in CIESM Workshop Monographs, ed. F. Briand, Monaco.

El-Gindy, A. H., and S. H. S. El-Din (1986), Water masses and circulation patterns in the deep layer of the eastern Mediterranean, *Oceanologica Acta*, *9*, 239–248.

Emery, W. J., and R. E. Thompson (2001), *Data Analysis Methods in Physical Oceanography*, 2 ed., Elsevier Science.

Gačić, M., G. L. E. Borzelli, G. Civitarese, V. Cardin, and S. Yari (2010), Can internal processes sustain reversals of the ocean upper circulation? The Ionian Sea example, *Geophys. Res. Lett.*, *37*, L09608, doi:10.1029/2010GL043216.

Gertman, I. F., N. Pinardi, Y. Popov, and A. Hecht (2006), Aegean Sea water masses during the early stages of the eastern Mediterranean Climatic Transient (1988– 1990), *Journal of Physical Oceanography*, *36*, 1841–1859.

Gündüz, M., and E. Özsoy (2005), Effects of the North Sea Caspian pattern on surface fluxes of Euro-Asian-Mediterranean seas, *Geophys. Res. Lett.*, *32*, 21, doi: 10.1029/2005GL024315.

Hecht A., and I. Gertman (2001), Physical features of the eastern Mediterranean resulting from the integration of POEM data with Russian Mediterranean cruises, *Deep Sea Research, Part I*, *48*/8, 1847–1876.

Hieke, W., P. Halbach, M. Türkay, and H. Weikert, eds. (1994), Mittelmeer 1993, Cruise No. 25, 12 May–20 August 1993, METEOR-Berichte, Universitaet Hamburg, 94–3, 256 pp.

Hoepffner, N., ed. (2006), *Marine and Coastal Dimension of Climate Change in Europe*, A report to the European Water Directors, European Commission, Directorate-General, Joint Research Center, Institute for Environment and Sustainability.

Josey, S. A. (2003), Changes in the heat and freshwater forcing of the eastern Mediterranean and their influence on deep-water formation, *J. Geophys. Res.—Oceans*, *108*, 3237, doi:10.1029/2003JC001778.

Josey, S. A., S. Somot, and M. Tsimplis (2011), Impacts of atmospheric modes of variability on Mediterranean Sea surface heat exchange, *J. Geophys. Res.*, *116*, C02032, doi:10.1029/2010JC006685.

Klein, B., W. Roether, D. Manca, D. Bregant, V. Beitzel, V. Kovacevic, and A. Luchetta (1999), The large deep-water transient in the eastern Mediterranean, *Deep Sea Res.*, *46*, 371–414.

Kontoyiannis, H., E. Balopoulos, A. Pavlidou, O. Gotsis-Skretas, G. Asimakopoulou, E. Papageorgiou, and A. Iona (2005), The hydrodynamics and biochemistry of the Cretan Straits (Antikithira and Kassos Straits) revisited in the period June/1997–May/1998, *Journal of Marine Systems*, *53*, 37–57.

Kovačević, V., B. B. Manca, L. Ursella, K. Schroeder, S. Cozzi, M. Burca, E. Mauri, R. Gerin, G. Notarstefano, and D. Deponte (2012), Water mass properties and dynamic conditions of the eastern Mediterranean in June 2007, *Progress in Oceanography*, *104*, 59–79.

Lacombe, H., and P. Tchernia (1960), Quelques traits genéraux de l'hydrologie Mediterranéene, *Cahiers Océanographiques*, 20(8), 528–547.

Lascaratos A., and K. Nittis, (1998), A high resolution 3-D numerical study of intermediate water formation in the Levantine Sea, *J. Geophys. Res.*, *13*, 18497–18511.

Lascaratos, A., W. Roether, K. Nittis, and B. Klein (1999), Recent changes in deep ocean formation and spreading in the eastern Mediterranean Sea, *Prog. Oceanogr.*, *44*, 5–36.

Lionello, P., P. Malanotte-Rizzoli, and R. Boscolo, eds. (2006), *Mediterranean Climate Variability*, Elsevier Science.

LIWEX Group (2003), The Levantine intermediate water experiment (LIWEX) group: Levantine basin—a laboratory for multiple water mass formation processes, *J. Geophys. Res.*, *198*, doi:10.1029/2002JC001643.

Malanotte-Rizzoli, P., and A. Hecht (1988), Large-scale properties of the eastern Mediterranean: A review, *Oceanol. Acta*, *11*, 323–335.

Malanotte-Rizzoli, P., B. Manca, M. d'Alcala, A. Theocharis, S. Brenner, G. Budillon, and E. Özsoy (1999), The eastern Mediterranean in the 80s and in the 90s: The big transition in the intermediate and deep circulations, *Dyn. Atmos. Oceans*, *29*, 365–395.

Malanotte-Rizzoli, P., B. Manca,, S. Marullo, M. d'Alcala, A. Theocharis, A. Bergamasco, G. Budillon, E. Sansone, G. Civitarese, F. Conversano, I. Gertman, B. Herut, N. Kress, S. Kioroglou, H. Kontoyannis, K. Nittis, B. Klein, A. Lascaratos, M. A. Latif, E. Özsoy, A. R. Robinson, R. Santoleri, D. Viezzoli, and V. Kovacevic (2003), The Levantine Intermediate Water Experiment (LIWEX) Group: Levantine Basin—laboratory for multiple water mass formation processes, *J. Geophys. Res.*, *108* (C9), 8101.

Malanotte-Rizzoli, P., and the Pan-Med Group (2012), Physical forcing and physical/biochemical variability of the Mediterranean Sea: A review of unresolved issues and directions of future research, Report of the Workshop Variability of the Eastern and Western Mediterranean circulation and Thermohaline Properties: Similarities and Differences, Rome, November 7–9, 2011, 48 pp.

Miller, A. R. (1974), Deep convection in the Aegean Sea, in *Processus de Formation des Eaux Oceaniques Profondes*, Colloques Internationaux du C. N. R. S. 215.

Nielsen, J. (1912), Hydrography of the Mediterranean and adjacent waters, in *Report of the Danish Oceanographic Expedition 1908–1910 to the Mediterranean and Adjacent Waters*, Vol. *1*, pp. 72–191.

Nittis, K., A. Lascaratos, and A. Theocharis (2003), Dense water formation in the Aegean Sea: Numerical simulations during the Eastern Mediterranean Transient, *J. Geophys. Res.*, *108* (C9), PBE 21, doi:10.1029/2002JC001352.

Ovchinnikov, I. M., and A. Plakhin (1984), Formation of the intermediate waters of the Mediterranean Sea in the Rhodes Cyclonic Gyre, *Oceanology*, *24*, 317–319.

Ovchinnikov, I. M., Y. I. Popov, and I. F. Gertman (1990), A study of the deep water formation in the eastern Mediterranean Sea, *Oceanology*, *30*, 6, 1039–1041.

Özsoy, E. (1999), Sensitivity to global change in temperate Euro-Asian seas (the Mediterranean, Black Sea and Caspian Sea): A review, in *The Eastern Mediterranean as a Laboratory Basin for the Assessment of Contrasting Ecosystems*, pp. 281–300, ed. P. Malanotte-Rizzoli and V. N. Eremeev, NATO Science Series 2, Environmental Security, 51, Kluwer Academic Publishers, Dordrecht.

Özsoy, E., A. Hecht, and Ü. Ünlüata (1989), Circulation and hydrography of the Levantine Basin: Results of POEM-coordinated experiments 1985–1986, *Prog. Oceanogr.*, *22*, 125–170.

Özsoy, E., A. Hecht, Ü. Ünlüata, S. Brenner, H. İ. Sur, J. Bishop, M. A. Latif, Z. Rozentraub, and T. Oğuz (1993), A synthesis of the Levantine Basin circulation and hydrography, 1985–1990, *Deep-Sea Res.*, *40*, 1075–1119.

Özsoy, E., and M. A. Latif (1996), *Climate Variability in the Eastern Mediterranean and the Great Aegean Outflow Anomaly*, International POEM-BC/MTP Symposium, Molitg les Bains, France, 1–2 July 1996, pp. 69–86.

Özsoy, E., Ü. Ünlüata, T. Oğuz, M. A. Latif, A. Hecht, S. Brenner, J. Bishop, and Z. Rozentroub (1991), A review of the Levantine Basin circulation and its variabilities during 1985–1988, *Dyn Atmos. Oceans*, *15*, 421–456.

Pinardi, N., and E. Masetti (2000), Variability of the large-scale general circulation of the Mediterranean Sea from observations and modelling: A review, *Palaeogeography, Palaeoclimatology, Palaeoecology*, *158*, 153–173.

Plakhin, Ye. A. (1972), Vertical winter circulation in the Mediterranean, *Oceanology*, *12*, 344–351.

POEM Group (A. R. Robinson, P. Malanotte Rizzoli, A. Hecht, A. Michelato, W. Roether, A. Theocharis, Ü. Ünlüata, N. Pinardi, A. Artegiani, A. Bergamasco, J. Bishop, S. Brenner, S. Christianidis, M. Gačić, D. Georgopoulos, M. Golnaraghi, M. Hausmann, H. G. Junghaus, A. Lascaratos, M. A. Latif, W. G. Leslie, C. J. Lozano, T. Oğuz, E. Özsoy, E. Papageorgiou, E. Paschini, Z. Rozentroub, E. Sansone, P. Scarazzato,R. Schlitzer, G. C. Spezie, E. Tziperman, G. Zodiatis, L. Athanassiadou, M. Gerges, M. Osman) (1992), General circulation of the eastern Mediterranean, *Earth-Sci. Rev.*, *32*, 285–309.

Pollak, M. I. (1951), The sources of deep-water in the eastern Mediterranean Sea, *Journal of Marine Research*, *10*, 128–152.

Press, W. H., S. A. Teukolsky, W. T. Vetterling, and B. P. Flannery (2007), *Numerical Recipes*, 3 ed., Cambridge University Press.

Roether, W., and R. Schlitzer (1991), Eastern Mediterranean deep-water renewal on the basis of chlorofluoromethane and tritium data, *Dyn. Atmos. Oceans*, *15* (3–5), 333–354.

Roether, W., B. Manca, B. Klein, D. Bregant, D. Georgopoulos, V. Beitzel, V. Kovacevic, and A. Lucchetta (1996), Recent changes in eastern Mediterranean deep waters. *Science*, *271*, 333–335.

Roether W., B. Klein, B. B. Manca, A. Theocharis, and S. Kioroglou (2007), Transient Eastern Mediterranean deep waters in response to the massive dense-water output of the Aegean Sea in the 1990s, *Progr. Oceanogr.*, *74*, 540–571, doi:10.1016/j.pocean.2007.03.001.

Samuel, S., K. Haines, S. Josey, and P. Myers (1999), Response of the Mediterranean sea thermohaline circulation to observed changes in the winter wind stress field in the period 1993, *J. Geophys. Res. Oceans*, *104*, 7771–7784.

Sayın, E., and Ş. T. Beşiktepe (2010), Temporal evolution of the water mass properties during the Eastern Mediterranean Transient (EMT) in the Aegean Sea, *J. Geophys. Res.*, *115*, C10025, doi:10.1029/2009JC005694.

Schlitzer, R., W. Roether, H. Oster, H. Junghans, H. Hausmann, H. Johannsen, and A. Michelato (1991), Chlorofluoromethane and oxygen in the eastern Mediterranean, *Deep Sea Res.*, *38*, 1531–1551.

Stratford, K., and K. Haines(2002), Modelling changes in Mediterranean thermohaline circulation 1987–1995 *J. Mar. Syst.*, *33*, 51–62.

Sur, H., E. Özsoy, and Ü. Ünlüata (1993), Simultaneous deep and intermediate depth convection in the Northern Levantine Sea, winter 1992, *Oceanol. Acta*, *16*, 33–43.

Theocharis, A., and D. Georgopoulos (1993), Dense water formation over the Samothraki and Lemnos plateaux in the North Aegean Sea (eastern Mediterranean Sea), *Continental Shelf Research*, *l3*(S-9), 919–939.

Theocharis, A., K. Nittis,, H. Kontoyiannis,, E. Papageorgiou, and E. Balopoulos (1999), Climatic changes in the Aegean influence the eastern Mediterranean thermohaline circulation (1986–1997), *Geophys. Res. Lett.*, *26*, 1617–1620.

Theocharis, A., B. Klein, K. Nittis, and W. Roether (2002), Evolution and status of the Eastern Mediterranean Transient (1997–1999), *Journal of Marine Systems*, *33–34*, 91–116.

Vervatis, V., S. Sofianos, and A. Theocharis (2011), Distribution of the thermohaline characteristics in the Aegean Sea related to water mass formation processes (2005–2006 winter surveys), *J. Geophys. Res.*, *116*, C09034, doi:10.1029/2010JC006868.

Wu, P., K. Haines, and N. Pinardi (2000), Toward an understanding of deep-water renewal in the eastern Mediterranean, *J. Phys. Oceanogr.*, *30*, 443–458.

Wüst, G. (1961), On the vertical circulation of the Mediterranean Sea, *J. Geophys. Res.*, *66*, 3261–3271.

Zervakis, V., D. Georgopoulos, and P. Drakopoulos (2000), The role of North Aegean in triggering the recent eastern Mediterranean climatic changes, *J. Geophys. Res.*, *105*, 26103–26116.

Zervakis, V., and D. Georgopoulos (2002), Hydrology and circulation in the North Aegean (eastern Mediterranean) throughout 1997 and 1998, *Mediterranean Marine Science*, *3*, 5–19.

Zervakis, V., D. Georgopoulos, and P. G. Drakopoulos (2000), The role of the North Aegean in triggering the recent eastern Mediterranean climatic changes, *J. Geophys. Res.*, *105* (C11), 103–116.

Zervakis, V., E. Krasakopoulou, D. Georgopoulos, and E. Souvermezoglou (2003), *Vertical Diffusion and Oxygen Consumption during Stagnation Periods in the Deep North Aegean*, Deep-Sea Res., Part I, Oceanographic Research Papers. *50*, 53–71.

Zodiatis, G. (1992), On the seasonal variability of the water masses circulation in the NW Levantine Basin-Cretan Sea and flows through the eastern Cretan Arc Straits, *Ann. Geophysicae*, *10*, 12–24.

Zodiatis, G. (1993). Water mass circulation between the SE Ionian–W Cretan basins through the western Cretan Arc Straits, *Bollettino di Oceanologia Teorica ed applicata*, *11*, 1, 61–75.

8

An Internal Mechanism Driving the Alternation of the Eastern Mediterranean Dense/Deep Water Sources

Alexander Theocharis, George Krokos, Dimitris Velaoras, and Gerasimos Korres

8.1. INTRODUCTION

The eastern Mediterranean Sea (EMed) is composed by two large basins, the Ionian and the Levantine, connected by the Cretan Passage (2000–4500 m deep), and of two marginal seas, the Adriatic and the Aegean that act as dense-water sources, driving its thermohaline circulation (Figure 8.1).

The Adriatic Sea has an elongated shape along a NW-SE axis, with an extended shelf area in its north part and a deep depression in the south, namely the South Adriatic Pit, with 1200 m maximum depth. It communicates with the Ionian Sea through the Otranto Strait, 900 m deep presenting a rather U-shaped channel with smooth topography. The dense Adriatic water overflows the Otranto Sill and mixes with north Ionian waters, to finally form the Eastern Mediterranean Deep Water (EMDW).

The Aegean Sea presents a more complicated physiographic structure, as far as the bottom topography, the numerous islands, the complexity of the deep subbasins and internal sills among them, the plateaus, and the series of six straits, namely the Cretan Arc Straits system, through which it communicates with the open EMed, the Ionian to the west, and the Levantine to the east. These straits are characterized by high relief and sill depths ranging from 150 to 1000 m [*Theocharis et al.*, 1993; *Kontoyiannis et al.*, 1999]. The most important and deepest straits are the Antikithira Strait (sill depth ~700 m) to the west of Crete and the Kassos (sill depth ~1000 m) and Karpathos (sill depth ~850 m) to the east. The denser Aegean waters overflow the sills of the deep straits and sink toward the deep regions of the Hellenic Trench.

It is of importance that through the eastern straits of the Cretan Arc, there is an intense horizontal circulation pattern that sustains the communication between the Aegean and the Levantine. Branches of the Asia Minor Current [*POEM group*, 1992] intrude continuously into the Aegean supplying it with heat and salt. The different subbasins of the Aegean are connected with the nearly 400 m contour, which determines their water mass structure and the exchanges over the sills. The largest subbasin of the Aegean is the Cretan Sea in its southern part, with two depressions of 2200 and 2500 m deep, respectively, in the eastern part.

The EMed underwent an extreme thermohaline change in the late 1980s to early 1990s, the well-documented Eastern Mediterranean Transient (EMT) [*Roether et al.*, 1996; *Klein et al.*, 1999; *Theocharis et al*, 1999b; *Roether et al.*, 2007]. The abrupt shift of the main source of the EMDW from the Adriatic to the Aegean Sea and the effectiveness of the latter has revealed that the deep thermohaline circulation is not in a seemingly almost steady state as it was believed for several decades in the past [*Nielsen*, 1912; *Wüst*, 1961; *Roether and Well*, 2001].

The production of very dense waters started in the entire Aegean Sea from 1987 onward, bearing different characteristics in each Aegean subbasin. These dense waters were progressively accumulated in the south Aegean Sea, feeding first the deep regions of its western part, especially the Myrtoan Sea (max depth ~1000 m) and then the deep depressions of the eastern Cretan Sea (Figure 8.1). More specifically, in the period 1987–1991, there was a rather continuous formation of denser waters that raised the deep isopycnals of the South Aegean toward shallower layers

Hellenic Center for Marine Research, Institute of Oceanography, Anavyssos, Greece

The Mediterranean Sea: Temporal Variability and Spatial Patterns, Geophysical Monograph 202. First Edition.
Edited by Gian Luca Eusebi Borzelli, Miroslav Gačić, Piero Lionello, and Paola Malanotte-Rizzoli.

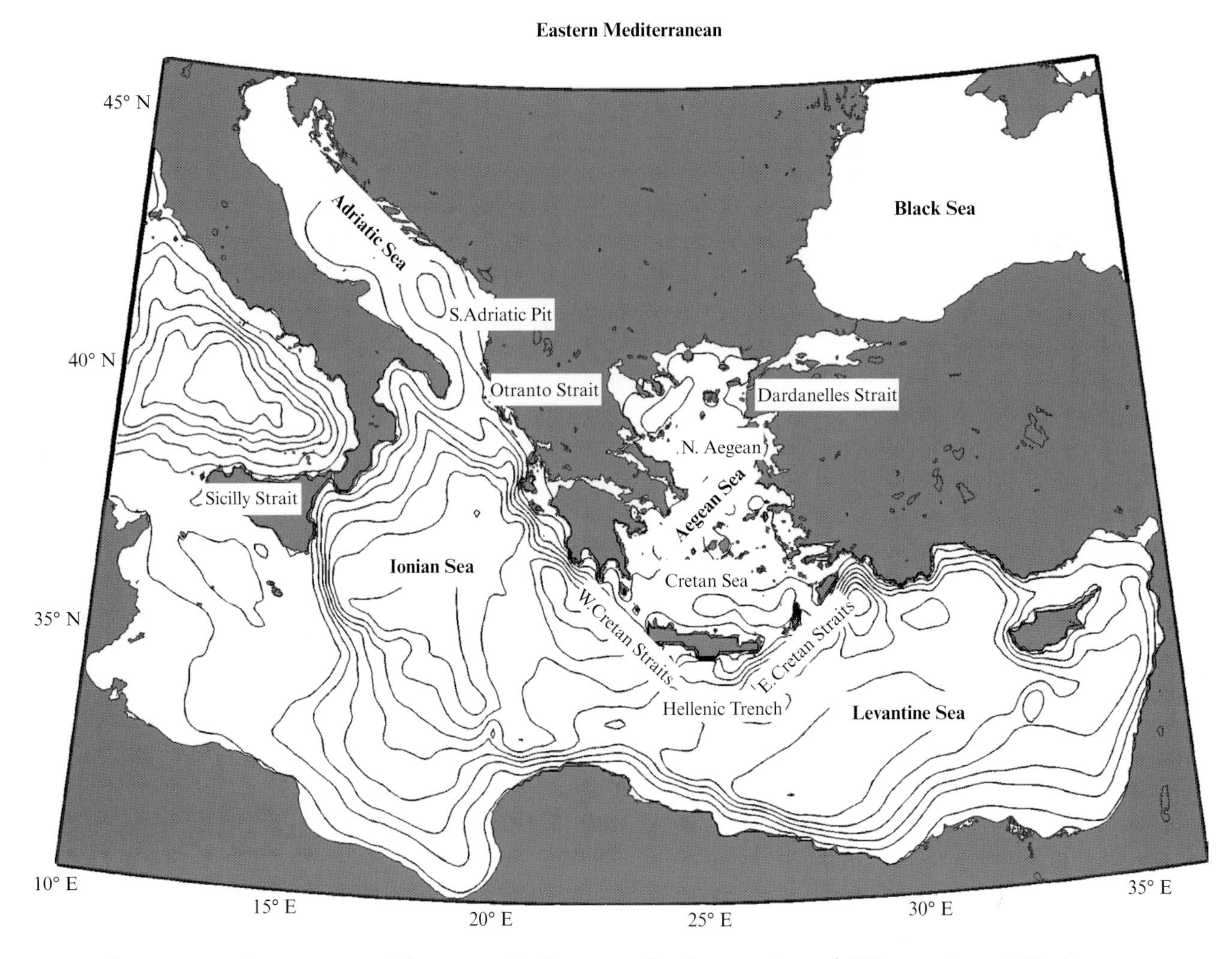

Figure 8.1. Bathymetric map of the eastern Mediterranean Sea (contour interval 500 m starting at 100 m).

[*Theocharis et al.*, 1999b]. At the same time, there was a continuous increase of temperature and salinity in the entire water column of the South Aegean. Finally, the exceptional cold air temperatures that occurred over the Aegean Sea during the winters of 1992–93, enhanced the massive production of even denser waters in the Cretan Sea, namely the Cretan Deep Water (CDW), that reached the unprecedented potential density values greater than 29.35 kg/m^3 [*Theocharis et al.*, 1999b]. These density values were much higher than that of the EMDW and even higher than that of the Adriatic Deep Water (ADW) [*Malanotte-Rizzoli et al.*, 1999]. The total amount produced within 7 years (1987–1993) was estimated up to 7 Sv [*Roether et al.*, 1996] of which 2.8 Sv outflowed from mid-1992 through 1994 [*Roether et al.*, 2007]. Thus, the average production rate per year was estimated in the order of 1 Sv, three times higher the rate of the mean production of the Adriatic Sea (0.3 Sv). Peak outflow possibly exceeded 3.5 Sv in 1993 [*Roether et al.*, 2007]. This huge amount of very dense water was discharged through the Cretan Arc Straits toward the Ionian and the Levantine seas, feeding first the deepest regions of the Hellenic Trench area, and then spreading into the entire EMed, causing an uplifting of the older EMDW of Adriatic origin and all water masses found above it, by several hundred meters. This new type of Eastern Mediterranean deep and bottom waters was warmer and more saline than the previously prevailing EMDW due to the different characteristics of the new source water (CDW). At the same time, the Aegean has produced an intermediate water mass, the so called Cretan Intermediate Water (CIW), with similar to the Levantine Intermediate Water (LIW) characteristics, that exited through the Cretan Arc Straits as well. Then following a westward pathway toward Sicily and Otranto straits, it has actually replaced the LIW at the western regions of the EMed [*Malanotte-Rizzoli et al.*, 1999].

The renewal time of the EMed deep and bottom waters is now estimated to be less than a decade [*Manca et al.*, 2002], while the respective time in the past ranged from 84 to 126 years [*Roether and Schlitzer*, 1991; *Schlitzer et al.*,

1991]. Therefore, numerous authors [*Roether et al.*, 1996, *Malanotte-Rizzoli et al.*, 1999; *Theocharis et al.*, 1999b; *Zervakis et al.*, 2000; *Demirov and Pinardi*, 2002; *Josey*, 2003, among others] proposed different factors and/or synergies, both atmospheric and oceanic, as responsible causes for the EMT, which is considered as a unique event in an almost steady thermohaline functioning of the EMed with one main source of Deep Water Formation (DWF) existing in the Adriatic. However, the Aegean Sea has been reported in the past [*Miller*, 1963, 1974; *Ovchinnikov and Plakhin*, 1965] as a secondary source of dense/deep waters. Dense waters with Aegean signature were sporadically observed outside the Cretan Arc Straits at depths of 500–1400 m before the 1990s.

It must be noted that the Aegean is considered an important contributor of salt and heat to the intermediate and deeper horizons of the EMed. The so-called Cretan Intermediate Water (CIW) and Cretan Deep Water (CDW) are formed in the Aegean at large by winter convection and/or shelf processes, and exit through the Cretan Arc Straits ventilating the intermediate and deep/bottom layers, respectively [*Georgopoulos et al.*, 1989; *Schlitzer et al.* 1991; *Roether et al.*, 1996; *Theocharis et al.*, 1993, 1999a]. Therefore, a plausible question arises, whether the EMT were really a unique episode within the last 100 years, or an extreme event in the frame of the internal variability of the EMed.

The examination of the last century hydrological datasets of the EMed gave evidence of another similar, but less intense than the EMT, event in the 1970s, which was salinity induced [*Lascaratos et al.*, 1999; *Theocharis et al.*, 2002a]. *Theocharis et al.*, [2002a] have given indications on the characteristics of the produced deep waters in the three basins—the Ionian, the Aegean, and the Levantine—but neither has provided a full interpretation on the DWF processes of this episode. Furthermore, *Beuvier et al.* [2010] supported this finding by a modeling study.

Here it is of importance to define what is meant by the term *similar to EMT* or *EMT-like* event. The EMT-like event can be described as a period of salinity and temperature preconditioning of the Aegean Sea, followed by dense water production, with densities ranging between those of LIW and EMDW and occasionally even higher, which overflow through the Cretan Arc sills. Then, these masses sink and spread in the eastern Mediterranean, contributing to the formation of a deep water mass with detectable signature. However, in an EMT-like event, the outflowing dense Aegean waters are not in adequate quantities or dense enough to replace already existing EMDW or reach the bottom layers, respectively. Therefore, the detected changes in the thermohaline circulation of the EMed during the 1990s could be characterized as a recurrent phenomenon with distinct variability in its intensity and consequences.

Recent studies [*Borzelli et al.*, 2009; *Gačić et al.*, 2010, 2011], taking into consideration the observed reversals in the Ionian Sea upper circulation after the EMT peak production period, have proposed a feedback mechanism between the redistribution of Ionian water masses related to variations of the thermohaline properties of the southern Adriatic and inversions of Ionian upper layer circulation. According to the aforementioned studies, these inversions driven by baroclinic vorticity production alternate the advection into the Adriatic of saline Aegean/Levantine masses or of fresher masses of Atlantic origin, changing the Adriatic thermohaline properties accordingly. *Gačić et al.* [2011] added that this mechanism can also explain the alternating predominance of the Adriatic and Cretan seas as dense-water sources in the EMed. Cyclonic upper Ionian circulation offers salty masses in the Adriatic, which make it more prone to winter convection, while the Aegean/Levantine seas are gaining buoyancy due to AW inflow. Conversely, anticyclonic upper Ionian circulation leads to decreased Aegean/Levantine buoyancy, while AW flowing into the Adriatic makes it more buoyant.

This study, based on numerical modeling and field observations, aims at proposing a different mechanism that drives the alternation of the two competitive dense-water sources, the Adriatic and the Aegean, at quasi-regular, almost decadal, time intervals. The proposed mechanism consists of a thermohaline pump that disturbs the EMed upper thermohaline cell, which results in the alternation of the DWF sources.

In order to examine the long-term variability of the thermohaline properties in the EMed, the circulation patterns and DWF processes, we revisit the datasets of the past century and up to 2010. At the same time, we investigate the ability of the hydrodynamic model to simulate the observed changes and we examine the relative importance of the involved forcing factors. The combined modeling and observational results show that the long-term variability cannot be exclusively explained by the atmospheric forcing.

The paper is organized as follows: section 2 deals with the various observational and atmospheric forcing datasets and the numerical model description. In section 3, the numerous results of this study are carefully examined and discussed. Finally, in section 4, we offer a summary of our findings and conclusions.

8.2. DATASETS AND MODEL DESCRIPTION

8.2.1. Data Series

In situ data used in this work originate from various datasets containing both CTD and bottle casts. Data referring to the Adriatic Sea (South Adriatic Pit) are derived from the MEDATLAS 2002 *in situ* database [*MEDAR Group*, 2002] covering a period from 1960 to 1996. Those referring to the Cretan Sea are a synthesis of various datasets, namely the MEDATLAS 2002 data, data from Soviet cruises as they appear in NOAA/NODC online datasets (http://www.nodc.noaa.gov/), data originating from the Hellenic Navy Hydrographic Service (HNHS) database and finally data from a number of cruises as found in the Hellenic National Oceanographic Data Center (HNODC) of Hellenic Center for Marine Research (HCMR). The Cretan Sea *in situ* dataset covers a period from 1960 to 2011, while the addition of data from all the aforementioned sources makes it perhaps the most comprehensive Cretan Sea dataset ever used. Casts that produced obvious spikes (mainly bottle casts) have been removed.

8.2.2. Model Description

For the present study, the 3-D, primitive equation, free surface, sigma coordinate Princeton Ocean Model (POM) is used [*Blumberg and Mellor*, 1987]. POM has been extensively used for applications in the Mediterranean Sea such as general circulation studies [*Zavatarelli and Mellor*, 1995], water mass formation studies, and circulation of its subbasins [*Lascaratos and Nittis*, 1998; *Demirov and Pinardi*, 2002; *Korres et al.*, 2002; *Nittis et al.*, 2003]. The current Mediterranean implementation follows the work by *Korres et al.* [2002] on a resolution of 1/10° x 1/10° in the horizontal and 25 sigma levels logarithmically distributed near surface and bottom in order to properly represent the dynamics of the boundary layers. The model bathymetry was obtained from the US Navy Digital Bathymetric Data Base (DBDB5). The Mediterranean model covers the geographical area 7° W to 36° E and 30.25° N to 45.75° N. Previous POM model implementations [*Zavatarelli and Mellor*, 1995; *Nittis et al.*, 2003; *Korres et al.*, 2002; *Kourafalou and Tsiaras*, 2007] have been shown to reproduce the main features of the Mediterranean/eastern Mediterranean circulation as well as processes that are considered of major importance for the overturning functioning.

8.2.3. Atmospheric Forcing

The atmospheric forcing that was used to drive the model simulation was derived from the high resolution (~0.5degrees) ARPERA dataset, produced upon downscaling of the ERA40 ECMWF reanalysis, using the version 4 of the global spectral ARPEGE-Climate model [*Déqué et al.*, 1994] as was described by *Somot et al.*, [2006]. The driving by the ERA40 reanalysis was performed through a spectral nudging technique described in *Herrmann and Somot*, [2008]. ARPERA uses a variable resolution configuration with a 2.5-stretching factor, a stretching pole located in the centre of the Tyrrhenian Sea and a resolution of ~50 km over the Mediterranean basin.

The dataset covers a period of 40 years (1960–2000) with 6-hour temporal resolution. The validation of the atmospheric forcing has been carried out indirectly through its efficiency on driving 3-D Mediterranean Sea hydrodynamic simulations. The reader is referred to a case study of the Gulf of Lions deep convection of the 1986–1987 winters where the use of ARPERA dataset proved to produce more accurate simulations of DWF processes than the use of ERA40 (*Herrmann and Somot*, 2008], mainly due to increased spatial variability and a better representation of extreme events. Other studies involve the evolution of the Mediterranean heat content and sea level over the 1961–2000 period [*Somot and Colin*, 2008; *Tsimplis et al.*, 2008, 2009].

Bulk formulas are used to determine the momentum, heat, and water fluxes at the air-sea interface. Solar and downward Longwave radiation are directly provided from the atmospheric forcing dataset, while the upward Longwave radiation is computed according to Stefan Boltzmann law using the sea surface temperature of the model. Latent and sensible heat fluxes are calculated through bulk aerodynamic formulas [*Gill*, 1982]. The turbulent exchange coefficients are estimated using the "Kondo scheme" [*Kondo*, 1975] in terms of air-sea temperature difference and wind speed at a single level (10 m) above the sea surface. For the transformation of air temperature and relative humidity from the available 2 m to 10 m, COARE [*Fairall*, 2003] formula is applied. The freshwater flux E-P is imposed as a freshwater flux at the sea surface. The evaporation rate is calculated from the evaporative heat flux.

For the present study satellite, modeled and *in situ* data were used to further evaluate the resulting fluxes and subsequently apply corrections concerning wind, precipitation, and SST.

The mean annual budgets were computed for the period 1960–2001 and correspond to -5.24 W/m^2 for the heat flux and 0.52 m/yr for the freshwater flux, which agree with accepted climatological values for the Mediterranean Sea as found in the literature [*Bethoux*, 1979].

8.2.4. River Discharge Data

Most major Mediterranean rivers (24) were accounted in a monthly resolution, chosen upon mean discharge significance and covering the whole basin. River outflow values correspond to observed and/or reconstructed data (RivDis database, SESAME, National datasets, graphs in publications, etc).

8.2.5. Boundary Conditions

8.2.5.1. Temperature and salinity climatology For use in open boundary conditions (i.e., Gibraltar) as well as in relaxation schemes, an annually varying seasonal climatology was constructed by combining the seasonal climatology of MEDAR MEDATLAS [*MEDAR Group*, 2002] and an interannual gridded dataset for the Mediterranean Sea covering the period 1945–2002. The latter is the 5-year moving temporal Gaussian window product described in *Rixen et al.* [2005], which was produced by interpolation of all the available measurements of the MEDATLAS 2002 *in situ* database on a 0.2 x 0.2 grid using a variational inverse method. The combination of the two datasets toward an annual varying seasonal climatology was created by imposing annual anomalies of the interannual dataset over the seasonal climatology of MEDAR MEDATLAS.

8.2.5.2. Relaxation schemes Salinity and temperature were weakly relaxed (on a timescale of 250 days) toward the interannually varying seasonal climatology with a vertical weighting function so that the damping is in the deep layers only. Surface temperature was also relaxed toward the SST fields contained in the ARPERA dataset, for consistency with the atmospheric fluxes via surface thermal flux, applied on a timescale of 250days.

8.2.5.3. Gibraltar Strait The model open boundary at its western edge is located in the Atlantic Ocean at -7° E and it is the boundary that controls the Gibraltar inflow/outflow. Open boundary conditions were set as follows:

1. Zero gradient condition for the free surface elevation
2. *Flather* [1976] boundary condition for the barotropic velocity normal to the open boundary
3. Sommerfeld radiation for the internal (baroclinic) velocities
4. Temperature and salinity at the open boundaries are advected upstream. When there is inflow through the open boundary, these fields are prescribed from the interannually varying seasonal climatology presented in 2.5.1.

8.2.5.4. Dardanelles Strait parameterization The other open boundary of the model is located at the Dardanelles Strait, which controls the salt exchange between the Aegean and the Marmara/Black Sea. At the surface, low salinity waters flow from the Marmara Sea into the Aegean, while saltier Aegean waters flow as an undercurrent into the Marmara Sea.

For the parameterization of Dardanelles inflow/outflow, an imposed open boundary condition is applied, following *Kanarska and Maderich* [2008]. Inflow of brackish waters of Black Sea origin occupying the upper 25-m layer and outflow of Aegean Sea waters occupying the bottom layer are prescribed. The inflow to the Aegean is set equal to 0.032 Sv whereas the outflow to the Marmara Sea is set equal to 0.022 Sv. The amplitude of the seasonal modulation is set to 0.005 Sv and it is assumed that the inflow is reaching maximum values during the end of March while the outflow during the end of July. The salinity of the inflowing waters is assumed equal to 28.5 psu with a seasonal modulation of 2 psu, and minimum values reached in mid-July. Temperature of inflowing waters follows *Beşiktepe et al.* [1994] seasonal values.

8.2.6. Model Run

The model was first integrated for 20 years perpetually forced with 1960's atmospheric forcing until a steady state seasonal cycle was reached considering the total kinetic energy as long as the stabilization of the hydrological variables in an annual basis. The final simulation was initialized from this state and covered the whole period 1960–2000.

8.3. RESULTS AND DISCUSSION

The Adriatic and the Aegean seas present similarities and differences as described previously in the introduction. Both areas exhibit multiple forms of DWF processes, involving shelf as well as open ocean convection. They both function as dense water formation regions for the EMed, with different water hydrological characteristics and efficiency in water production rates. Their functioning resembles complex marginal seas, which, depending on their state, in terms of salinity and temperature preconditioning and the imposed atmospheric forcing, result in different behavior regarding their contribution in the overturning circulation. The nature of DWF process has been shown to influence strongly the resulting formation and outflow of deep waters and subsequently the respective inflow at the upper layers [*Maxworthy*, 1997]. Interestingly, the Adriatic Sea mainly produces deep to bottom waters for the EMed, while the Aegean presents a broader spectrum of water mass characteristics, intermediate with almost LIW characteristics to very dense bottom waters [*Georgopoulos et al.*, 1989; *Schlitzer et al.*, 1991, *Theocharis et al.*, 1999a].

Two types of mechanisms of DWF are identified in the Adriatic Sea. Part of deep water is formed through typical shelf convection in the North Adriatic, where production is controlled by the large river runoff and the severity of the winter period. The North Adriatic provides dense water constantly, but not in large quantities, contributing through mixing in the South Adriatic region [*Zore-Armanda*, 1963, *Vilibić and Orlic*, 2002]. The major part of the deep-water outflow originates in the South Adriatic Pit [*Mantziafou and Lascaratos*, 2004].

The case of the Aegean Sea is more complicated, as convection events occur in various subbasins with different morphological and hydrological characteristics. DWF by shelf convection processes has been documented in the various plateaus of the Aegean (Samothraki, Limnos, Cyclades, etc.), while open ocean convection has been observed in the North Aegean, at the Myrtoan Sea in the southwestern Aegean and, most important, in the Cretan Sea [*Miller*, 1974; *Theocharis and Georgopoulos*, 1993; *Theocharis et al.*, 1999a; *Zervakis et al.*, 2000; *Lykoussis et al.*, 2002; *Gertman et al.*, 2006; *Sayin et al.*, 2011].

Apart from the DWF mechanism involved (open sea and/or shelf formation), most of the studies on DWF relate the intensity of convection and the volume of dense water produced, with the deep outflow through the strait(s) that is counterbalanced by a respective inflow from the upper layers [*Chu*, 1991; *Maxworthy*, 1997; *Marshall and Schott*, 1999; Spall, 2004]. Although this relation is not straightforward, as there are many factors that influence the exchanges through the straits (such as the shape and size of the production area, the respective morphological and dynamic characteristics of the neighboring basins, and the depth of the convection), the exchanged volumes satisfactorily represent the intensity of DWF processes in the two basins.

Temporal evolution (annual means) of the volume of water exchanged at different density values through the strait(s) for the study period 1960–2000, computed through the model hindcast simulation, is presented in Figure 8.2. The results show a remarkable variability both in the net inflow/outflow as well as in the densities of water exchanged. The net flows at Otranto Strait alter from intense to negligible exchange values, while the Aegean presents a more complicated functioning. Furthermore, we observe a continuous positive trend in the densities and the amount of outflowing deep waters from the Aegean and a difference in the volume and densities of water exchanged between the two active periods (1960s and 1980s) for the Adriatic (Figure 8.2), with greater values in the second period. The observational data series analysis shows the same trend in the deep waters of the Aegean at levels greater than or equal to 1000 m as well as for the Adriatic deep water at the South Adriatic Pit, at depths greater than or equal to 900 m (Figure 8.3).

More specifically, in Figure 8.2, we observe the large outflow of high-density Aegean water, the so-called Cretan Deep Water (CDW), during the EMT period (1988–1995), which coincides with the observed decay phase of the Adriatic from 1987 onward that followed the previous intense active phase during 1980–1987. The above mentioned EMT period is well documented in the literature based on data analyses [*Roether et al.*, 1996; *Theocharis et al.*, 1999a, 1999b; *Malanotte-Rizzoli et al.*, 1999; *Klein et al.*, 1999] and is reproduced by model results. However, it is evident that a "similar" event occurred in the 1970s when again the decay phase (1972–78) that followed an intense period of DWF in the Adriatic (1967–72) was accompanied by an intense DWF period in the Aegean (1971–79), which was also pointed out and discussed in earlier studies [*Theocharis et al.*, 2002a; *Beuvier et al.*, 2010].

It is thus revealed by the model that both dense-water sources present cycles of intense DWF periods, which are followed by periods of less-intense activity in each source area creating an anticorrelated pattern between them. The above results are consistent with the observations (Figure 8.3) that show both the oscillating anticorrelated pattern of the deep-water characteristics of the sources and consequently of the DWF activity and finally the potential alternation of the DWF sources in the EMed. After 1992, the Aegean DW densities appear higher than the Adriatic DW (Figure 8.3). Importantly, although available data for the South Adriatic DW end up in 1996, according to data presented by *Cardin et al.* [2011], the anticorrelated oscillation persists up to 2009, with the South Adriatic DW becoming denser than the Aegean DW after 2006.

However, short-term intense outflows from both basins exist (1967–69, 1982–83 for the Aegean and 1992–93 for the Adriatic) that do not match the oscillating pattern, which, as we will discuss later, can be attributed to the different functioning of the two basins reenforced by the prevalence of extreme atmospheric forcing conditions for certain time periods.

Although the purpose of this work is far from a detailed analysis on the DWF processes, we hereafter discuss some interesting aspects of the two source areas functioning. The effects of different types of DWF processes and the depth of the convection that is reached during winters may be used to explain their different behavior. In the case of moderate convection events with shallow penetration depth, the outflow will depend on the baroclinic gradient between the source area and the adjacent open sea, constrained by rotation [*Marshall and Schott*, 1997]. As convection acts largely to mix the water vertically, it does not contribute significantly to a net vertical volume flux [*Spall*, 2004, 2010]. Indeed, observations have shown that when convection in the Adriatic does not cease completely but is confined in shallower depths [*Klein et al.*, 2000], the simulated net exchanges exhibit a dramatic reduction. However, when convection reaches the bottom below the sill depth of the strait, dense water will overflow toward the deeper basins out of the strait(s) and will behave as a gravity boundary current that may be further stabilized and controlled by topography [*Spall*, 2004]. Therefore, deep (relatively to the sill depth) convection increases dramatically the outflow from the DW source area and consequently increases the upper layer respective inflow.

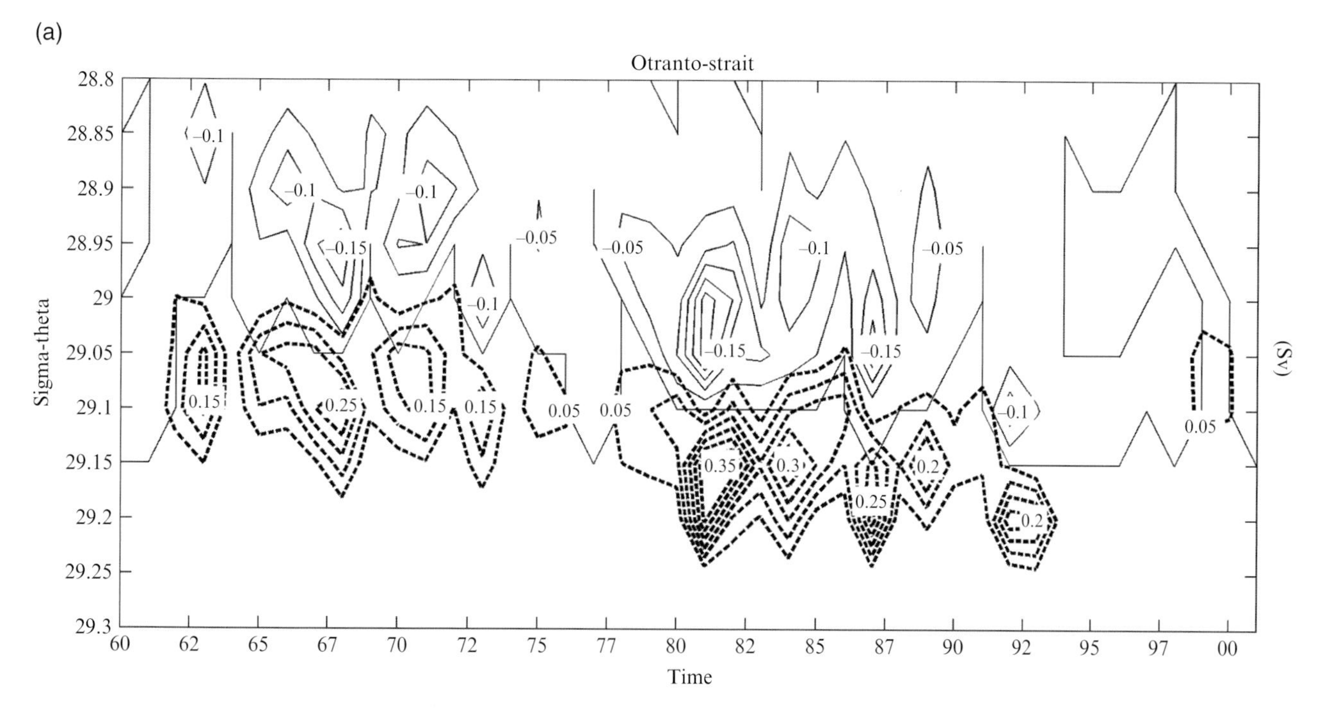

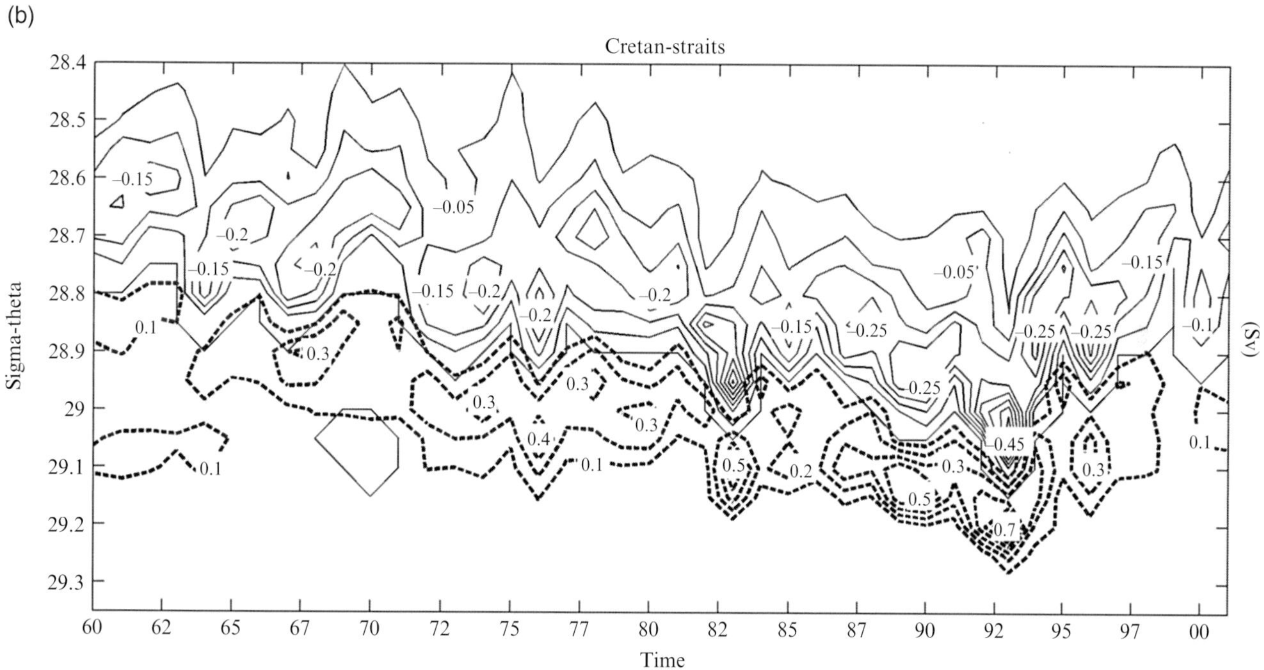

Figure 8.2. Temporal evolution (annual means) for the period 1960–2000, of the net volume (Sv) of water exchanged at different density values through the straits of the two DW source areas, (a) the Otranto Strait for the Adriatic and (b) the Cretan Arc Straits for the Aegean. Dashed lines represent outflows and solid lines inflows.

Observations also reveal that outflows through the Otranto Strait occur mainly at the bottom layers [*Zore-Armanda*, 1969]. Thus, the resulting net exchanges do not exhibit a linear dependence on the intensity of the convection, but are strongly influenced by the depth of the convection in relation to the sill depth at their straits. As we discuss later, this affects strongly the two source areas' contribution in the overturning process, and their interaction with the adjacent basins.

The same reasoning for the relation between the intensity of the exchanges and bottom reaching convection may be used to explain the more complex functioning of

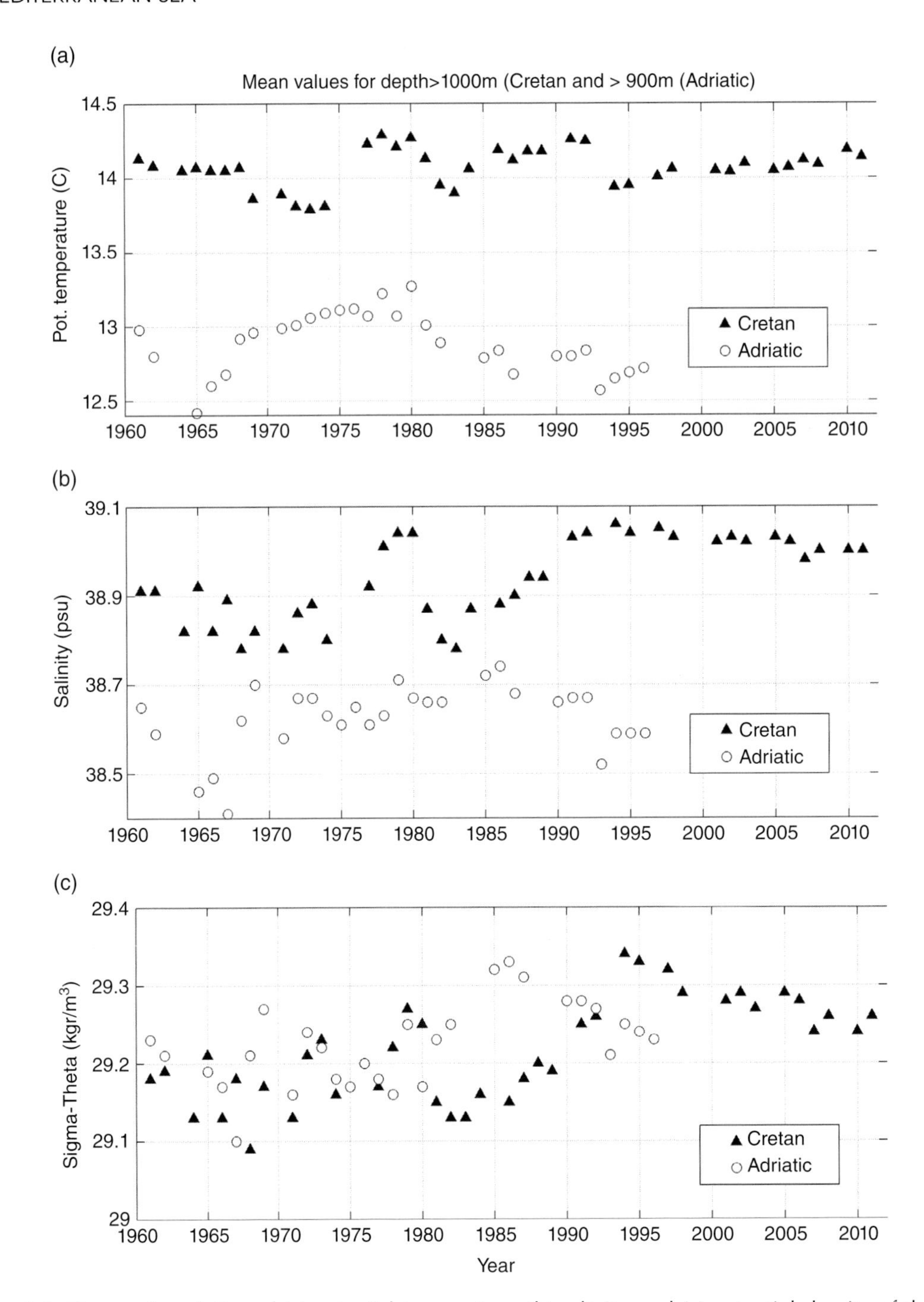

Figure 8.3. Temporal evolution of (a) potential temperature, (b) salinity, and (c) potential density of the deep layers at the South Aegean (Cretan) Sea and South Adriatic Pit from field observations. Yearly mean values are presented. The depth threshold is 1000 m for the Cretan Sea and 900 m for the Adriatic Sea.

the Aegean. As mentioned above, DWF in the Aegean occurs in various subbasins, which are connected with sill depths shallower than 400 m. Since these depths can be reached during winter convection events, they can provide the necessary outflow to sustain the almost constant simulated and observed inflow from the Levantine Basin in the upper layers. However, when the Aegean is in its preconditioned phase and/or experiences extreme atmospheric forcing, the depth of the convection taking place in the Cretan Sea can reach the sill depths and create strong outflows. In this case, the Cretan Sea due to its great size (relative to the other Aegean subbasins) dominates the DW production and becomes the main contributor of the observed water exchanges. These periods are characterized by deeper outflows, as the Cretan Sea is the deepest subbasin of the Aegean and communicates with the open sea with the deepest sill depths.

Furthermore, in case of deep outflow, water masses can be advected not only at the surface but also from depths greater than the warm and saline intermediate water (LIW) horizons. Persistent DWF activity and DW outflow through the Cretan Arc Straits will lead to the formation of a distinct layer of minimum temperature and salinity in the Aegean. This has a double effect toward the decay of DWF efficiency of the DW source. First, it speeds up the freshening of the water column; second, it increases stratification, preventing gradually winter deep convection [*Theocharis et al.*, 1999a]. This can be considered as a typical structural characteristic for both post-EMT and EMT-like periods. This is the case of the less-saline Transitional Mediterranean Water (TMW) intrusion during the EMT period, which was an additional factor toward the weakening of DWF processes in the Cretan Sea, as it was also shown by observations [*Theocharis et al.*, 1999b; *Vervatis et al.*, 2011]. This structure is also evident in the 1970s event (not shown), where fresher water masses intruded below the intermediate layers of the Cretan Sea, hindering the convection processes. Finally, the same hydrological feature was repeated in the 1960s [*Miller et al.*, 1970], possibly indicating another EMT-like event that occurred in the 1950s.

8.3.1. Evolution of the Hydrological Characteristics

The Aegean became the main source of the dense waters in the EMed after a long salinity-driven preconditioning phase (1986–1991), while the following winters (1992 and 1993) the atmospheric forcing (characterized by very low air temperatures) played the key role in the massive production of the so-called very dense Cretan Deep Water (CDW), which in turn outflowed through the Cretan Arc Straits and finally filled the deep and bottom layers of the entire EMed. The above salinity-driven period was attributed first to the E-P changes of 1988–1993 over the Levantine and the Aegean areas [*Lascaratos et al.*, 1999, *Theocharis et al.*, 1999b; *Josey*, 2003] and also to the observed increased (by 1–4 times) salt transport from the Levantine into the Aegean through the eastern Cretan Arc Straits [*Theocharis et al.*, 1999b]. The above preconditioning and intense formation phases were followed by a long enough decay phase (1995–2007), as it was indicated by *Theocharis et al.*, [2002b] and shown in Figure 8.4. This fact revealed the transient nature of the EMT event. Finally, in Figure 8.4, it is evident that since 2007 the salinity has increased in the South Aegean.

Model results show that the temporal evolution of the hydrological characteristics in the entire water column of the two source areas reveals that periods of intense DWF activity are accompanied by increased temperature and salinity values in each active basin, respectively (Figure 8.5). This gradual salinity increase precedes the period of the maximum DWF activity, denoting the salinity preconditioning of each source area. On the contrary, during the decay phase, salinity decreases, and as a result weakens the DWF ability respectively.

In more detail, the EMT period (1988–1995) is characterized by an increase in salinity and temperature in the Aegean Sea and vice versa for the Adriatic, almost in the entire water column (Figure 8.5). A similar situation is observed during almost the entire decade of the 1970s and probably before 1962 as well. The intervals in between, during the 1960s and 1980s, correspond to the Adriatic intense preconditioning and intense DWF periods. Therefore, the above evolution exhibits anticorrelated oscillations of both salinity and temperature between the two basins. *Gačić et al.*, [2011] have shown the existence of such anticorrelated oscillation for the period 1986–2009 for the Ionian on the one hand and the Levantine and Aegean on the other, using field observations.

The model results obtained here concerning the anticorrelated salinity and temperature oscillations between the two source areas are confirmed by field observations (Figures 8.3 and 8.4). More specifically, the preconditioning of the Aegean Sea for the first DWF event of the 1970s is evident from 1971–78, while for the EMT event, it ranges from 1982–1992 (Figure 8.4). Interestingly, many authors reported that the EMT event was primarily salinity-driven during its preconditioning phase, 1986–1992, based on the available POEM program data [*Roether et al.*, 1996; *Malanotte-Rizzoli et al.*, 1999; *Theocharis et al.*, 1999b]. For the Adriatic Sea after its intense DWF period in the late 1960s, mean salinity has decreased during 1968–1972 and remained relatively low up to 1978, starting to increase again in 1979 onward signifying the next intense DWF period of the 1980s (Figures 8.3 and 8.4).

8.3.2. Evolution of the Atmospheric Forcing

The atmospheric variability has been considered by many authors as the main driving mechanism for the EMT event [i.e., *Demirov and Pinardi*, 2002; *Beuvier et al.*, 2010; *Josey et al.*, 2011]. Here, we examine the long-term evolution of various key atmospheric parameters over the source areas. In Figure 8.6, we present the time evolution of air temperature, wind intensity, and precipitation over each basin, which are the main factors controlling air-sea interaction. All parameters are extracted from the ARPERA dataset, which was used to force the hindcast model experiment. To better present their variability in relation to the observed anticorrelated

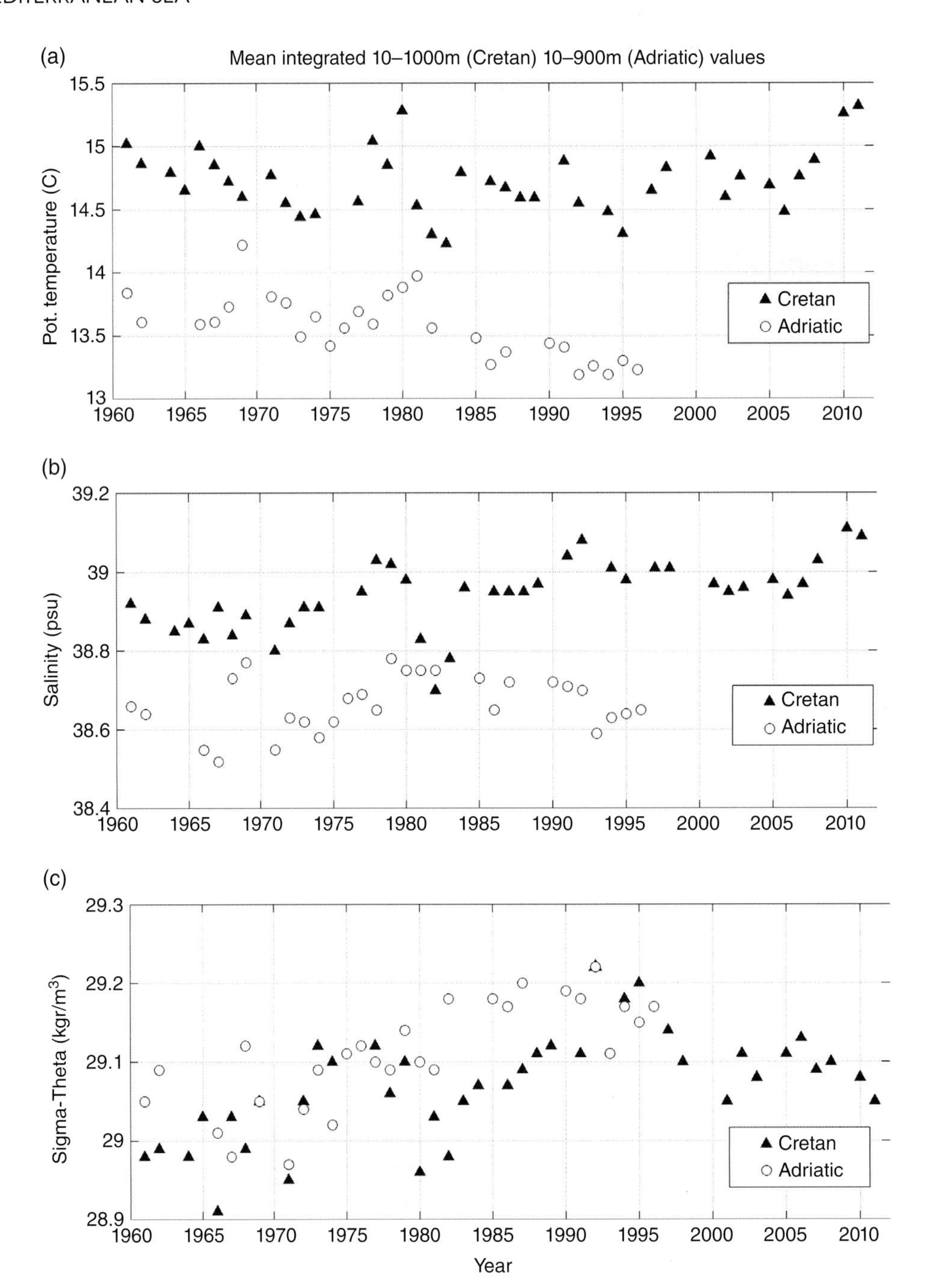

Figure 8.4. Temporal evolution of (a) potential temperature, (b) salinity, and (c) potential density at the South Aegean (Cretan) Sea and South Adriatic Pit from field observations. Yearly mean integrated values are presented. The integration layer is 10 to 1000 m for the Cretan Sea and 10 to 900 m for the Adriatic Sea.

decadal oscillations of the two source areas' hydrological characteristics and DWF activity, a 5-year moving average fit is also plotted.

During the 40-year study period of 1960–2000, air temperature is characterized by a negative trend in the first 20 years, which is followed by a positive trend for the remaining period (Figure 8.6a). At the same time, the air temperature time series over the two source areas are highly correlated. Intense DWF in the Aegean generally coincides with local air temperature minima on an annual basis. Indeed, EMT was triggered by extremely low winter air temperatures, which are strongly evident even in the annual mean values (Figure 8.6a). However, the activation of the Aegean can also be triggered even when it is not accompanied by increased salinity, indicating the susceptibility of the Aegean Sea to intense atmospheric forcing, mainly due to its generally higher salt content relative to the Adriatic (i.e., as can be seen in Figure 8.2 for 1983). As for the precipitation, a noticeable reduction is revealed for the period 1980–1990 (Figure 8.6b), which coincides with the observed salinity increase in the same period in the Aegean and may explain its positive trend, while the

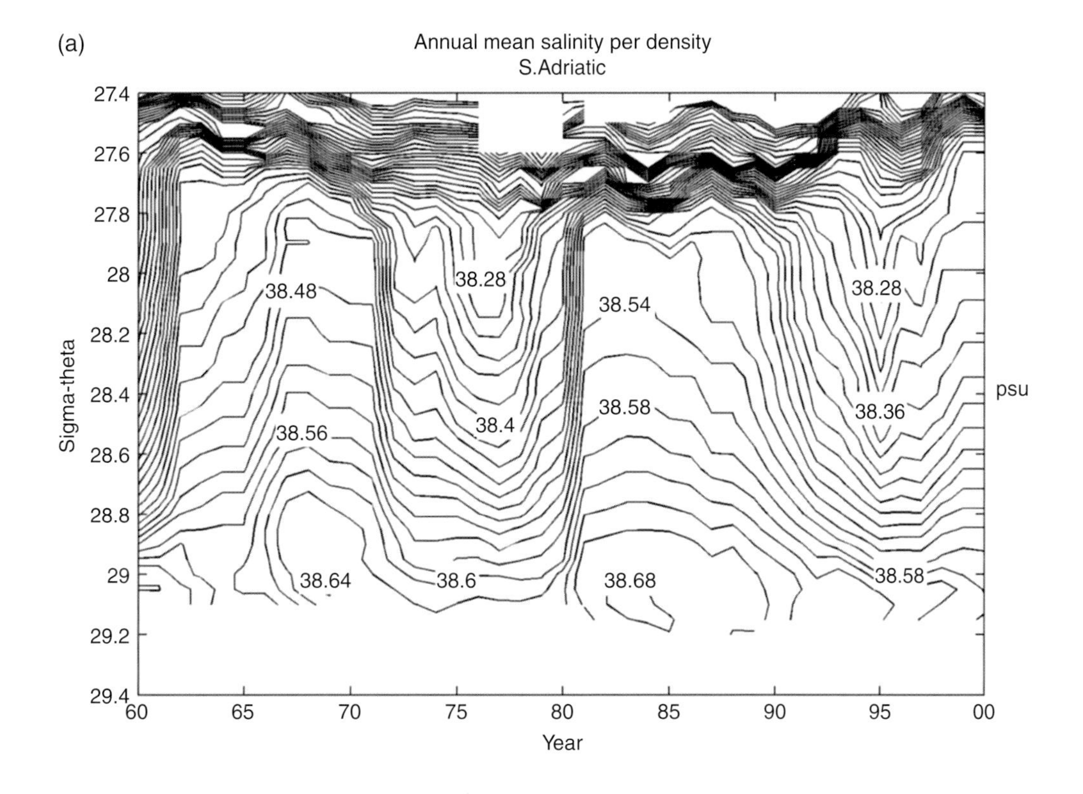

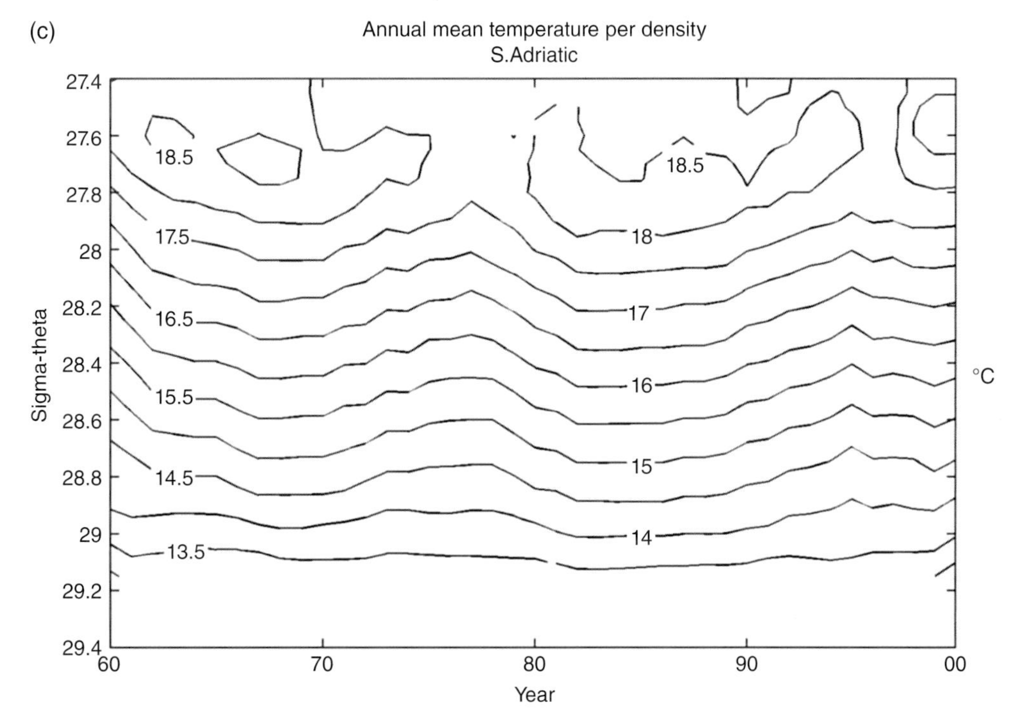

Figure 8.5. Temporal evolution (annual means) for the period 1960–2000, of salinity and temperature for the South Adriatic (a and c) and South Aegean seas (b and d) at different densities.

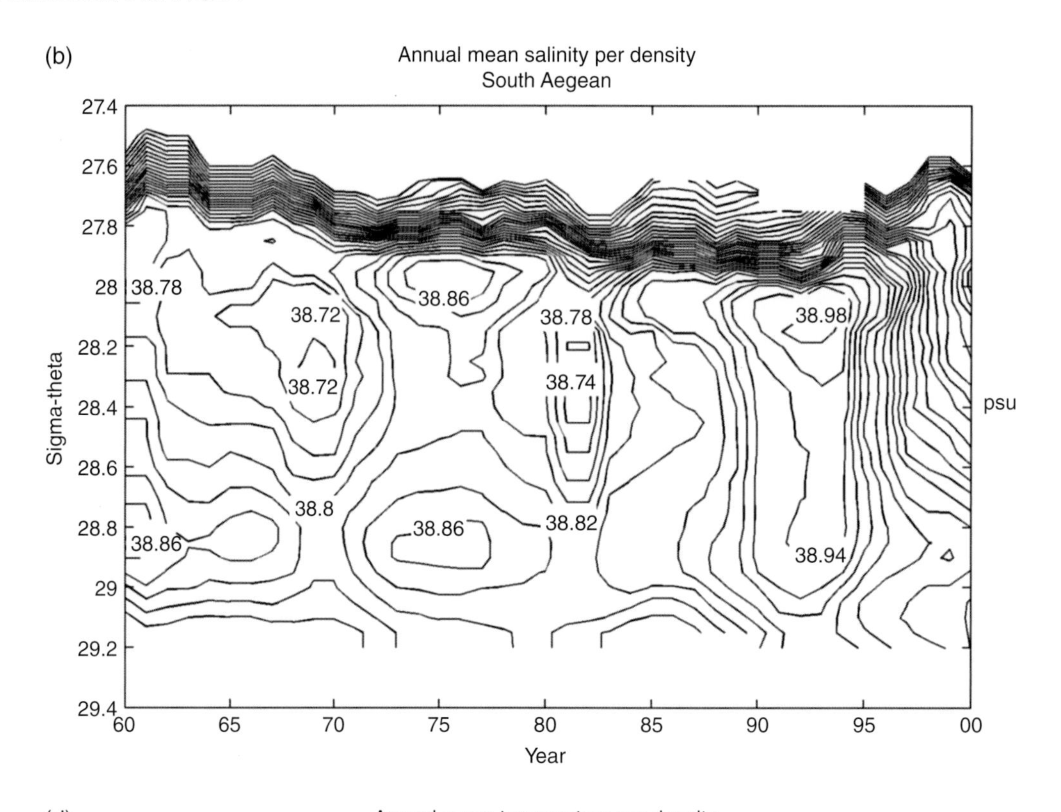

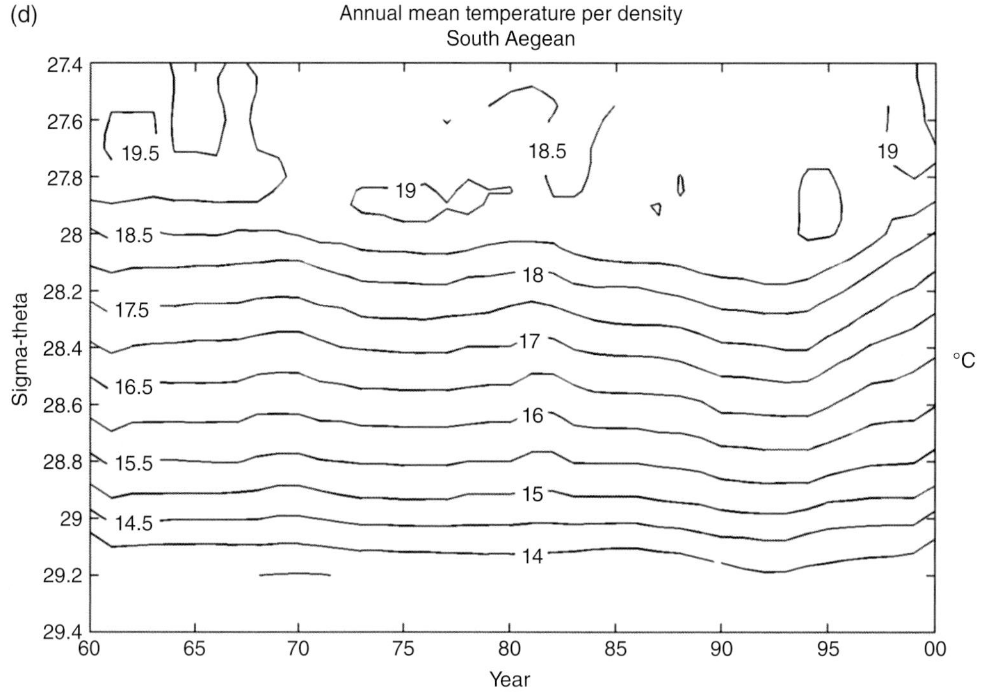

Figure 8.5. (continued)

Adriatic shows an opposite behavior. Furthermore, no similar reduction is found in the 1960s that could be related to the salinity increase in the Aegean. Wind patterns exhibit persistent annual but no significant long-term variations (Figure 8.6c).

Therefore, it is found that besides the fact that there is significant (long- and/or short-term) variability in the atmospheric parameters over the two sources areas, there is no clear evidence for the existence of an anticorrelated long-term atmospheric pattern between them, in contrast to the evolution of their hydrological characteristics presented above.

Conversely, the computed air-sea temperature difference (Figure 8.7a), which is a key factor for the determination

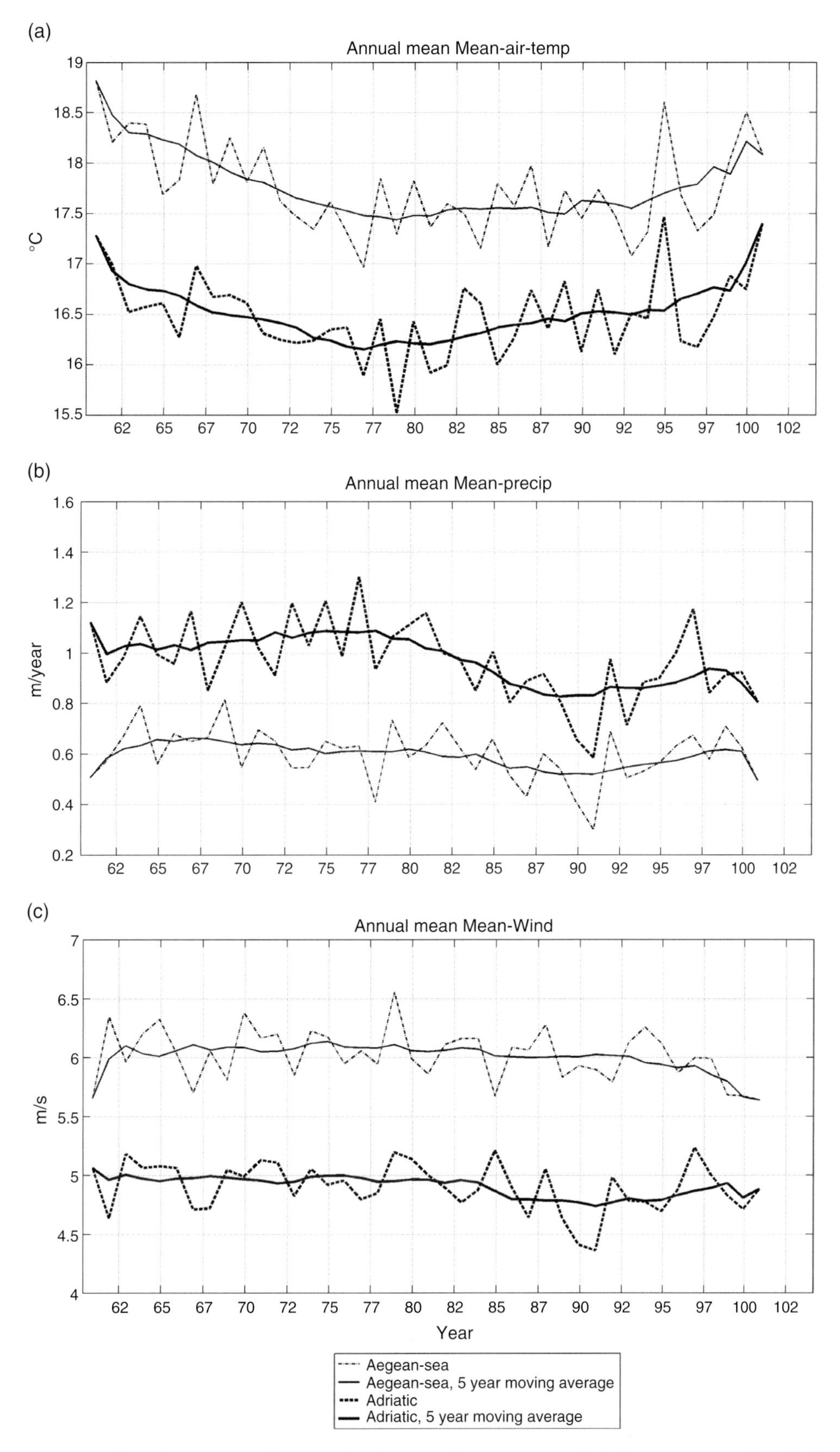

Figure 8.6. Temporal evolution (annual means) for the period 1960–2000 of (a) air temperature, (b) wind intensity, and (c) precipitation over the Adriatic (dashed lines) and Aegean seas (dashed dotted lines). Solid lines represent 5-year moving averages.

of the evaporative and consequently of the total heat fluxes, is characterized by an oscillating behavior over the two source areas. This has significant implications in the resulting total heat flux over each basin, resulting in decadal anticorrelated oscillations, corresponding to DWF (production/outflow) patterns (Figure 8.7c). The relation of these oscillations to the long-term variability of sea-surface temperature can be attributed to the increased heat content that is observed in the source areas during their intense production periods (Figures 8.5c and 8.5d).

Therefore, although atmospheric forcing unambiguously plays an important role in triggering DWF, with the Aegean Sea showing the greatest susceptibility, its variability does not coincide with the oscillations of the hydrological characteristics and the periods of intense DWF. Inversely, the long-term air-sea interaction is influenced by the hydrological variability through increased thermal and evaporative fluxes.

The anticorrelated behavior of the hydrological characteristics of the two source areas further suggests that this may be attributed to lateral redistribution of hydrological properties. In the EMed, heat and salt are accumulated due to increased evaporation as well as higher temperatures in the southern EMed and mainly in the Levantine. The variations in salinity and temperature in the two areas, the Aegean and Adriatic, may reflect changes of the water mass pathways.

8.3.3. Salinity Lateral Redistribution

The salinity variations in the upper and intermediate layers of the Levantine and the Ionian seas before and during the EMT event (1986–1995) were attributed to changes in the water mass pathways of the AW and LIW/CIW, which redistributed the salt in the entire EMed [*Malanotte-Rizzoli et al.*, 1997, 1999]. Numerical studies and observations have documented that the Ionian upper layer basinwide circulation undergoes reversals on decadal timescales passing from anticyclonic to cyclonic mode and vice versa [*Pinardi and Navarra*, 1993; *Pinardi et al.*, 1997; *Demirov and Pinardi*, 2002; *Larnicol et al.*, 2002; *Pujol and Larnicol*, 2005; *Vigo et al.*, 2005]. Observations showed [*Malanotte-Rizzoli et al.*, 1997, 1999] that during the preconditioning phase of the EMT (1987–1991), the northeastward branch of the Atlantic Ionian Stream (AIS) follows a large anticyclonic meander in the Ionian interior that was expanded northward in 1991 compared to the previous pattern of 1987. This replaced the cyclonic circulation that was established before 1987 in the northern part of the Ionian causing freshening of the upper layers. These waters reach Otranto Strait and finally enter the Adriatic Sea. In parallel, the AW eastward flow is progressively reduced and at the same time, the intermediate waters LIW/CIW recirculate in the Levantine, thus increasing their salinity and consequently the salinity of the Aegean Sea, due to the direct communication between these basins.

Furthermore, since 1997 another reversal of the circulation from anticyclonic to cyclonic occurred [*Manca*, 2000; *Larnicol et al.*, 2002; *Theocharis et al.*, 2002b; *Pujol and Larnicol*, 2005]. *Borzelli et al.*, [2009] proposed that the reversal was associated with the vorticity transfer due to redistribution of deep-water masses in the final stage of the EMT and showed that it was not sustained by the wind, as suggested by *Demirov and Pinardi* [2002]. *Gačić et al.* [2010, 2011] introduced the concept of the North Ionian Gyre (NIG) and confirmed that the Ionian upper-layer, basinwide circulation underwent a reversal in 1997 and demonstrated another reversal in 2006, suggesting the existence of a decadal timescale in the reversals due to interaction between the Adriatic and the Ionian. This phenomenon was named the Adriatic-Ionian Bimodal Oscillation System (BiOS), driven by a feedback mechanism that functions due to the differences in salinity between the high-salinity waters originating from the EMed (Levantine and/or Aegean) and the lower salinity AW.

In this study, we reproduced the evolution of the AW and LIW/CIW pathways in the period 1960–2000 using the model hindcast simulation. The AW pathway can be well represented by the minimum of salinity in the upper 250 m, while the LIW/CIW pathways by the salinity maximum in the intermediate layers (200–600 m). There is a good agreement comparing the model results with the published maps of the pathway changes based on observations for the years 1987 and 1991 (not shown) [*Malanotte-Rizzoli et al.*, 1999].

The period 1987–1991 is considered as a salinity preconditioning phase for the Aegean. However, the pathways' structure of the AW and LIW/CIW substantially differs between 1987 and 1991, as these years represent the starting and nearly final stage of the preconditioning phase respectively [*Theocharis et al.*, 1999b]. In 1987, the AW, after an anticyclonic meandering in the central Ionian, was still free to flow eastward through the Cretan Passage, while in 1991 we observe an important northward shift of the AIS supplying the North Ionian with much lower salinity waters as well as a reduced eastward AW flow. Additionally, an accompanying increase in the salinity of both the upper and the LIW horizons in the Levantine Basin occurred due to recirculation of the already saline water masses (LIW/CIW). These high-salinity waters, in turn, entered directly into the Aegean Sea through the eastern Cretan Straits. This was an additional internal process that contributed to the increase of salinity of the Aegean during the preconditioning phase (1987–1991) of EMT.

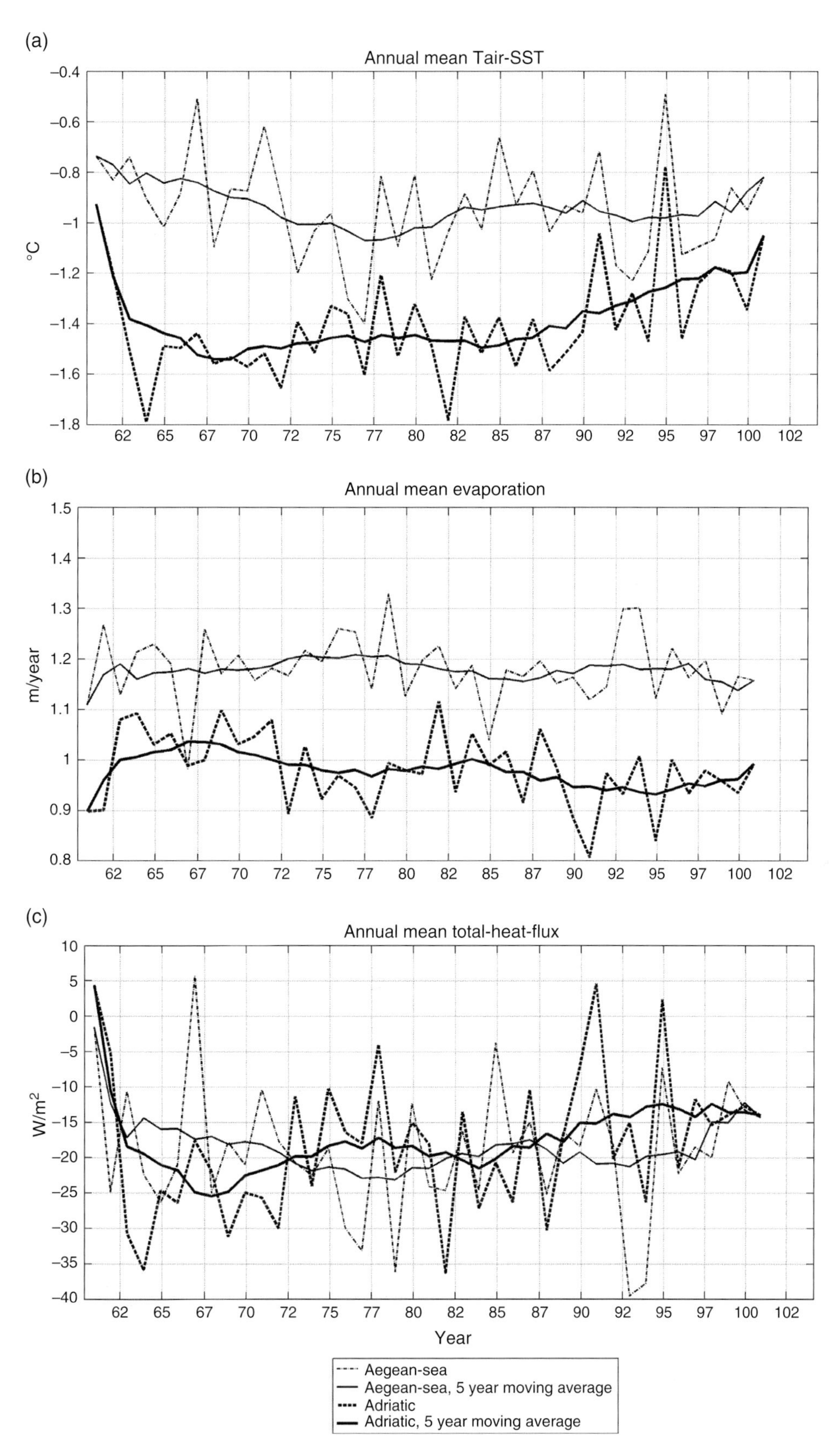

Figure 8.7. Temporal evolution (annual means) for the period 1960–2000, of (a) air temperature and sea surface temperature difference, (b) evaporative fluxes, and (c) total heat fluxes over the Adriatic (dashed lines) and Aegean seas (dashed dotted lines). Solid lines represent 5-year moving averages.

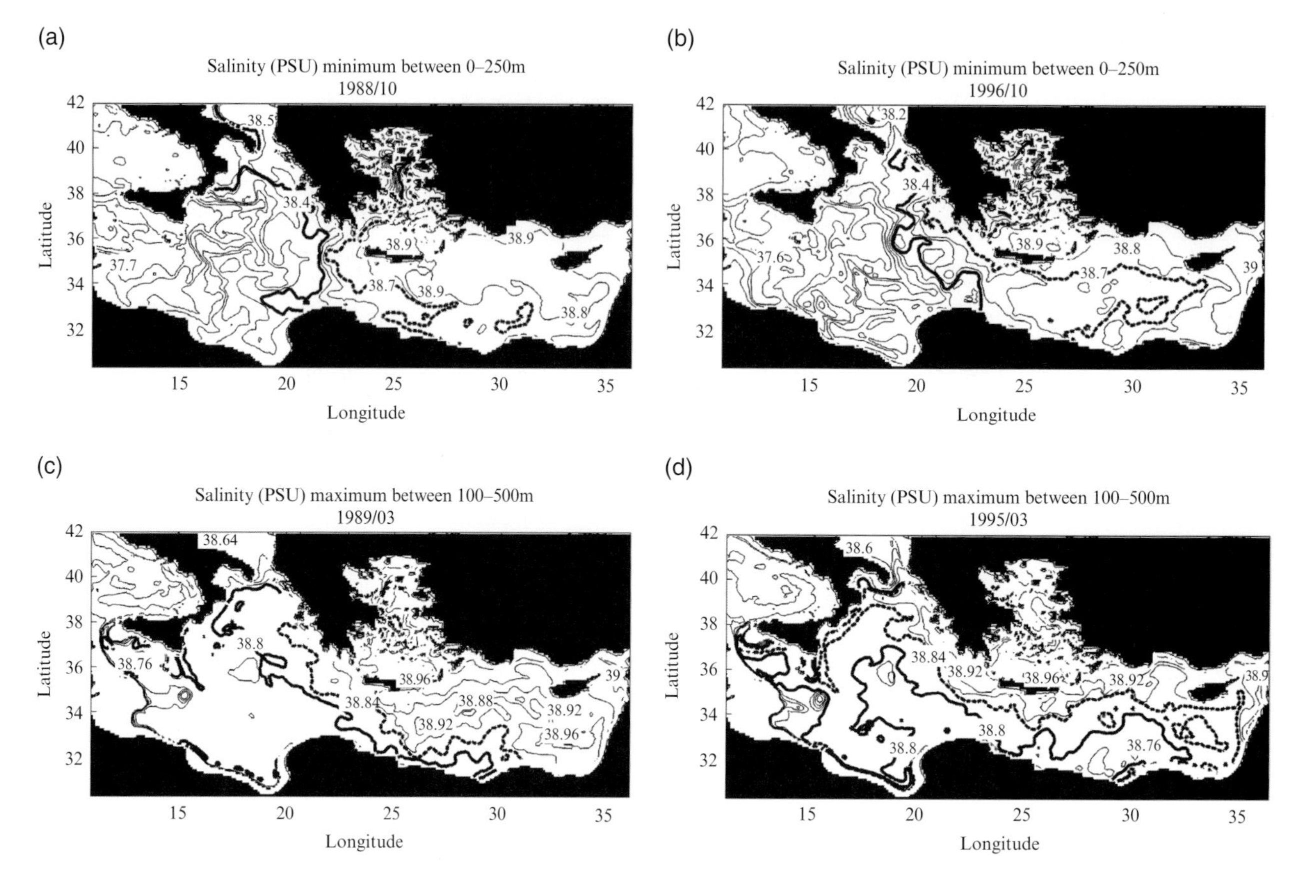

Figure 8.8. Horizontal distribution of salinity minimum (AW pathway) in the upper 250 m (upper panel a and b) and of salinity maximum (LIW/CIW pathway) in the intermediate layer 100–500 m (lower panel c and d) for the preconditioning (Oct/1988 – March/1989) and decay phase (Oct/1996–March/1995) of the EMT event.

In this work, we show the water mass pathway distribution in 1988–1989 and 1995–1996 (Figure 8.8), as these years mark the early stage of the preconditioning of the Aegean and the decay phase of EMT respectively, clearly indicating the changes in the entire cycle of the EMT. In 1988, we can already detect differences from 1987, as the northward shift and confinement of the AW in the Ionian is more pronounced, indicating also that the thermohaline changes related to the EMT have started before 1991 in agreement with the views presented by *Malanotte-Rizzoli et al.* [1999] based on data interpretation. This is manifested by the important Aegean outflow through the Cretan Arc Straits that was measured since 1989 (up to 1 Sv) but with densities not exceeding 29.18 kg/m^3 (much lower than the extreme value 29.4 kg/m^3) combined with the increased inflow from the eastern Cretan Straits [*Theocharis et al.*, 1999b]. In 1996, we show the restoration of the AW eastward flow, which marks the decay phase of the Aegean as the main DW source in the EMed. In the same Figures 8.8c and 8.8d, we show the corresponding changes in the intermediate water masses pathways. In 1988, as in 1991, the LIW/CIW recirculate in the Levantine due to the lack of AW inflow thus increasing their already high salinity, while in 1995 (post-peak period for the deep Aegean outflow) we show the restoration of LIW/CIW general westward cyclonic circulation toward Otranto. Therefore, as the decay phase for the Aegean is initiated by the restoration of the eastward AW flow, freshening the Levantine and subsequently the Aegean, the intermediate water mass acting as "salt carrier" moves geostrophically in the Ionian. As the upper thermohaline cell is restored, the salt carrier can reach the Otranto Strait reenhancing the salinity preconditioning of the Adriatic. Indeed, a post-EMT freshening of the surface/ subsurface/ intermediate water masses in the Aegean has been also verified by observations [*Theocharis et al.*, 2002b; *Velaoras and Lascaratos*, 2010].

In Figure 8.9, we present model simulation salinity maps corresponding to the 1970 period, showing that the same changes in the circulation patterns were responsible for the salinity lateral redistribution in the entire EMed. We observe again the northward shift of the AW and its confinement in the Ionian during the

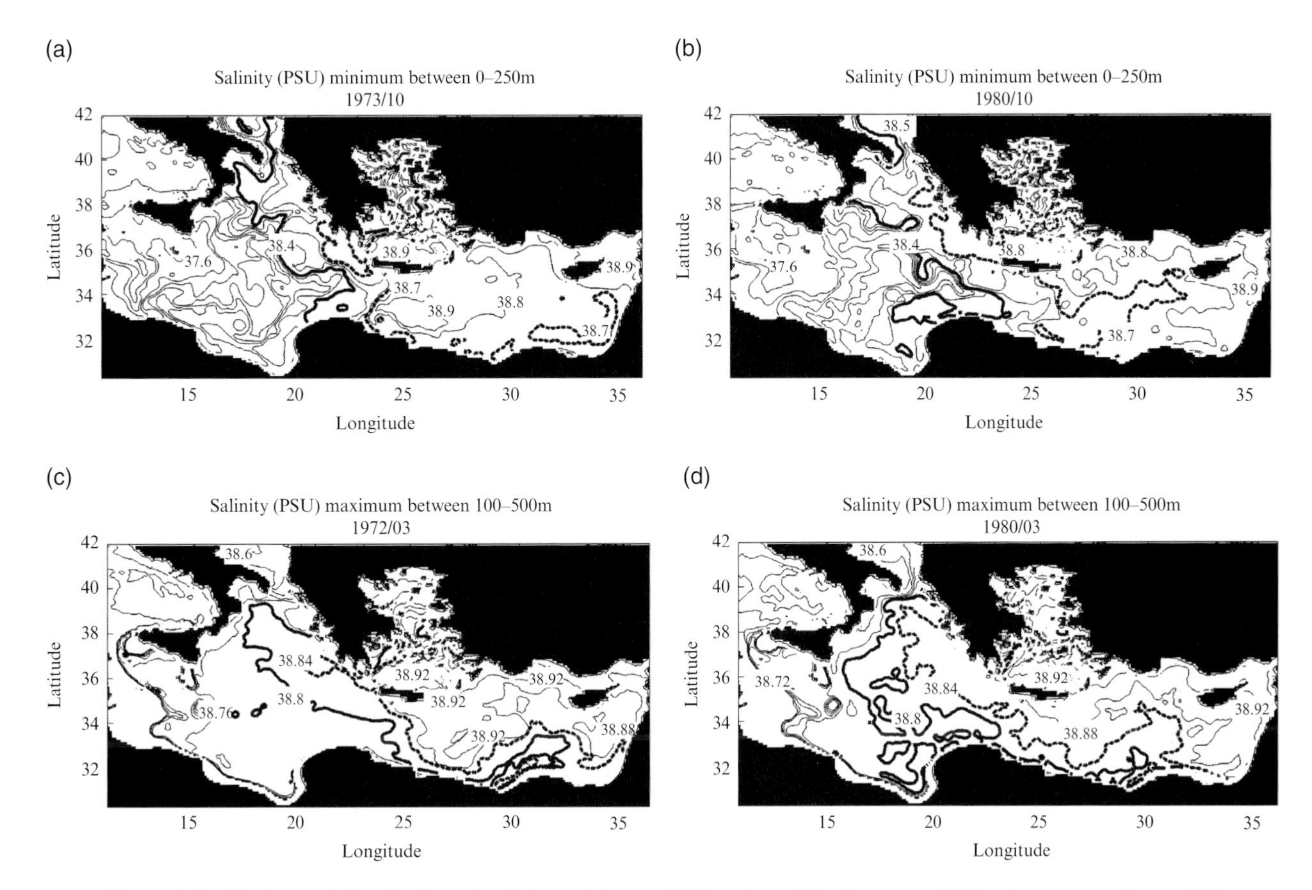

Figure 8.9. Horizontal distribution of salinity minimum (AW pathway) in the upper 250 m (upper panel a and b) and of salinity maximum (LIW/CIW pathway) in the intermediate layers 100–500 m (lower panel c and d) during the preconditioning (Oct/1973–March/1972) and decay phase (Oct/1980–March/1980) of the 1970s event.

preconditioning phase of the Aegean Sea, while the restoration of its eastward flow is taking place during the decay phase (Figures 8.9a and 8.9b). In Figures 8.9c and 8.9d, we observe the recirculation of LIW in the Levantine due to reduction of AW inflow as well as the restoration of its general cyclonic circulation toward Otranto during the decay phase of the Aegean Sea, respectively.

As can be seen from the upper-layer mean salinity evolution diagram (Figure 8.10), preconditioning for the Aegean means decay for the Adriatic due to the freshening caused by the AW and lack of inflow of intermediate salt-carrier water masses. Similarly decay for the Aegean means preconditioning for the Adriatic. The evolution of salinity in the Ionian exhibits also oscillations, revealing its intermediate role as a passage for the water masses. The salinity increase in the Ionian evolves when LIW/CIW exits the Aegean-Levantine region and precedes the salinity increase in the Adriatic. Similarly, the salinity decrease periods in the Ionian are related to the increase of salinity in the Levantine-Aegean and mark the progressive decrease in the Adriatic that follows.

8.3.4. The Role of Lateral Advection

The major question that arises concerns the forcing mechanism that creates these changes in the circulation patterns, and whether this mechanism can be attributed to external (atmospheric) or internal (oceanic) functioning.

As mentioned above, the changes in the circulation patterns that lead to the observed oscillations of the hydrological characteristics have been associated with changes in the outflows from the two source areas [*Borzelli et al.*, 2009; *Gačić et al.*, 2010]. Although this may act to sustain the observed circulation variability, little attention has been given to the large changes that have been observed in the inflow of water masses toward the two source areas.

We showed that both source areas have a significant outflow of dense water through their deepest layers during their active DWF phases. Following the principle of continuity of volume, we conclude that such deep outflow must be balanced by upper layer inflow, which is also evident when examining the exchanges through their strait(s) (Figure 8.2). This can be considered as a "pumping" mechanism that advects upper-layer water masses toward

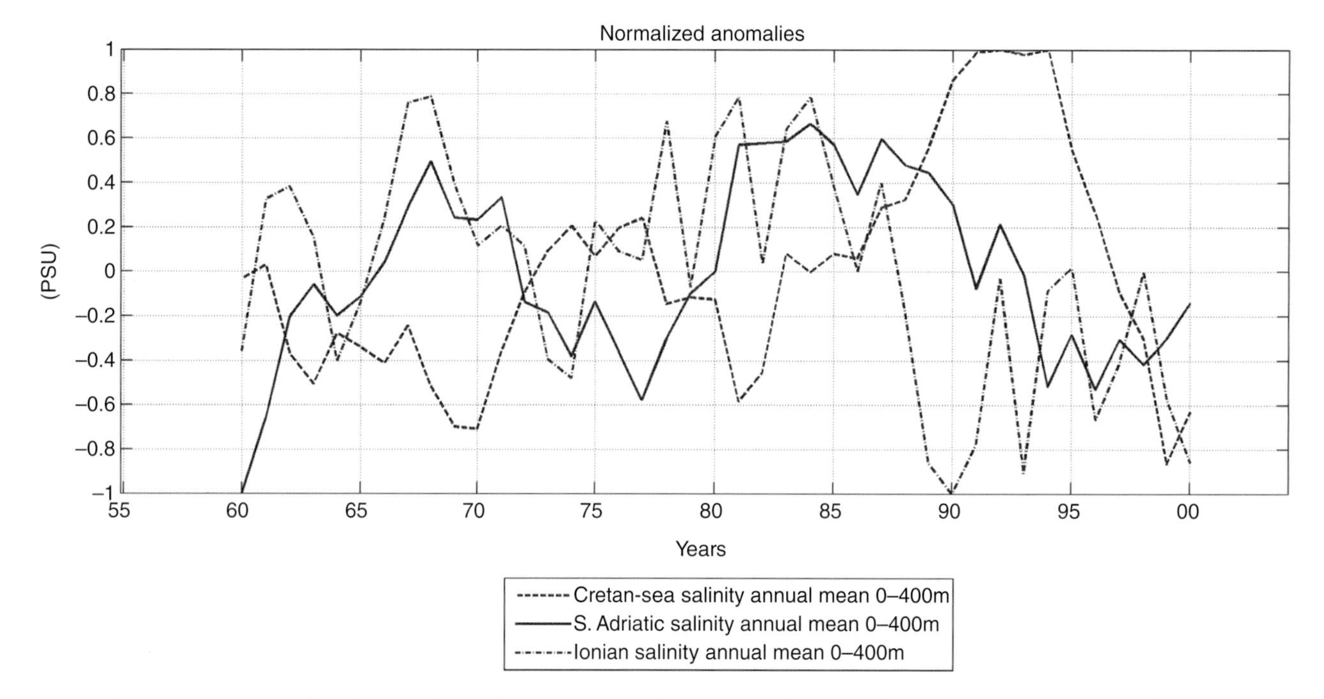

Figure 8.10. Normalized anomalies of the temporal evolution (annual means) for the period 1960–2000 of mean salinity in the upper layers (0–400 m) for the South Aegean (Cretan) (dashed line), Ionian (dashed dotted line), and South Adriatic (solid line) seas.

the active area. The connection between meridional transport and DWF intensity has also been suggested by *Pisacane et al.* [2006] in their study on long-term variability of the EMed circulation and properties using perpetual atmospheric forcing. Moreover, the progressively increasing inflow of the upper layer (0–200 m) Levantine waters (LSW, LIW) toward the Aegean through the Cretan Straits during the EMT period [*Theocharis et al.*, 1999b], as it was noted above, gives us an excellent working example of such a pumping mechanism.

In Figure 8.11, we give estimates from model results for the upper-layer net transport across different transects in the EMed as a response to the deep outflows from the two source areas. We consider the annual mean net transports in the upper layers (0–400 m) in three successive west-east transects in the Ionian Sea at latitudes 37° N, 38° N, 39° N, compared with the outflow of dense/deep water from the Adriatic through the Otranto Strait (Figure 8.11b). Furthermore, we show the transports in the upper layers (0–200 m) entering the Levantine-Aegean along the transect at 22° E, which is compared to the outflow from the deeper layers of the Cretan Straits (Figure 8.11a). These intercomparisons intend to give insight on the long-term evolution of volume transports in the upper/intermediate layers and to show their relation to the variability and magnitude of water exchanged during the convection events.

The response of the upper layer net flow to the overturning circulation induced by the intensity of DWF reveals that the Ionian Sea upper-layer water masses are strongly influenced by lateral advection from the two source areas. The anticorrelated transport patterns toward the two source areas is mainly manifested on long-term timescales as well as on an annual basis, and reflects their competitive functioning. The long-term effect on the net flow can then be used to explain the observed changes in the circulation patterns. As discussed above, the general circulation of the upper layers of the EMed is strongly influenced by the inflow of the relatively less saline AW through the Sicily Strait, which follows an anticyclonic meander in the Ionian interior and continues eastward toward the Levantine. The upper-layer thermohaline cell of the EMed closes with the return flow of the intermediate water masses originating in the Levantine and Cretan seas. However, during periods when the DWF intensifies in either of the two source areas, the influence of lateral advection in the upper layers toward the active area may progressively alter the water masses pathways, leading to significant changes of the circulation patterns. These changes are mainly manifested with the reversal of the circulation in the North Ionian. The anticyclonic mode reflects the northward expansion of the AIS meander due to the intensified advection toward the Adriatic, with an accompanying reduction of AW volumes flowing to the east. The latter weakens the upper-layer thermohaline cell, gradually reducing the return westward flow of the intermediate water masses toward the Ionian Basin due to water mass conservation.

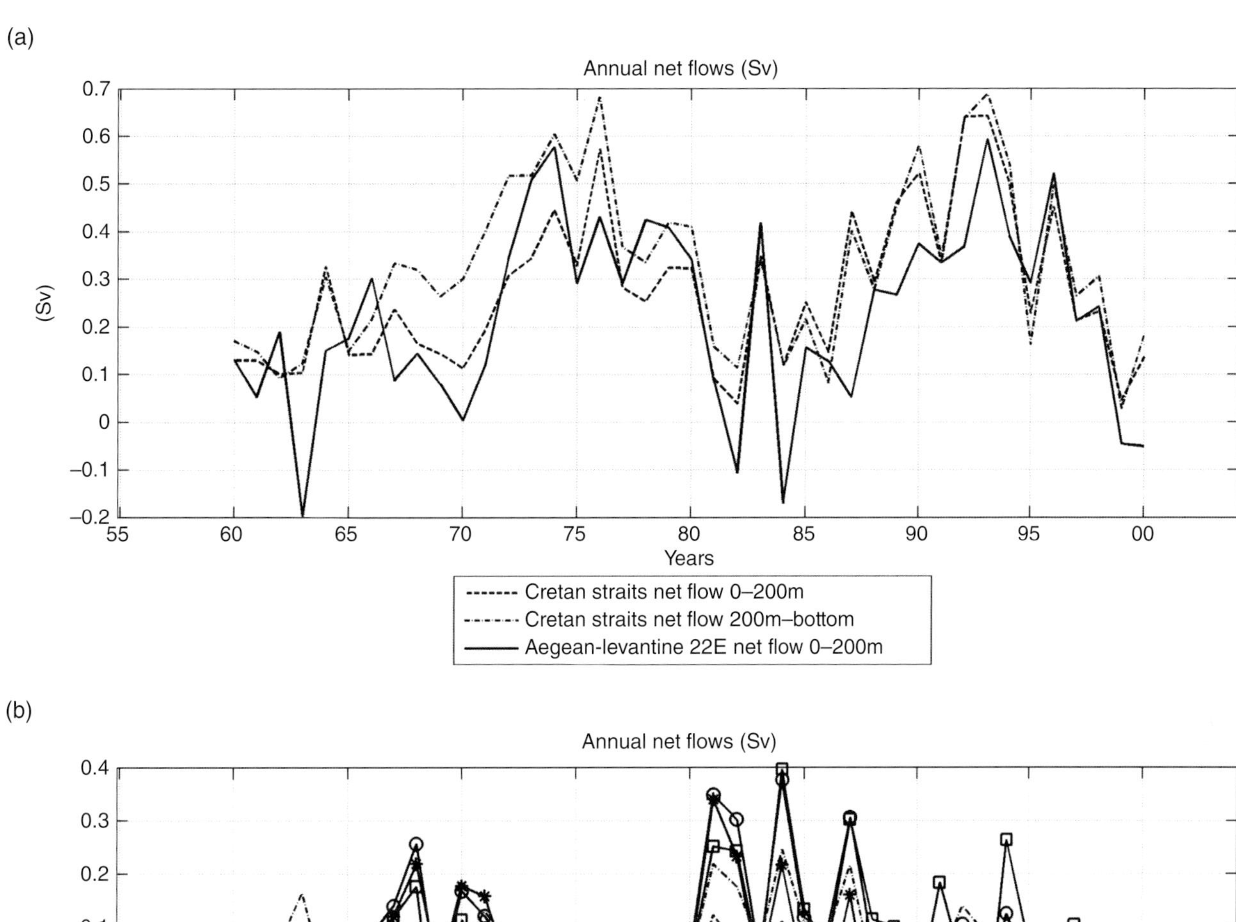

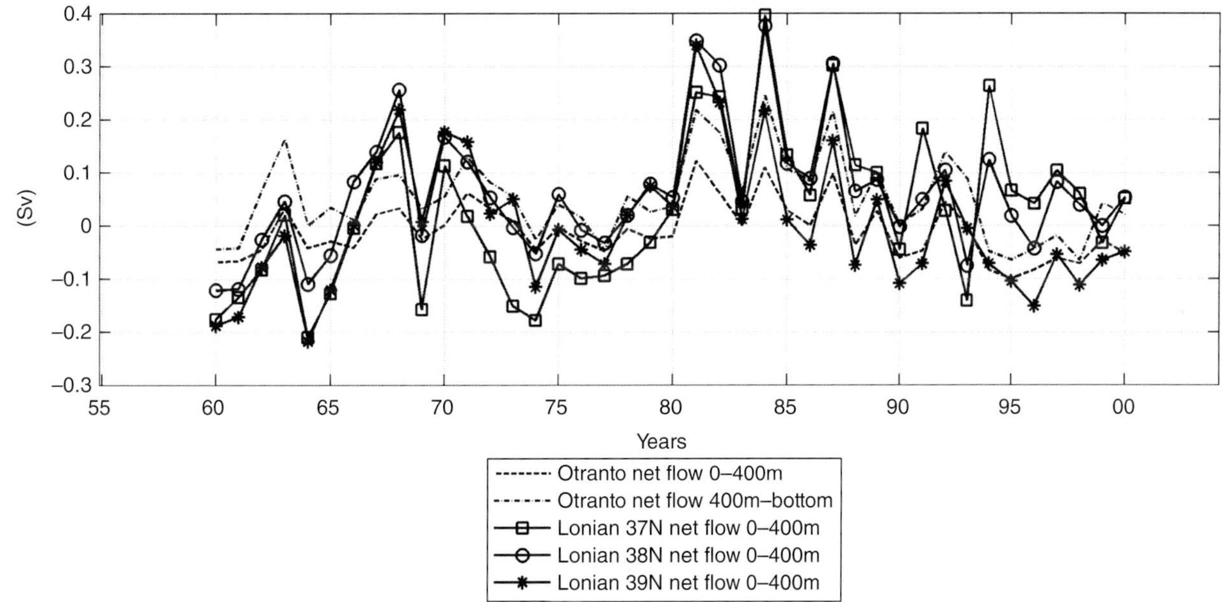

Figure 8.11. (a) Annual net transport in the upper 0–200 m in the Levantine (solid line) along 22° E and the corresponding net flows (0–200, 200-bottom) through the Cretan Arc Straits, (b) annual net transport in the upper 0–400 m in the Ionian along sections 37° N, 38° N, and 39° N, and the corresponding net flows (0–400, 400-bottom) through the Otranto Strait.

These combined changes gradually lead to the freshening of the Adriatic Sea and reduce its efficiency for DWF and its effect on lateral advection. At the same time, the disturbance of the upper EMed themohaline cell causes recirculation of the saline water masses in the Levantine progressively leading to preconditioning of the neighboring Aegean Sea, which enhances its DWF efficiency. Subsequently, the Aegean gains the primary role in advecting the respective upper-layer water masses. This results in the restoration of the eastward flow of AW toward the Levantine and the southward constriction of its anticyclonic meander in the Ionian, which is followed

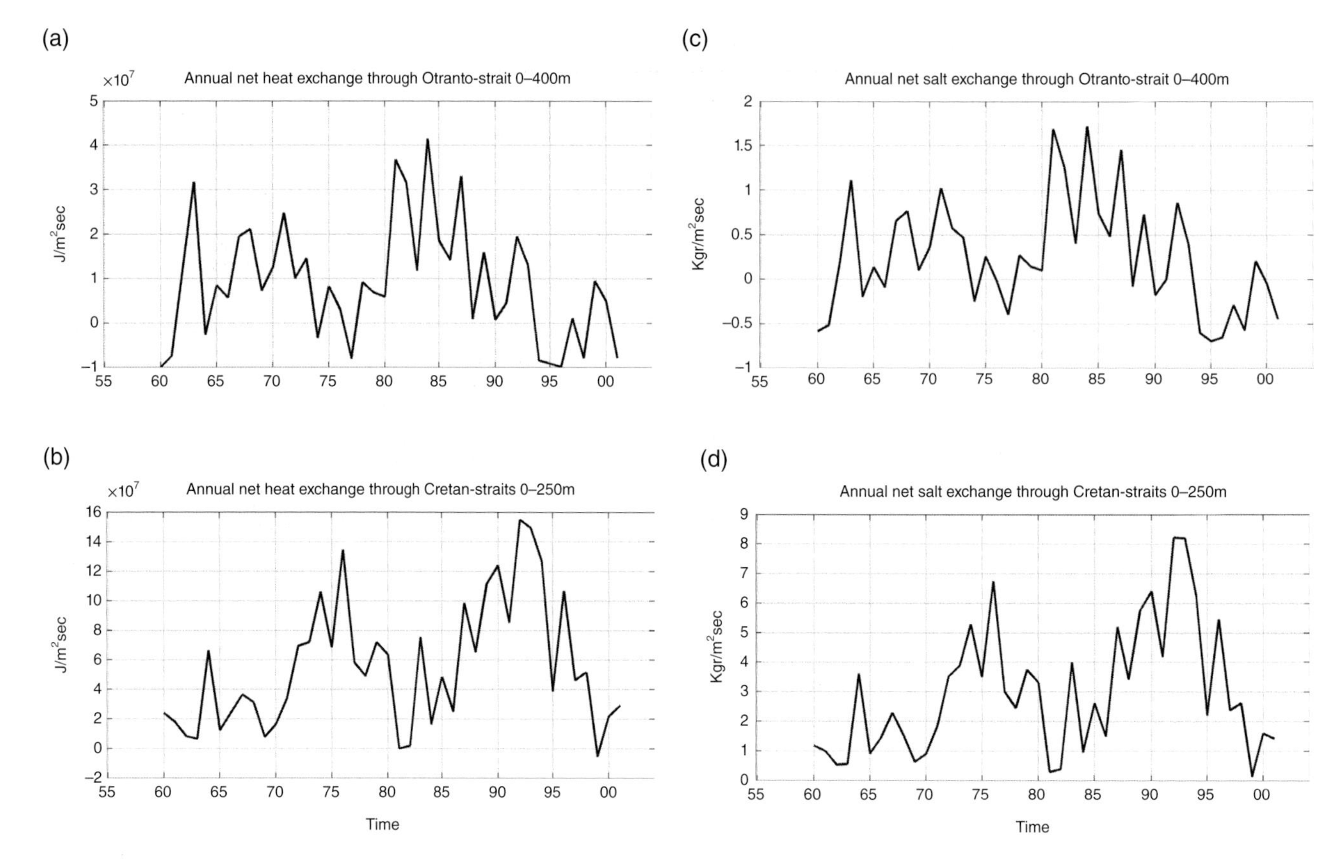

Figure 8.12. Annual net heat (left panels) and salt (right panels) exchange through Otranto Strait (a and b) and Cretan Arc Straits (c and d).

by the reestablishment of the westward flow of surface and intermediate water masses from the Levantine/Aegean region. The general cyclonic circulation of LIW/CIW in the Ionian governed by the Coriolis effect accompanied by the subsequent restriction of the anticyclonic flow pattern of the AW due to its southward shift, consists the cyclonic mode of the North Ionian circulation.

According to this concept, the combined effect of the spatial variability of the anticyclonic meander of the AIS and the subsequent variability in the westward flow of the Levantine/Aegean waters are portrayed as reversals of the North Ionian circulation.

8.3.5. Flows through Straits—Salinity and Heat

In order to quantify the effects of lateral advection toward the convection areas and its apparent relation with the observed oscillations of DWF in the two source areas, we computed the heat and salt fluxes at their straits that are due to volume exchanges representing the net mean flow. For the computations of lateral fluxes we use

$$H_c = \int_A U \rho c_p T dA \quad S_c = \int_A U \rho S dA$$

Where U is the mean horizontal velocity per unit area and depth, dA is the unit area, c_p is the specific heat capacity of sea water, and *rho* is the potential density.

Annual means of the heat (H_c) and salt (S_c) transport were selected in order to focus on their interannual variability. Due to the difference of the two source areas in their overturning functioning, as discussed above, the lateral fluxes are computed in the first 400 m for the Adriatic and the first 200 m for the Aegean Sea, since they represent approximately the average depth of inflowing water masses.

The resulting lateral fluxes show that increased heat (Figures 8.12a, 12b) and salt (Figures 8.12c, 12d) transport is connected with periods of intense DWF in both source areas, and exhibits the same anticorrelated oscillation pattern.

The resulting net salt flux computed at the straits (Figures 8.12c, 8.12d) during the preconditioning phase of the source areas can be attributed both to the increased salinity in the available water masses off their straits and to the increased inflowing volume as well (reflecting velocity anomalies times the mean salt content). The decay of DWF activity is initiated by the decrease of the salt content of the source area in accordance with the changes in the main water masses pathways and is then followed by the reduction of the volume of inflowing upper/ intermediate layer water masses due to the respective decrease of the overturning functioning of the source area.

Similar behavior can be observed for the lateral heat fluxes (Figures 8.12a, 8.12b). Periods of intense DWF activity is also correlated with increased heat content due to lateral advection of warm water masses (the LIW and/or CIW). This is initially contradictory to the functioning of DWF since it provides positive buoyancy, which reduces deep convection by antagonizing the surface buoyancy loss and increases the stratification of the water column. Indeed, years with maximum inflow are followed by years of moderate convection events that result in lower lateral advection. However, increased lateral heat inflow is correlated with increased evaporation (Figure 8.7b) and larger thermal fluxes to the atmosphere (Figure 8.7c).

Convection events that follow an increase in heat content result in vertical redistribution of heat from the intermediate layers toward the surface, sustaining higher SST and increased heat exchange. This was also shown when examining the air-sea temperature difference and the respective evaporative and heat fluxes (Figure 8.7b). *Josey* [2011], using different atmospheric datasets that cover the 1949–2000 period, showed that periods of intense DWF in the Aegean are strongly correlated with increased air-sea heat fluxes, although in his study this is related only to the atmospheric forcing variability. Furthermore, recent studies on the EMT [*Romanski et al.*, 2012] connect the intense atmospheric forcing with enhanced cyclone activity over the Aegean Sea. Thus, although the initial influence of thermal preconditioning induces stratification and balances heat loss to reduce the depth of the convection, the resulting increased air-sea interaction acts as a secondary mechanism to further enhance salinity preconditioning through the corresponding increase in evaporation. The above reasoning implies that the observed oscillations may constitute a coupled ocean-atmosphere interaction mechanism, with a positive feedback owing to the internal redistribution mechanism.

8.4. SUMMARY AND CONCLUSIONS

The Adriatic and the Aegean seas alternatively act as two competitive and distinct deep/dense water source areas for the eastern Mediterranean on an almost decadal timescale, although they present several similarities and differences in morphology, hydrology, and dynamics, as well as in the prevailing meteorological conditions.

The present work, based on numerical modelling and observations for the period 1960–2000, deals with the interdependence among Deep/Dense Water Formation activity in the two source areas, the variability of the circulation as well as the salinity and temperature oscillations occurring in all basins of the EMed.

It is revealed that during the 40-year period, both dense water sources present cycles of intense DWF periods, which are followed by periods of less-intense activity creating an anticorrelated, competitive oscillating pattern between them. The latter concerns the hydrological properties and structure of the entire water column, the characteristics of dense waters formed in each basin, and the outflowing and the respective inflowing water volumes. Furthermore, a continuous positive trend in the densities and the amount of outflowing deep waters from the Aegean during the entire period as well as difference in water density between the two active periods for the Adriatic in the 1960s and 1980s are evident.

From the examination of the water mass hydrological characteristics, it is revealed that the cycles of intense DWF periods can be attributed to enhanced salt content. The salinity increase precedes the period of the maximum DWF activity denoting the salinity preconditioning of each source area. On the contrary, during the decay phase, salinity decreases weakening the DWF efficiency respectively.

As far as the atmospheric influence is concerned, although there is significant long- and short-term variability in the atmospheric parameters, this variability is not in an anticorrelated state between the two basins and does not exhibit oscillations similar to those of the hydrological parameters. Conversely, oscillations are evident in the total heat and evaporative fluxes that depend on the air and sea surface temperature difference, which present decadal anticorrelated oscillating patterns over each basin corresponding to DWF variability. This reveals the dependence on the long-term evolution of sea surface temperature that can be attributed to the increased heat content observed in each source area during the periods of intense DW production. However, intense atmospheric forcing is a prerequisite for triggering DWF and may result in significant outflows from both basins even for periods that do not match the oscillating pattern, showing its important role as a short term factor for DWF processes.

The above findings along with the anticorrelated behavior of the two source areas led us to examine the internal functioning, related to the circulation variability and the lateral redistribution of water properties, as being the main mechanism for the alteration of the DWF source in the EMed.

The massive outflow of dense water through the deepest layers of the straits during the active phases of each basin is balanced by inflow from the upper layers. This can be considered as a pumping mechanism that advects upper-layer available water masses toward the active area.

According to this mechanism, the DWF process taking place in either active source area gradually modifies the main water mass pathways significantly disturbing the upper thermohaline cell of the EMed.

Particularly in the case of the Adriatic during the active period, the AW meander (AIS) is shifted further northward in the Ionian Sea significantly reducing its contribution to the eastward flow. At the same time, the LIW/CIW masses recirculate in the Levantine Sea greatly reducing their westward flow due to water mass conservation. The above major changes in the water mass pathways reflect the weakening of the upper thermohaline cell of the EMed. This leads to a gradual increase in salinity of the surface/intermediate waters in the Levantine basin, which precondition the Aegean Sea. In the opposite case, during the active phase of the Aegean, the AW is gradually advected eastward and at the same time the westward flow of the LIW/CIW is restored, thus reestablishing the thermohaline cell of the EMed. Therefore, the proposed NIG and BiOS system and the associated reversals of circulation in the north Ionian are the result of the above-mentioned functioning (mechanism).

The increased AW volume received by the active source area progressively leads to its freshening and the subsequent reduction of the DWF ability/intensity. This process preconditions the inactive source area through the corresponding changes in the pathways of LIW and/or CIW. These intermediate water masses act as salt and heat carriers, directly resulting in salinity preconditioning due to the increased salt flux and enhancing air-sea interaction due to the increased heat flux, which is a secondary mechanism to further enhance salinity preconditioning through the corresponding increase in evaporation. Therefore, the proposed mechanism affects both the preconditioning and the decay phases of the two source areas.

Importantly, in the case of the Aegean Sea active periods (1972–1980 and 1987–1995), strong inflow of water occurs also from horizons deeper than LIW depths, the so-called TMW, thus providing additional to the surface, less-saline, and cooler waters. TMW then forms a salinity and temperature minimum layer at intermediate depths, which is an additional factor toward freshening of the Aegean water column and can be considered a characteristic, indicative vertical structure in the South Aegean that develops after periods of intense DWF and outflows, such as post-EMT (after 1990) or post-EMT-like (after 1950 and 1970) events

Finally, it is revealed that the EMT event corresponds to a period of resonance of atmospheric variability with internal oscillation modes. It can thus be considered an extreme event in the frame of the internal variability of the EMed and can be attributed to the combined effect of long-term preconditioning of the Aegean due to internal oscillations of hydrological characteristics, further enhanced by a positive trend in their evolution, followed by a short period of extreme atmospheric forcing. The above reasoning implies that the observed oscillations may constitute a coupled ocean-atmosphere interaction mechanism. Therefore, further investigation is required on the combined effects and interactions of internal processes and external atmospheric forcing.

Acknowledgments. This work has been supported by the MEDECOS Project (ERA-NET–MarinERA, EU FP6). All data providers are deeply acknowledged. The authors would like to express their gratitude to all scientists, technicians, and crews of the various oceanographic vessels for their continuous efforts throughout the last century in collecting and processing all field data used in this work. Special thanks to the HNHS and the HNODC for making their data available. We are also grateful to M. Rixen for the provision of the interannual gridded dataset for the Mediterranean, S. Sommot for providing the atmospheric forcing dataset through SESAME project, and Kostas Tsiaras for his overall help.

REFERENCES

Beşiktepe, Ş, H. I. Sur, E. Ozsoy, M. A. Latif, T. Oguz, and U. Unluata (1994), Circulation and hydrography of the Sea of Marmara, *Progress in Oceanography*, *34*, 285–334.

Bethoux, J. P. (1979), Budgets of the Mediterranean Sea: Their dependence on the local climate and on the characteristics of the Atlantic waters, *Oceanologica Acta*, *2*, 157–163.

Beuvier, J., F. Sevault, M. Herrmann, H. Kontoyiannis, W. Ludwig, M. Rixen, E. Stanev, K. Béranger, and S. Somot (2010), Modeling the Mediterranean Sea interannual variability during 1961–2000: Focus on the Eastern Mediterranean Transient, *Journal of Geophysical Research*, *115*, C08017, 27, doi: 10.1029/2009JC005950.

Blumberg, A. F., and G. L. Mellor (1987), A description of a three-dimensional coastal ocean circulation model, pp. 1–16, in *Three-Dimensional Coastal Ocean Circulation Models*, ed. N. S. Heaps, AGU, Washington, D.C.

Borzelli, G. L. E., M. Gačić, V. Cardin, and G. Civitarese (2009), Eastern Mediterranean transient and reversal of the Ionian Sea circulation, *Geophysical Research Letters*, *36*, L15108, doi: 10.1029/2009GL039261.

Cardin V., M. Bensi, and M. Pacciaroni (2011), Variability of water mass properties in the last two decades in the South Adriatic Sea with emphasis on the period 2006–2009, *Continental Shelf Research*, *31*, 9, 951–965, doi: /10.1016/j.csr.2011.03.002.

Chu, P. C. (1991), Geophysics of deep convection and deep water formation in oceans, pp. 3–16 in *Deep Convection and Deep Water Formation in the Oceans*, ed. P. C. Chu and J. C. Gascard, Elsevier Science Publishers B.V., Amsterdam.

Demirov, E., and N. Pinardi (2002), Simulation of the Mediterranean Sea circulation from 1979 to 1993: Part I, The interannual variability, *Journal of Marine Systems*, *33–34*, 23–50, doi: 10.1016/S0924-7963(02)00051-9.

Déqué M., C. Dreveton, A. Braun, and D. Cariolle (1994), The ARPEGE/IFS atmosphere model: A contribution to the French community climate modeling, *Climate Dynamics*, *10*, 4–5, 249–266, doi: 10.1007/BF00208992.

Fairall, C. W., E. F. Bradley, J. E. Hare, A. A. Grachev, and J. B. Edson (2003), Bulk parameterization of air-sea fluxes: Updates and verification for the COARE algorithm, *Journal of Climate*, *16*, 571–591, doi: 10.1175/1520-0442(2003)016<0571:BPOAS F>2.0.CO;2 .

Flather, R. A. (1976), A tidal model of the northwest European continental shelf, *Memo. Soc. Roy. Sci. Liege*, *6* (10), 141–164.

Gačić, M., G. Civitarese, G. L. Eusebi Borzelli, V. Kovacevic, P.-M. Poulain, A. Theocharis, M. Menna, A. Catucci, and N. Zarokanellos (2011), On the relationship between the decadal oscillations of the northern Ionian Sea and the salinity distributions in the eastern Mediterranean, *Journal of Geophysical Research*, *116*, C12002, doi: 10.1029/2011JC007280.

Gačić M., G. L. Eusebi Borzelli, G. Civitarese, V. Cardin, and S. Yari (2010), Can internal processes sustain reversals of the ocean upper circulation? The Ionian Sea example, *Geophysical Research Letters*, *37*, L09608, doi: 10.1029/2010GL043216.

Georgopoulos, D., A. Theocharis, and G. Zodiatis (1989), Intermediate water formation in the Cretan Sea (South Aegean Sea), *Oceanologica Acta*, *12*, 4, 353–359.

Gertman, I., N. Pinardi, Y. Popov, and A. Hecht (2006), Aegean sea water masses during the early stages of the eastern Mediterranean Climatic Transient (1988–90), *Journal of Physical Oceanography*, *36*, 1841–1859, doi: 10.1175/JPO2940.1.

Gill, A. (1982), *Atmosphere-Ocean Dynamic*, International Geophysics Series, vol. *30*, Academic Press, London, p. 662.

Herrmann, M. J., and S. Somot (2008), Relevance of ERA40 dynamical downscaling for modeling deep convection in the Mediterranean Sea, *Geophysical Research Letters*, *35*, L04607, doi: 10.1029/2007GL032442.

Josey, S. A. (2003), Changes in the heat and freshwater forcing of the Eastern Mediterranean and their influence on deep water formation, *Journal of Geophysical Research*, *108*, C7, 3237, doi: 10.1029/2003JC001778.

Josey, S. A., S. Somot, and M. Tsimplis (2011), Impacts of atmospheric modes of variability on Mediterranean Sea surface heat exchange, *Journal Geophysical Research*, *116*, C02032, doi:10.1029/2010JC006685.

Kanarska, Y., and V. Maderich (2008), Modeling of seasonal exchange flows through the Dardanelles Strait, *Estuarine, Coastal and Shelf Science*, *79*, 3, 449–458, doi: 10.1016/j.ecss.2008.04.019

Klein, B., W. Roether, B. Manca, and A. Theocharis (2000), The evolution of the Eastern Mediterranean Climatic Transient during the last decade: The tracer viewpoint, pp. 21–26 in *The Eastern Mediterranean Climatic Transient: Its Origin, Evolution and Impact on the Ecosystem*, CIESM Workshop Series., vol. *10*, ed. F. Briand, Mediterranean Science Committee, Trieste.

Klein, B., W. Roether, B. Manca, D. Bregant, V. Beitzel, V. Kovacevic, and A. Luchetta (1999), The large deep water transient in the Eastern Mediterranean, *Deep-Sea Research Part I*, *46*, 371–414.

Kondo, J. (1975), Air-sea bulk transfer coefficients in diabatic conditions, *Boundary-Layer Meteorology*, *9*, 91–112.

Kontoyiannis, H., A. Theocharis, E. Balopoulos, S. Kioroglou, V. Papadopoulos, M. Collins, A. F. Velegrakis, and A. Iona, (1999), Water fluxes through the Cretan Arc Straits, Eastern Mediterranean Sea: March 1994 to June 1995, *Progress in Oceanography*, *44*, 511–529, doi: 10.1016/S0079-6611(99)00044-0.

Korres, G., A. Lascaratos, E. Hatziapostolou, and P. Katsafados (2002), Towards an ocean forecasting system for the Aegean Sea, *The Global Atmosphere and Ocean System*, *8*, 2–3, 191–218, doi: 10.1080/1023673029000003534.

Kourafalou, V., and K. Tsiaras (2007), A nested circulation model for the North Aegean Sea. *Ocean Science*, *3*, 1, 1–16., doi: 10.5194/os-3-1-2007.

Larnicol, G., N. Ayoub, and P. Y. Le Traon (2002), Major changes in Mediterranean Sea level variability from 7 years of TOPEX/Poseidon and ERS-1/2 data, *Journal of Marine Systems*, *33–34*, 63–89, doi: 10.1016/S0924-7963(02)00053-2.

Lascaratos, A., and K. Nittis (1998), A high-resolution three-dimensional numerical study of intermediate water formation in the Levantine Sea, *Journal of Geophysical Research*, *103*, C9, 18497–18511.

Lascaratos, A., W. Roether, K. Nittis, and B. Klein (1999), Recent changes in the deep water formation and spreading in the Eastern Mediterranean Sea, *Progress in Oceanography*, *44*, 1–3, 5–36.

Lykoussis, V., G. Chronis, A. Tselepides, N. B. Price, A. Theocharis, I. Siokou-Frangou, F. Van Wambeke, R. Danovaro, S. Stavrakakis, G. Duineveld, D. Georgopoulos, L. Ignatiades, A. Souvermezoglou, and F. Voutsinou-Taliadouri (2002), Major outputs of the recent multidisciplinary biogeochemical researches undertaken in the Aegean Sea, *Journal of Marine Systems*, *33–34*, 313–334, doi: 10.1016/S0924-7963(02)00064-7.

Malanotte-Rizzoli, P., B. B. Manca, M. Ribera D'Alcala, A. Theocharis, A. Bergamasco, D. Bregant, G. Budillon, G. Civitarese, D. Georgopoulos, A. Michelato, E. Sansone, P. Scarazzato, and E. Souvermezoglou (1997), A synthesis of the Ionian hydrography, circulation and water mass pathways during POEM-Phase I, *Progress in Oceanography*, *39*, 153–204.

Malanotte-Rizzoli, P., B. B. Manca, M. Ribera D'Alcala, A. Theocharis, S. Brenner, G. Budillon, and E. Ozsoy (1999), The eastern Mediterranean in the 80s and in the 90s: The big transition in the intermediate and deep circulations, *Dynamics of Atmospheres and Oceans*, *29*, 365–395.

Manca, B. B. (2000), Recent changes in dynamics of the eastern Mediterranean affecting the water characteristics of the adjacent basins, pp. 21–26, in *The Eastern Mediterranean Climatic Transient: Its Origin, Evolution and Impact on the Ecosystem*, CIESM Workshop Series, vol. 10, ed. F. Briand, Mediterranean Science Committee, Trieste.

Manca, B. B., L. Ursella, and P. Scarazzato (2002), New development of eastern Mediterranean circulation based on hydrological observations and current measurements, *Marine Ecology*, *23*, Supp. 1, 237–257, doi: 10.1111/j.1439-0485.2002.tb00023.x.

Mantziafou, A., and A. Lascaratos (2004), An eddy resolving numerical study of the general circulation and deep-water formation in the Adriatic Sea, *Deep-Sea Research: Part I: Oceanographic Research Papers*, *51*, 7, 921–952, doi: 10.1016/j.dsr.2004.03.006.

Marshall, J., and F. Schott (1999), Open-ocean convection: Observations, theory, and models, *Reviews in Geophysics*, *37*, 1–64.

Maxworthy, T. (1997), A frictionally and hydraulically constrained model of the convectively driven mean flow in partially enclosed seas, *Deep Sea Research*, *44*, 8, 1339–1354.

MEDAR Group (2002), MEDAR/MEDATLAS 2002 Database, Cruise inventory, observed and analyzed data of temperature and bio-chemical parameters, 4 CDROM.

Miller, A. R. (1963), Physical oceanography of the Mediterranean Sea: A discourse, *Rapp. Comm. Int. Mer Med.*, *17*, 3, 857–871.

Miller, A. R. (1974), Deep convection in the Aegean, in *Processus de formation des eaux oceaniques profondes, Colloques Internationaux du CNRS*, *215*, 156–163.

Miller, A. R., P. Tchernia, H. Charnock, and D. A. McGill (1970), *Mediterranean Sea Atlas of Temperature, Salinity, Oxygen Profiles and Data from Cruises R/V Atlantis and R/N Chain with Distribution of Nutrient Chemical Properties*, the Woods Hole Oceanographic Institution Atlas Series, vol. *3*, pp. 190.

Nielsen, J. (1912), Hydrology of the Mediterranean and adjacent waters, in *Report of the Danish Oceanographic Expedition 1908–1910 to the Mediterranean and Adjacent Waters*, ed. J. Schmidt, pp. 77–192, Andr. Fred. Host & Son, Copenhagen.

Nittis, K., A. Lascaratos, and A. Theocharis (2003), Dense water formation in the Aegean Sea: Numerical simulations during the Eastern Mediterranean Transient, *Journal of Geophysical Research*, *108*, C9, 8120, doi:10.1029/2002JC001352.

Ovchinnikov, I., and Y. Plakhin (1965), Formation of Mediterranean deep water masses, *Oceanology*, *5* (4), 40–47.

Pinardi, N., and A. Navarra (1993), Baroclinic wind adjustment processes in the Mediterranean Sea, *Deep Sea Research II*, *40*, 6, 1299–1326.

Pinardi N., G. Korres, A. Lascaratos, V. Roussenov, and E. Stanev (1997), Numerical simulation of the interannual variability of the Mediterranean Sea upper ocean circulation, *Geophysical Research Letters*, *24*, 425–428.

Pisacane G., V. Artale, S. Calmanti, and V. Rupolo (2006), Decadal oscillations in the Mediterranean Sea: A result of the overturning circulation variability in the eastern basin? *Climate Research*, *31*, 257–271.

POEM Group (Robinson, A.R., P. Malanotte-Rizzoli, A. Hecht, A. Michelato, W. Roether, A. Theocharis, U. Unluata, N. Pinardi, A. Artegiani, A. Bergamasco, J. Bishop, S. Brenner, S. Christianidis, M. Gačić, D. Georgopoulos, M. Golnaraghi, M. Hausmann, H.-G. Junghaus, A. Lascaratos, M. A. Latif, W. G. Leslie, C. J. Lozano, T. Oguz, E. Ozsoy, E. Papageorgiou, E. Paschini, Z. Rozentroub, E. Sansone, P. Scarazzato, R. Schlitzer, G.-C. Spezie, E. Tziperman, G. Zodiatis, L. Athanassiadou, M. Gerges , and M. Osman) (1992), General circulation of the eastern Mediterranean, *Earth-Science Review*, *32*, 285–309.

Pujol, M. I., and G. Larnicol (2005), Mediterranean Sea eddy kinetic energy variability from 11 years of altimetric data, *Journal of Marine Systems*, *58*, 121–142, doi: 10.1016/j.jmarsys.2005.07.005.

Rixen M., J. Beckers, S. Levitus, J. Antonov, T. Boyer, C. Maillard, M. Fichaut, E. Balopoulos, S. Iona, H. Dooley, M. Garcia, B. Manca, A. Giorgetti, G. Manzella, N. Mikhailov, N. Pinardi, and M. Zavatarelli (2005), The western Mediterranean deep water: A proxy for climate change— art. no. L12608, *Geophysical Research Letters*, *32*, L12608, doi: 10.1029/2005GL022702.

Roether, W., and R. Schlitzer (1991), Eastern Mediterranean deep water renewal on the basis of chlorofluoromethane and tritium, *Dynamics of Atmospheres and Oceans*, *15*, 3–5, 333–354.

Roether, W., and R. Well (2001), Oxygen consumption in the Eastern Mediterranean. *Deep Sea Research Part I: Oceanographic Research Papers*, *48*, 6, 1535–1551, doi: 10.1016/S0967-0637(00)00102-3.

Roether, W., B. B. Manca, B. Klein, D. Bregant, D. Georgopoulos, V. Beitzel, V. Kovacevic, and A. Luchetta (1996), Recent changes in eastern Mediterranean deep waters, *Science*, *271*, 333–335.

Roether, W., B. Klein, B. B. Manca, A. Theocharis, and S. Kioroglou (2007), Transient eastern Mediterranean deep waters in response to the massive dense-water output of the Aegean Sea in the 1990s, *Progress in Oceanography*, *74*, 540–571, doi: 10.1016/j.pocean.2007.03.001.

Romanski, J., A. Romanou, M. Bauer, and G. Tselioudis (2012), Atmospheric forcing of the eastern Mediterranean transient by mid-latitude cyclones, *Geophysical Research Letters*, *39*, L03703, doi: 10.1029/2011GL050298.

Sayın E., C. Eronat, Ş. Uçkaç, and Ş. T. Beşiktepe (2011), Hydrography of the eastern part of the Aegean Sea during the Eastern Mediterranean Transient (EMT), *Journal of Marine Systems*, *88*, 502–515, doi: 10.1016/j.jmarsys.2011.06.005

Schlitzer, R., W. Roether, H. Oster, H-G. Junghans, M. Hausman, H. Johannsen, and A. Michelato (1991), Chlorofluoromethane and oxygen in the eastern Mediterranean, *Deep-Sea Research Part , Oceanographic Research Papers*, *38*, 12, 1531–1551.

Somot S., and J. Colin (2008), First step towards a multi-decadal high-resolution Mediterranean sea reanalysis using dynamical downscaling of ERA40, *Research Activities in Atmospheric and Oceanic Modeling*, CAS/JSC Working group on numerical experimentation, Report 38, http://collaboration.cmc.ec.gc.ca/science/wgne/index.html.

Somot, S., F. Sevault, and M. Déqué (2006), Transient climate change scenario simulation of the Mediterranean Sea for the twenty-first century using a high-resolution ocean circulation model. *Climate Dynamics*, *27*, 851–879, doi: 10.1007/s00382-006-0167-z.

Spall M. A. (2004), Boundary currents and water mass transformation in marginal seas, *Journal of Physical Oceanography*, *34*, 1197–1213.

Spall, M. A. (2010), Dynamics of downwelling in an eddy-resolving convective basin, *Journal of Physical Oceanography*, *40*, 2341–2347, doi: 10.1175/2010JPO4465.1.

Theocharis, A., A. Lascaratos, and S. Sofianos (2002a), Variability in the sea water properties in the Ionian, Cretan and Levantine Seas during the last century, pp. 71–74, in *Tracking Long Term Hydrological Changes in the Mediterranean*

Sea, CIESM Workshop Series., vol. *16*, ed. F. Briand, Mediterranean Science Committee, Monaco.

Theocharis, A., and D. Georgopoulos (1993), Dense water formation over the Samothraki and Limnos Plateaux in the north Aegean Sea (eastern Mediterranean Sea), *Continental Shelf Research*, *13* (8/9), 919–939.

Theocharis, A., B. Klein, K. Nittis, and W. Roether (2002b), Evolution and status of the eastern Mediterranean transient (1997–1999), *Journal of Marine Systems*, *33–34*, 91–116, doi: 10.1016/S0924-7963(02)00054-4

Theocharis, A., D. Georgopoulos, A. Lascaratos, and K. Nittis (1993a), Water masses and circulation in the central region of the eastern Mediterranean: Eastern Ionian, south Aegean, and northwest Levantine, 1986–1987, *Deep-Sea Research Part II: Topical Studies in Oceanography*, *40*, (6), 1121–1142.

Theocharis, A., E. Balopoulos, S. Kioroglou, H. Kontoyiannis, and A. Iona (1999a), A synthesis of the circulation and hydrography of the south Aegean Sea and the Straits of the Cretan Arc (March 1994–January 1995), *Progress in Oceanography*, *44*, 469–509.

Theocharis, A., K. Nittis, H. Kontoyiannis, E. Papageorgiou, and S. Balopoulos (1999b), Climatic changes in the Aegean Sea influence the eastern Mediterranean thermohaline circulation (1986–1987), *Geophysical Research Letters*, *26* (11), 1617–1620.

Tsimplis, M. N., A. G. P. Shaw, A. Pascual, M. Marcos, M. Pasaric, and L. Fenoglio-Marc (2008), Can we reconstruct the 20th century sea level variability in the Mediterranean Sea on the basis of recent altimetric measurements? in *Remote Sensing of the European Seas*, ed. V. Barale, and M. Gade, pp. 307–318, Springer, Berlin.

Tsimplis, M. N., M. Marcos, B. Pérez, P. Challenor, M. J. Garcia-Fernandez, and F. Raicich (2009), On the effect of the sampling frequency of sea level measurements on return period estimate of extremes—Southern European examples, *Continental Shelf Research*, *29*, 18, 2214–2221, doi: 10.1016/j.csr.2009.08.015.

Velaoras D., and A. Lascaratos (2010), North-central Aegean Sea surface and intermediate water masses and their role in triggering the eastern Mediterranean Transient, *Journal of Marine Systems*, *83*, 58–66, doi: 10.1016/j.jmarsys.2010.07.001.

Vervatis D. V., S. S. Sofianos, and A. Theocharis (2011), Distribution of the thermohaline characteristics in the Aegean Sea related to water mass formation processes (2005–2006 winter surveys), *Journal of Geophysical Research*, *116*, C09034, doi:10.1029/2010JC006868.

Vigo, I., D. Garcia, and B. F. Chao (2005), Change of sea level trend in the Mediterranean and Black seas, *Journal of Marine Research*, *63*, 6, 1085–1100, doi: 10.1357/002224005775247607.

Vilibić I., and M. Orlić (2002), Adriatic water masses, their rates of formation and transport through the Otranto Strait, *Deep Sea Research Part I: Oceanographic Research Papers*, *49*, 8, 1321–1340, doi: 10.1016/S0967-0637(02)00028-6.

Wüst, G. (1961), On the vertical circulation of the Mediterranean Sea, *Journal of Geophysical Research*, *66*, 3261–3271.

Zavatarelli, M., and G. L. Mellor (1995), A numerical study of the Mediterranean Sea circulation, *Journal of Physical Oceanography*, *25*, 1384–1414.

Zervakis, V., D. Georgopoulos, and P. Drakopoulos (2000), The role of the North Aegean in triggering the recent eastern Mediterranean climatic changes, *Journal of Geophysical Research*, *105*, C11, 26103–26116.

Zore-Armanda, M. (1963), Les masse d'eau de la Mer Adriatique, *Acta Adriatica*, *10*/3, 5–88.

Zore-Armanda, M. (1969), Water exchanges between the Adriatic Sea and the eastern Mediterranean, *Deep Sea Research*, *16*, 171–178.

9

Thermohaline Variability and Mesoscale Dynamics Observed at the Deep-Ocean Observatory E2M3A in the Southern Adriatic Sea

Manuel Bensi[1], Vanessa Cardin[1], and Angelo Rubino[2]

9.1. INTRODUCTION

In the Mediterranean Sea, strong surface cooling and evaporation due to winter spells of cold and dry air trigger Deep Water Formation (DWF). This phenomenon is crucial in shaping the basin thermohaline circulation because it enables a strong connection between ocean surface and deep layers. It takes place in different regions of the Mediterranean Sea and contributes to the driving of thermohaline cells, which are considered smaller-scale analogues of the world ocean thermohaline cells [*Malanotte-Rizzoli et al.*, 1997].

Convection that takes place in the Levantine Basin leads to the formation of a large open thermohaline cell characterized by the transformation of surface Atlantic Water into Levantine Intermediate Water (LIW). Convection that takes place in the Gulf of Lion and in the Adriatic Sea or in the Aegean Sea, in contrast, leads to the formation of two secondary closed cells characterized by the transformation of surface and intermediate waters into, respectively, Western and Eastern Mediterranean Deep Water [*Lascaratos et al.*, 1999].

In the Adriatic Sea, the Bora predominates. It is a dry and cold wind from the NE, which blows in intense episodic bursts [*Oddo et al.*, 2005] and whose spatial distribution is strongly influenced by the orography of the eastern Adriatic land margins [*Vilibić*, 2003; *Orlić et al.*, 1994]. It is responsible for strong sea-surface cooling. The latter phenomenon is associated with intense buoyancy flux at the air-sea interface, which triggers episodes of dense water formation both in the Northern and in the Southern Adriatic. The volume contribution of the dense waters produced in the northern subbasin to the Adriatic outflow through the Strait of Otranto ranges between 4% and 40% and it shows very high annual variability [see, e.g., *Cardin and Gačić*, 2003; *Vilibić*, 2003; *Supić and Vilibić*, 2006; *Mantziafou and Lascaratos*, 2008; *Cardin et al.*, 2011; *Rubino et al.*, 2012]. In the Southern Adriatic Pit (see Figure 9.1), whose maximum depth is ~1250 m, a semipermanent cyclonic gyre [*Gačić et al.*, 1997] causes the doming of a high salinity layer, typically located between 300 and 600 m depth and formed of Levantine waters coming from the Ionian Sea. Under such favorable conditions, winter episodes of strong heat losses (which can exceed 1000 Wm^{-2}) give birth to convective events with typical temporal scales of several days and spatial scales of 20 km (*Gačić et al.* [2001] and references therein).

The newly produced dense water, mixed with waters produced in the northern Adriatic, forms the Adriatic Deep Water, which overflows through the western bottom layers of the Strait of Otranto, eventually sinking in the abyssal plain of the Ionian Sea through bottom-arrested currents [*Hainbucher et al.*, 2006; *Rubino and Hainbucher*, 2007; *Rubino et al.*, 2012; *Bensi et al.*, 2012]. It contributes to the formation of the Eastern Mediterranean Deep Water [*Wüst*, 1961; *Hopkins*, 1978, 1985; *Schlitzer et al.*, 1991; *Roether et al.*, 1996].

Due to the harsh atmospheric and oceanic conditions typical of convective events, direct observations of

[1] *Istituto Nazionale di Oceanografia e di Geofisica Sperimentale (OGS), Sgonico (Trieste), Italy*

[2] *Dipartimento di Scienze Ambientali, Informatica e Statistica, Universita' Ca' Foscari di Venezia, Venezia, Italy*

The Mediterranean Sea: Temporal Variability and Spatial Patterns, Geophysical Monograph 202. First Edition.
Edited by Gian Luca Eusebi Borzelli, Miroslav Gačić, Piero Lionello, and Paola Malanotte-Rizzoli.

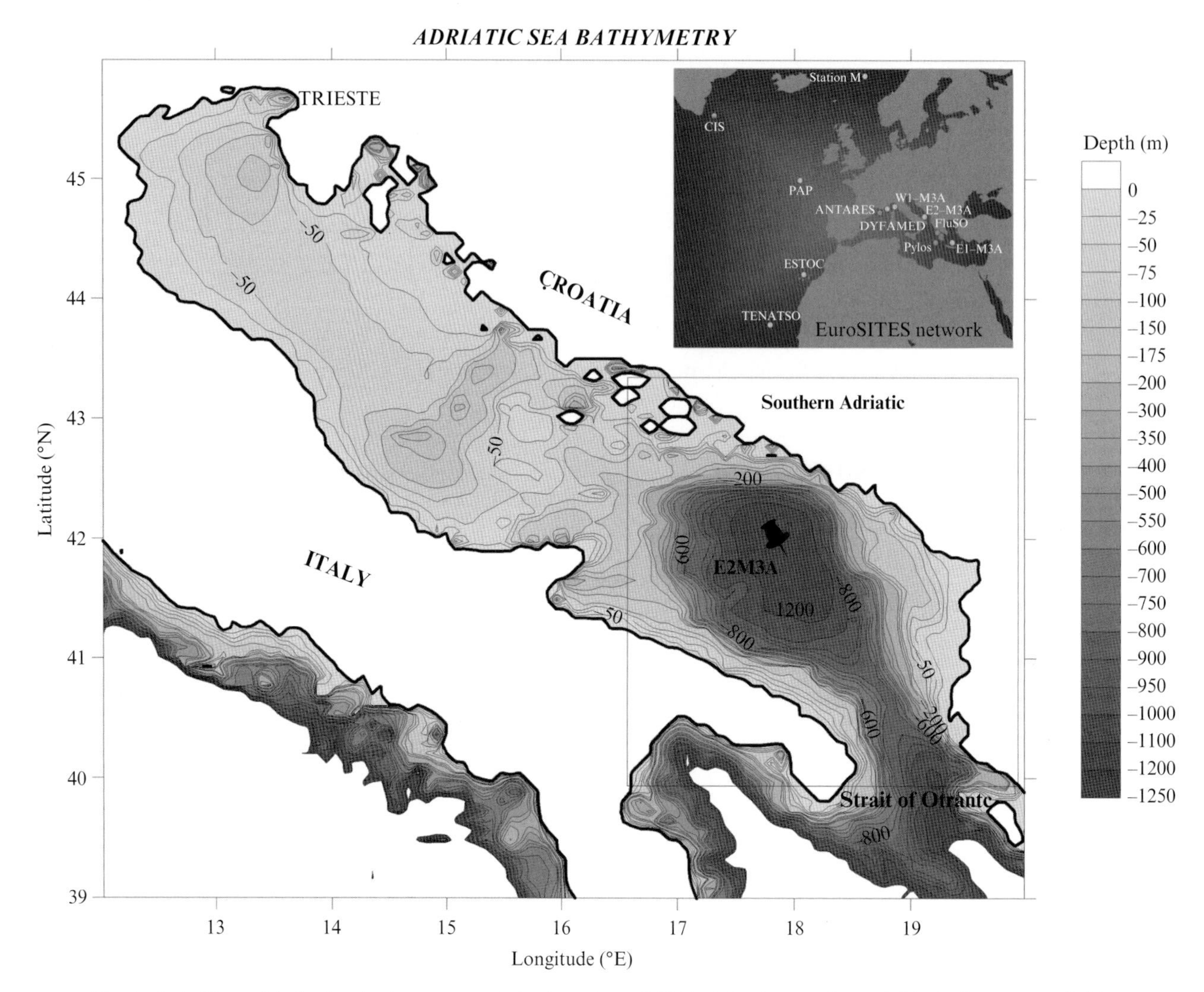

Figure 9.1. The Adriatic Sea bathymetry and the location of E2M3A in the southern Adriatic (large panel). The small panel shows the EuroSITES oceanographic buoy networks.

open-ocean convection are scarce [*Marshall and Schott*, 1999]. Moreover, long time series describing the evolution of the water column before and after major convective events are also rare. Indeed, moored instruments, which can provide continuous measurements throughout the water column, are needed to appreciate the details of the convective preconditioning, active phase, and exporting [*Marshall and Schott*, 1999]. Observations from such instruments, when available, can even unveil aspects of the small-scale couplings between physical, biological, and chemical processes active during convection [*Dur et al.*, 2007].

Since 2006, the southern Adriatic has been constantly monitored by means of a deep-ocean observatory (site E2M3A) located in its central part at 41° 50′N, 17° 45′E (Figure. 9.1). The deep-ocean observatory E2M3A was deployed in November 2006 during the Italian project VECTOR (VulnErabilità delle Coste e degli ecosistemi marini italiani ai cambiamenti climaTici e loro ruolO nei cicli del caRbonio mediterraneo), and enhanced after 2008 within the framework of the FP7-EuroSITES (European Ocean Observatories Network) project. The time series obtained are the longest ones available so far in this region.

In this paper, we present high-frequency oceanographic data (temperature, salinity, vertical and horizontal currents, and heat fluxes at the sea-air interface) collected at E2M3A between November 2006 and September 2010. We have used these data to investigate the thermohaline variability occurring in the southern Adriatic during this period. The analysis of the data is complemented by an analysis of oceanographic cruise data collected in the region. As a result, the temporal variability of the water masses in the southern Adriatic and an accurate view of the deep convection phases during four consecutive winters are described in detail. Particular attention is given to the observation of mesoscale eddies at E2M3A and to their possible role in heat and salt exchanges between intermediate and deep layers.

The paper is organized as follows: in section 2, the datasets used and the methods are described. Section 3 is focused on the thermohaline variability in the study region, and the evidence of deep convection events as well as the effect induced by the passage of mesoscale eddies on the water column are discussed. Finally, in section 4 an overall discussion is presented and conclusions are drawn.

9.2. DATASETS AND METHODS

The dataset used in this study comprises conductivity-temperature-depth (CTD) and current time series collected at E2M3A (Figure. 9.1) between 16 November 2006 and 16 September 2010, as well as different CTD profiles collected in the southern Adriatic during several oceanographic campaigns.

During the whole study period, four maintenance cruises were carried out. The deep mooring provided high-frequency sampling data to resolve episodic events and rapid processes. The payload of the observation site consisted of conductivity-temperature (CT; SBE37 MicroCAT) and CTD sensors (SBE16plusV2 SEACAT) at different nominal depths (350 m, 550 m, 750 m, 1000 m, and 1200 m). Furthermore, an upward-facing acoustic Doppler current profiler (ADCP) RDI 150 kHz (located at ~300 m) and a recording current meter (RCM 11) Aanderaa (located next to the bottom, ~1200 m) were deployed to provide current measurements at E2M3A.

The deep-ocean observatory underwent a change of configuration: the first configuration (Figure. 9.2a) operated between November 2006 and October 2009 and the second one (Figure. 9.2b) between October 2009 and September 2010. During the second configuration phase, the observatory was shifted ~44 km south of the first mooring position. In order to maintain all instruments at the same depth and to render the observations in the two phases more comparable, it was deployed at a similar depth (~1200 m). All the instruments used in this experiment were checked and calibrated before and after each deployment at the CTO (Oceanographic Calibration Center) of the Istituto Nazionale di Oceanografia e di Geofisica Sperimentale, OGS, Trieste. The obtained dataset consists of temperature (T), salinity (S), current velocity, and direction.

Temperature and salinity were measured using only CT sensors, with a sampling interval of 15 min, during the first phase. During the second phase, two CTD sensors with a sampling interval of 30 min were used in place of two of the CTs, at 350-m and 750-m depth (Figure. 9.2). The ADCP measured current profiles (u, v, and w components) every 15 min during the first phase and every 60 min during the second phase; in both cases the bin length was 5 m. In order to get rid of spurious vertical velocities, the signal from the diurnal vertical migration of zooplankton, which can reach several cm s^{-1} [*Marshall and Schott*, 1999], was removed from the ADCP time series. It was identified by its daily clockwork regularity. The Aanderaa RCM 11, moored a few meters above the ocean floor, measured the velocity and the direction of currents together with the *in situ* temperature every 30 min. Accuracies of the E2M3A instruments are given in Table 9.1.

All time series were processed applying a postprocessing routines package created with MATLAB to obtain cleaned and despiked data. The final corrected dataset was composed of average hourly data. A low-pass filter with a cutoff period at 33 hours [*Flagg et al.*, 1976] was applied to the hourly data in order to obtain subinertial nontidal flow.

The E2M3A data were validated by means of *in situ* measurements (eight cruises between November 2006 and July 2010) carried out in the southern Adriatic during the VECTOR, SESAME (Southern European Seas: Assessing and Modelling Ecosystem changes), and MSM (Maria S. Merian) cruises. All CTD downcast profiles (carried out by means of a SeaBird SBE19 plus) were corrected and the data were averaged every 1 db with overall accuracies within 0.002°C for temperature, 0.003 for salinity, and 2% of saturation for dissolved oxygen (DO). DO data obtained from CTD casts were calibrated comparing them with Winkler data obtained from discrete samples. A regression analysis using a polynomial of degree 2 enabled fitting of the downcast profiles of the CTD with the Winkler measurements obtained from water sample analyses [*Bensi and Kückler*, 2009].

Correlation coefficients calculated between E2M3A data and CTD profiles collected in the approaches to the mooring location indicate that the oceanographic phenomena occurring at the E2M3A site can accurately represent those typical of the open-ocean southern Adriatic. Indeed, at a distance of about 45 km from the mooring position, along the Bari-Dubrovnik section, the correlation between the data collected at E2M3A and CTD profiles was ~98%–99% for temperature and ~94%–96% for salinity. Increasing the distance up to 70 km decreased these correlations to 95% and 88% for temperature and salinity, respectively.

Air-sea heat fluxes were calculated according to *Cardin and Gačić* [2003] using meteorological data from the ECMWF (Centre for Medium-Range Weather Forecasts) operational dataset. Mean sea level pressure, total cloud cover, wind speed at 10 m above the mean sea level, air temperature, dew-point temperature at 2 m above the mean sea level, and skin temperature were used to calculate heat flux components as solar radiation (Q_S), long-wave

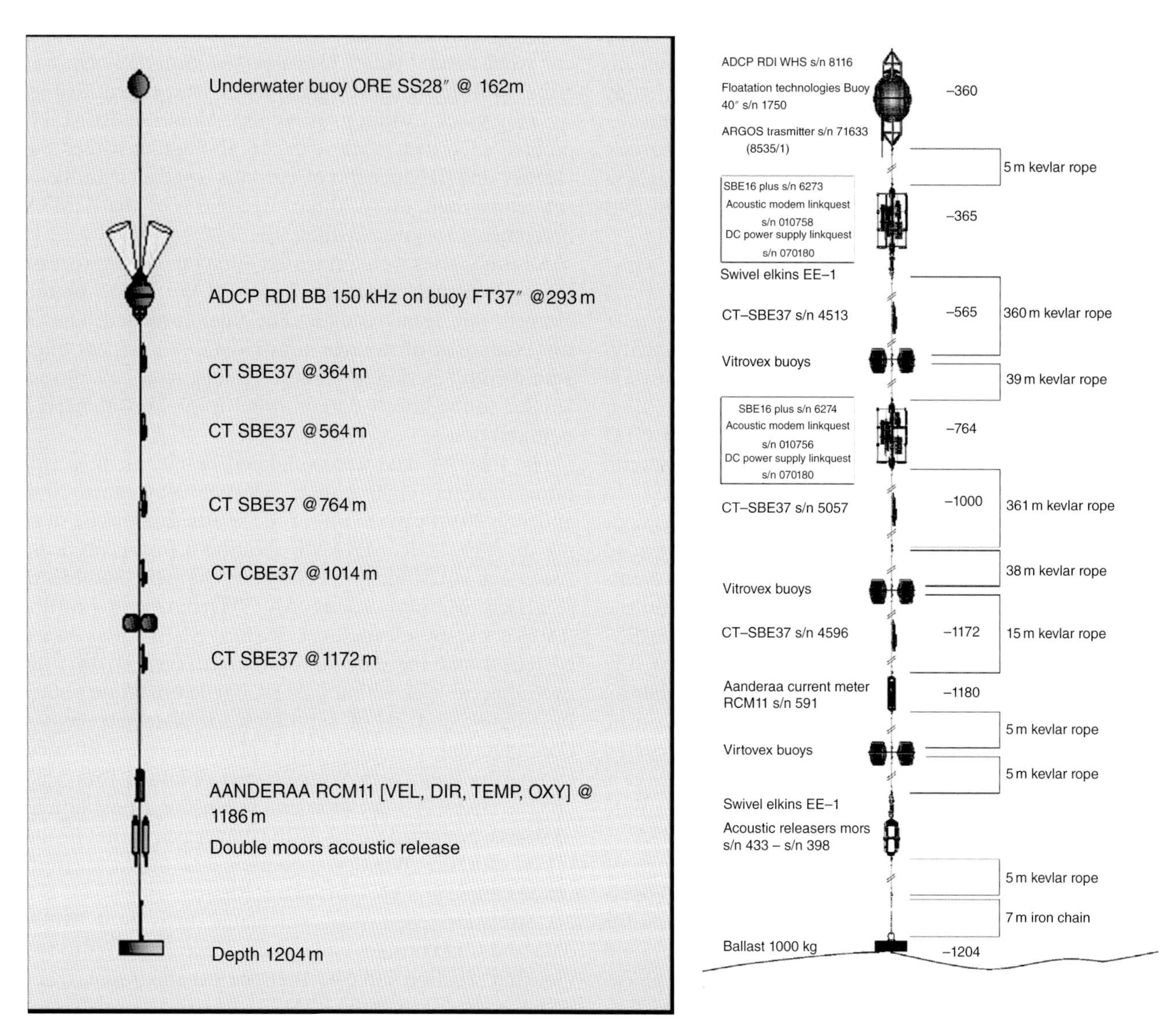

Figure 9.2. E2M3A Oceanographic buoy configurations: the first configuration (a) was used from November 2006 to October 2009 at 41° 50′ N, 17° 45′ E, while the second configuration (b) from October 2009 to September 2010 was at 41° 31.6′N, 18° 04.6′E.

Table 9.1. Accuracies of oceanographic instruments used at E2M3A.

Accuracy	CT (SBE37)	CTD (SBE16)	ADCP	RCM
Temperature (°C)	±0.002	±0.005	±0.01	±0.05°C
Conductivity (S m-1)	±0.0003	±0.0005	n/a	±0.2% of range
Dissolved Oxygen (ml l-1)	n/a	±2% of true value	n/a	n/a
Light transmission (%)	n/a	±0.003% of true value*	n/a	n/a
Current speed (cm s-1)	n/a	n/a	±1% ± 5 mm/s	±0.3
Current direction (deg)	n/a	n/a	±2°	±0.35°

*Gathered from the sensitivity of the optical part by N. Medeot and R. Nair (personal communication).

back radiation (Q_B), and latent (Q_L) and sensible (Q_S) heat. The net air-sea heat exchange over the observation site was calculated adding the above-mentioned components; negative values represent upward heat flux (heat loss). Daily net heat fluxes were determined at 42.50° N, 17.50° E by adding the local solar radiation to the daily mean long-wave and turbulent fluxes [*Cardin and Gačić*, 2003].

The total heat content (THC) and total salt content (TSC) within the near-surface layer (50–250 m), an intermediate layer (350–1000 m), and a near-bottom layer (1000–1200 m) were calculated from mooring data using the following equations [*Hecht et al.*, 1985]:

$$THC = \int_{-z1}^{-z2} \rho(z) Cp T(z) dz;$$

$$TSC = \int_{-z1}^{-z2} \rho(z) S(z) dz;$$

where z_1 and z_2 are the maximum and the minimum water depths, T the *in situ* temperature, ρ the *in situ* density and Cp the heat capacity of seawater. Heat and salt contents were calculated by vertically integrating the observed temperatures and salinities between each time series. The stability of the water column is expressed by the Brunt-Väisälä frequency $N = \sqrt{-\frac{g}{\rho_0} \cdot \left(\frac{d\rho_\theta}{dz}\right)}$, where ρ_θ (T, S) is the potential density and ρ_0 is the average potential density between the two seawater parcels. At E2M3A, N was obtained by calculating the difference between the potential densities at the extremes of each layer (as recorded by the moored CT or CTD), with the reference pressure calculated at the average layer depth.

The surface layers of the ocean undergo a regular cycle of convection and restratification in response to the annual cycle of buoyancy fluxes at the sea surface. If we consider thermal and haline expansion coefficients (α and β, respectively), the surface buoyancy flux (in $m^2\ s^{-3}$) can be expressed as:

$$B_0 = B_q + B_p \Rightarrow B_0 = \frac{g\alpha Q_0}{\rho c} + \frac{g\beta Q_L S}{\rho L},$$

where B_q is the thermal buoyancy, B_p the haline buoyancy (due to fresh water), Q_0 the net heat flux and Q_L its contribution to the evaporation, g the gravity, c the specific heat of water, S the surface salinity, and P the precipitation [*Schott et al.*, 1993]. Positive buoyancy flux indicates the surface becoming lighter as a result of heating and/or freshening. The maximum depth H reached by a convective plume is proportional to the magnitude of the surface buoyancy flux [*Marshall and Schott*, 1999]: $H = \frac{\sqrt{2\int B_0 dt}}{N}$, where dt is the time interval that characterizes the event and N is assumed constant [*Marshall and Schott*, 1999].

9.3. RESULTS

9.3.1. Thermohaline Variability in the Southern Adriatic between 2006 and 2010

To evaluate the effects of atmospheric forcing in convective events, heat fluxes were calculated at E2M3A for the whole study period (Figure 9.3). In the southern Adriatic, heat loss at the sea surface is principally enhanced by strong northerly or northeasterly winds; their duration and intensity deeply influence the resulting events (see Figure 9.3a). However, the relationship between maximum convection depth and surface heat flux is not straightforward [*Lilly et al.*, 2003]; for instance, convective preconditioning, on which the strength of the resulting winter convection crucially depends, is a complex phenomenon involving different oceanic and atmospheric events over a broad temporal scale (see, e.g., *Marshall and Schott* [1999]). Overall, winter 2007 was mild (maximum heat flux values of about −350 Wm^{-2} were observed at the end of January). In contrast, the following winters were much harsher (record values of ~−800 Wm^{-2} by mid-February 2008, ~−600 Wm^{-2} at the end of January 2009, and ~−450 Wm^{-2} in January 2010 were registered).

This variability is consistent with that observed in the time series of T and S at E2M3A (Figure 9.4): deep convection was virtually absent during winter 2006/2007 but it was strong during winters 2007/2008 and 2008/2009 and also present during winter 2009/2010.

Following *Cardin et al.* [2011] and referring to the mooring configurations depicted in Figure 9.2, in the following analyses the water column is considered a two-layer system in order to better describe its thermohaline variability on annual and interannual scales. The upper layer includes the CTs and CTDs located at 350-m, 550-m, and 750-m water depth ("intermediate layer"), while the deeper part includes the CT sensors located at 1000 m and 1200 m ("deep layer").

In the intermediate layer, the time series at 350-m depth clearly shows an increase in temperature and salinity between the end of 2006 and the end of 2007 (Figure 9.4a,b). This feature could be considered the manifestation of the end of a warming and salinification phase that started in 2003 in the southern Adriatic (see also *Cardin et al.* [2011]). Indeed, a progressive decrease in temperature and salinity was evident down to 550 m starting from winter 2008, associated with intense convective events (among the strongest observed in the area). Moreover, the salinity time series at 750-m depth also showed an abrupt decrease in winter/spring 2008, while during the

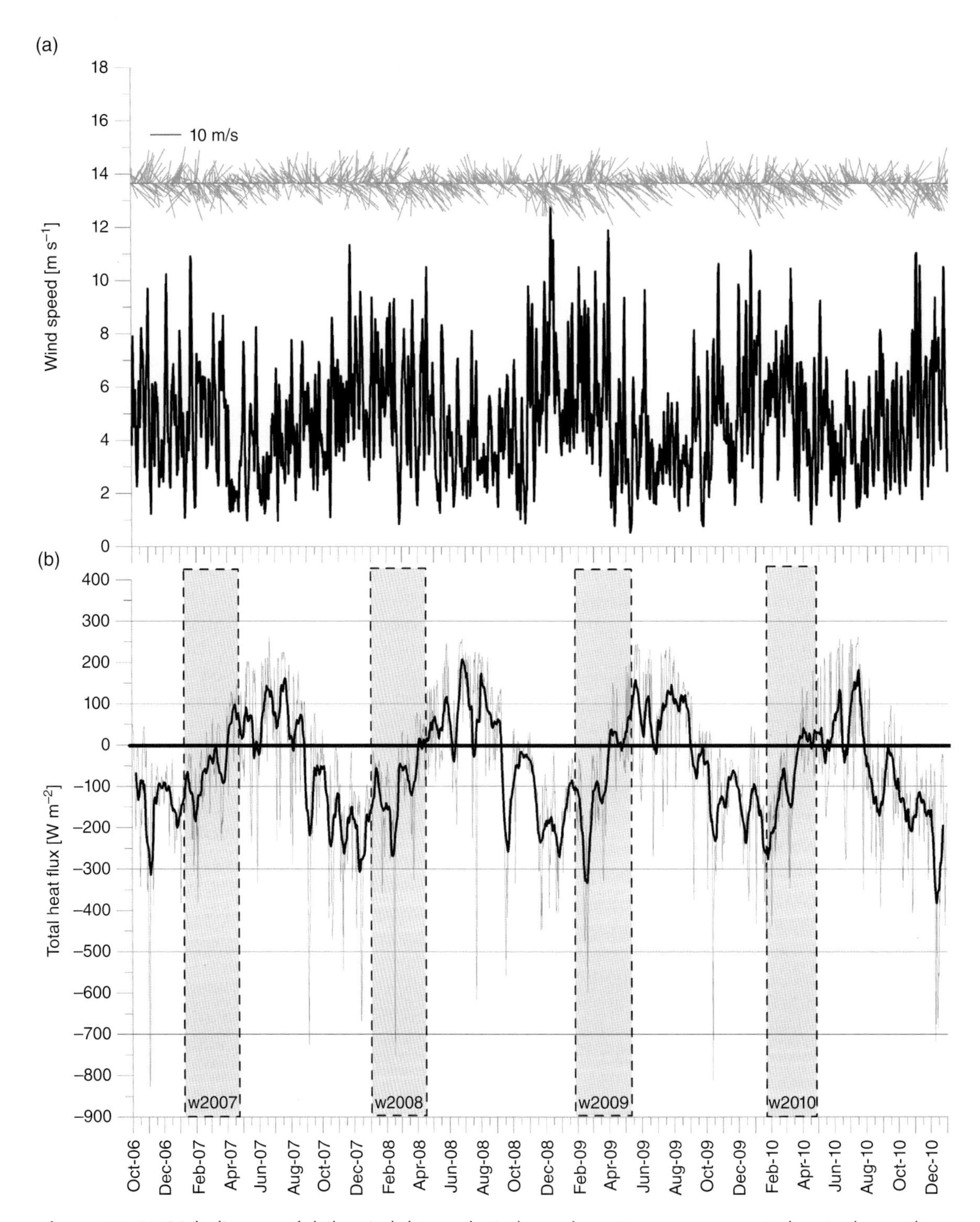

Figure 9.3. (a) Stick diagram of daily wind data and wind speed moving average over 3 days in the southern Adriatic (at 42.50° N, 17.50° E) for the period 2006–2010 obtained from ECMWF operational dataset; (b) daily total surface heat fluxes (grey line indicates observed values; black line indicates moving averages over 15 days). Shaded regions indicate winter periods (Jan–Apr).

following years it was characterized by a slight increase. The intrusion of low-salinity water observed from April 2008 at 750-m depth seems to be attributable to lateral advection. Due to the paucity of data, here we could only hypothesize that this signal was related to the arrival of North Adriatic Deep Water (NAdDW) formed during winter 2008 by convection on the northern Adriatic shelf. This assumption is based on the evidence of intense

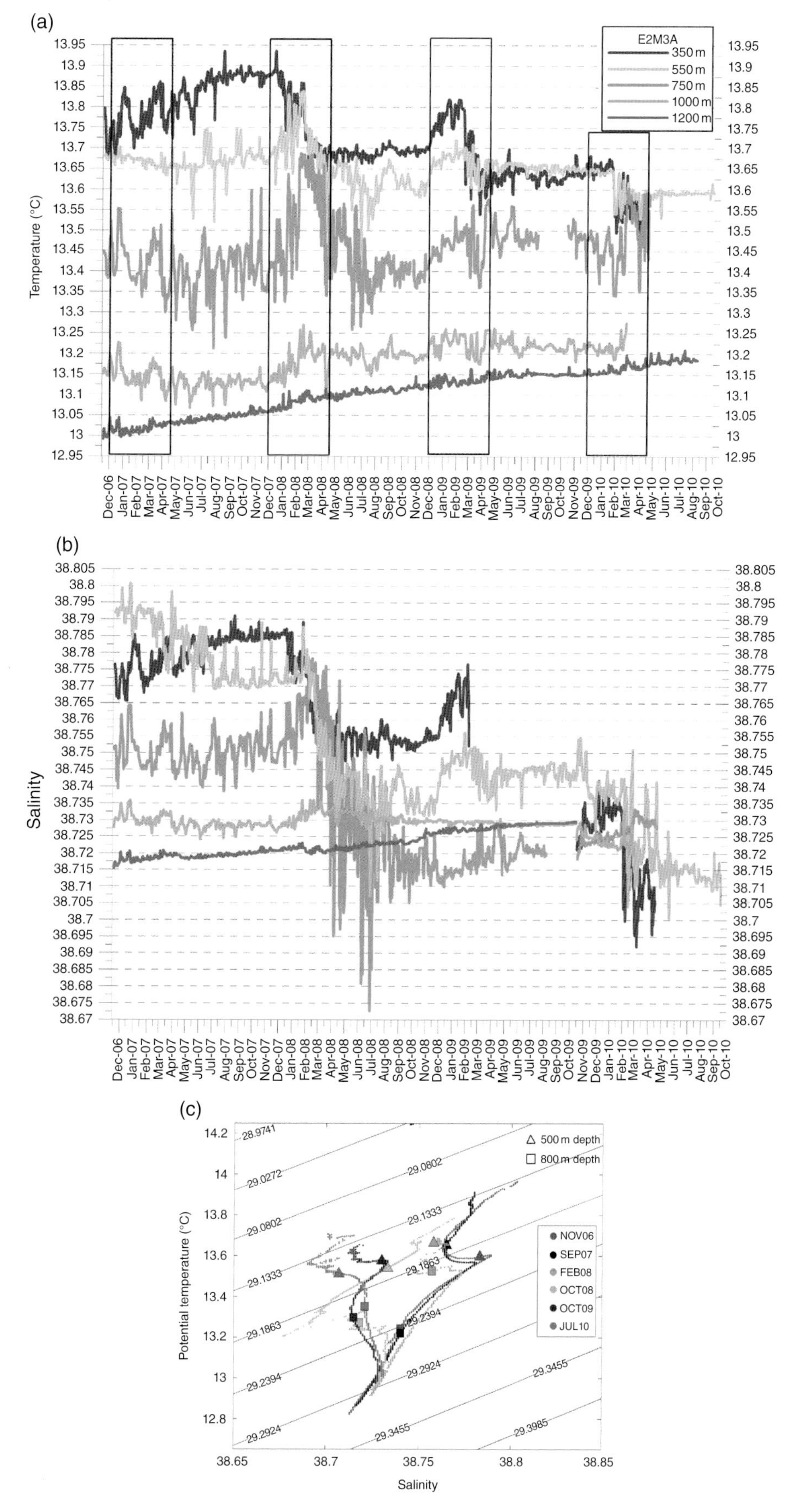

Figure 9.4. Time series of *in situ* (a) temperature and (b) salinity recorded by the CT and CTD sensors installed at E2M3A. Data were despiked and filtered with a 33-h Hamming filter. Winter periods are indicated in panel a. Potential temperature (θ)–Salinity (S) diagram of CTD profiles carried out during the oceanographic cruises in the proximity of the E2M3A (panel (c). For color detail, please see color plate section.

NAdDW production in January and February 2008 [*Cardin et al.*, 2011] in the northern Adriatic.

Gačić et al. [2010] report the entrance into the Adriatic of a noticeable quantity of Modified Atlantic Water, which presumably started in 2007 and which partly replaced waters of Levantine origin usually entering the basin through the Strait of Otranto.

The sudden temperature and salinity decreases noticed after each winter convection phase (2008/2009/2010) confirms the effectiveness of DWF processes, which act on the water column, transferring colder and fresher surface waters toward deeper levels. The combination of DWF and lateral advection processes was responsible for the decrease in temperature down to 550-m depth and in salinity down to 750-m depth after winter 2008. Interestingly, salinity reached values lower than the underlying layer after autumn 2009 (Figure. 9.4b), leading to the disappearance of the LIW core usually located at ~300–400 m (see Figure 9.4c). This fact can be related to the switch of the surface circulation in the northern Ionian from cyclonic to anticyclonic occurring between 2006 and 2007 [*Gačić et al.*, 2010; 2011], which could have reduced or even impeded the entrance of LIW in the Adriatic Sea.

The deep layer showed a tendency opposed to that observed in the intermediate layer. In particular, sensors moored at 1200-m depth revealed a constant temperature increase (linear trend: ~0.05°C y^{-1}) during the whole period (Figure 9.4a). Salinity also showed a similar positive trend (linear trend: ~0.004 y^{-1}) with values increasing from 38.717 to 38.729 between November 2006 and October 2009 (Figure 9.4b). The conductivity sensor at 1200-m depth stopped working after November 2009 but, very probably, the positive trend in the bottom layer continued during 2010, in accordance with CTD profiles carried out in July 2010 (during the MSM15/4 cruise). The effect of vertical mixing induced by deep convection was partly visible, for example, during February 2008 in the time series at 1000-m depth, as a temporary increase in temperature and salinity.

The positive trends observed close to the bottom could also be a consequence of the arrival of dense waters of North Adriatic origin, which usually fill the deepest layers of the southern Adriatic after having sunk through submarine canyons [*Cardin et al.*, 2011; *Rubino et al.*, 2012]. The influence of the LIW, particularly abundant during those years, together with the reduced North Adriatic river runoff observed in the period 2005–2007 [*Cozzi and Giani*, 2011], may have contributed to the salinity increase in the NAdDW formed during these years.

9.3.2. Heat and Salt Content Changes

Heat and salt content were calculated using E2M3A temperature and salinity data (see Figure 9.5). Our analyses reveal that the intermediate layer started losing heat and salt from winter 2008 when the strongest episode of DWF occurred, as described previously. An overall heat and salt loss in the intermediate layer (~ 0.02×10^{10} MJ m^{-2} and ~ 0.003×10^{7} kg m^{-2}, respectively) was accompanied by an overall increase in heat and salt content in the deeper layer (~ +0.01×10^{10} MJ m^{-2} and ~ 0.0012×10^{6} kg m^{-2}). Since no clear evidence of deep convection deeper than 1000 m was observed at E2M3A or from the data collected during the oceanographic cruises, we can exclude the possibility that heat and salt in the deep layer increased as a consequence of local vertical transport associated with DWF.

In order to extend the assessment of THC and TSC to the near surface (where no data from E2M3A exist), CTD measurements performed during eight dedicated cruises in the proximity of the mooring within the E2M3A operation time were also used (Figure 9.5a). The data reveal that in the layer between 50 m and 350 m, a negative trend in heat and salt content existed.

Such a trend is partly masked by the fact that this near-surface layer undergoes strong seasonal variability. So, for instance, the abrupt decrease in THC recorded in February 2008 (cruise IT1; see Figure 9.5a) was the consequence of the DWF events during that period in the area. As mentioned previously, the general negative trend observed in this layer can be associated with the establishment of an anticyclonic circulation in the northern Ionian basin [*Gačić et al.*, 2010], which contrasts with the flow of waters of Levantine origin into the Adriatic Sea. As a consequence, relatively fresh cold waters accumulated in the basin upper layer.

9.3.3. Dense Water Formation Episodes Observed at the E2M3A Site

Both background cyclonic circulation and preconditioning are able to reduce the stability of the water column within the southern Adriatic gyre [*Killworth*, 1983; *Marshall and Schott*, 1999]. To determine the stability of the water column at E2M3A during winter periods, the squared Brunt-Väisälä frequency N^2 was calculated in the layers 350–550 m and 550–750 m. Minimum values of N^2 were reached during February 2008 (Figure. 9.6), in correspondence with energetic convective pulses. Afterward, N^2 underwent a constant increase in the layer 350–550 m (N^2 is centred at 440 m) associated with the restored stability of the water column, but the highest values recorded during 2007 were not reached again within our time series.

The surface buoyancy flux was calculated considering the salinity at the first sensor depth (350 m) under the assumption that, during convection events, the upper water layer is almost homogeneous. Values of ~2.0–2.5×10^{-7} $m^2 s^{-3}$ were observed for the convective event occurring on

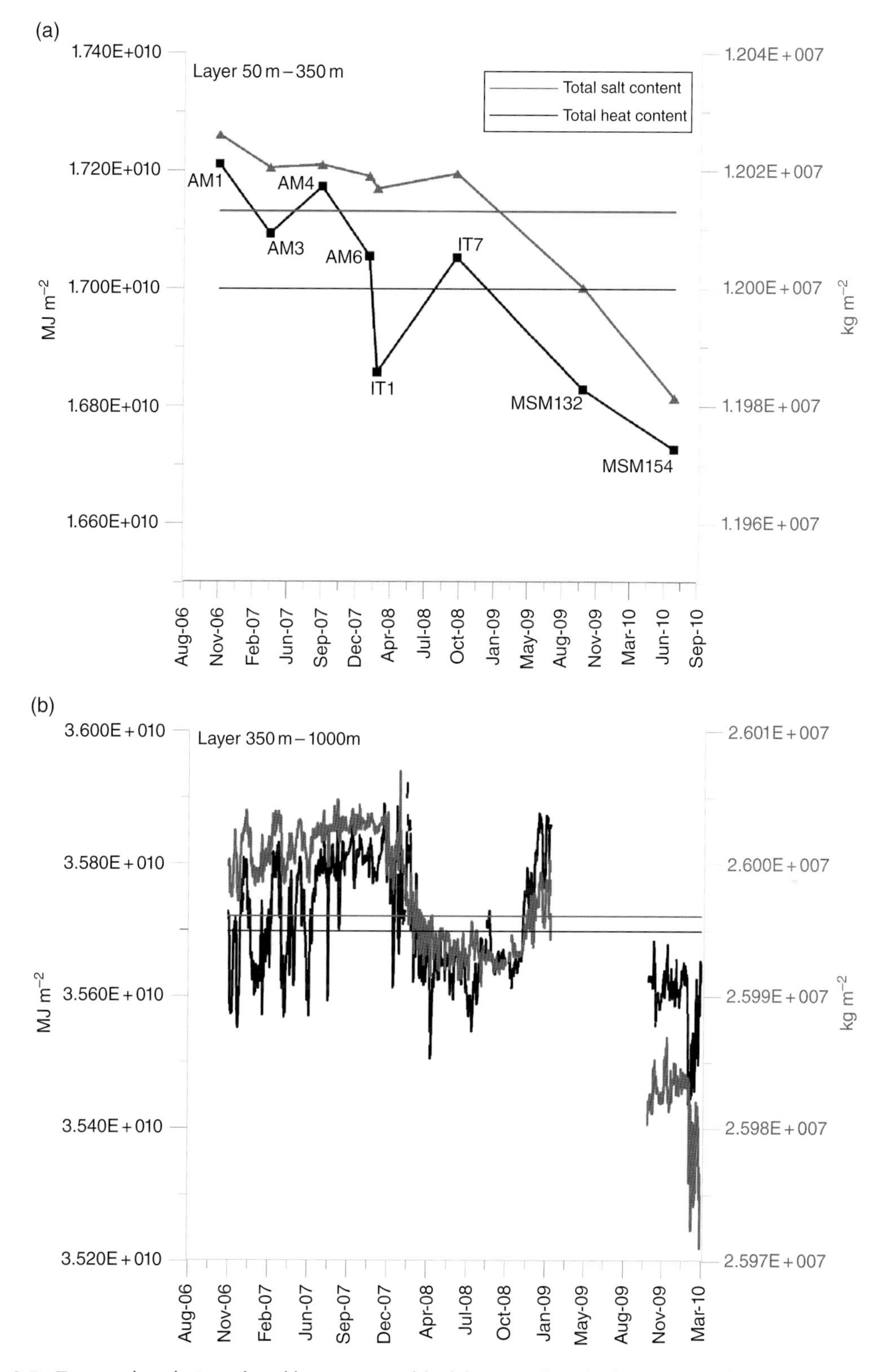

Figure 9.5. Temporal evolution of total heat content (black lines) and total salt content (red lines) at E2M3A. Panel (a) shows the results in the layer between 50 m and 350 m, as obtained from CTD casts in the proximity of E2M3A during its operational time; panels (b) and (c) show the results in the intermediate and deep layers, respectively. For color detail, please see color plate section.

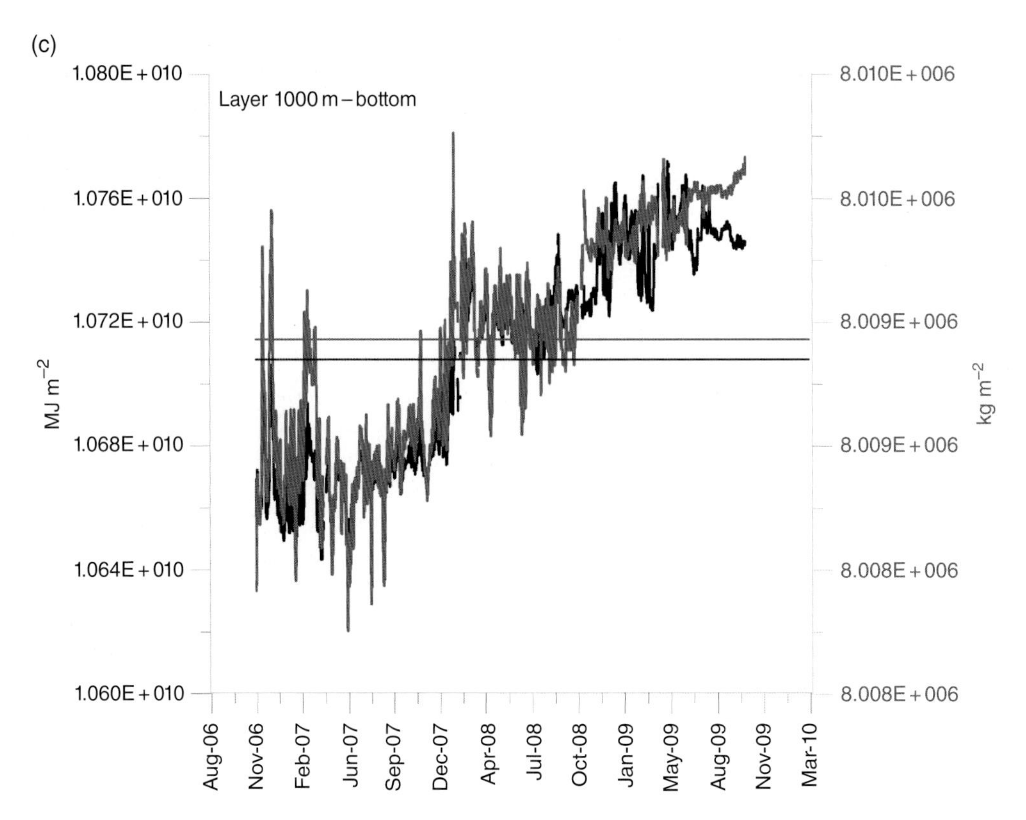

Figure 9.5. (continued)

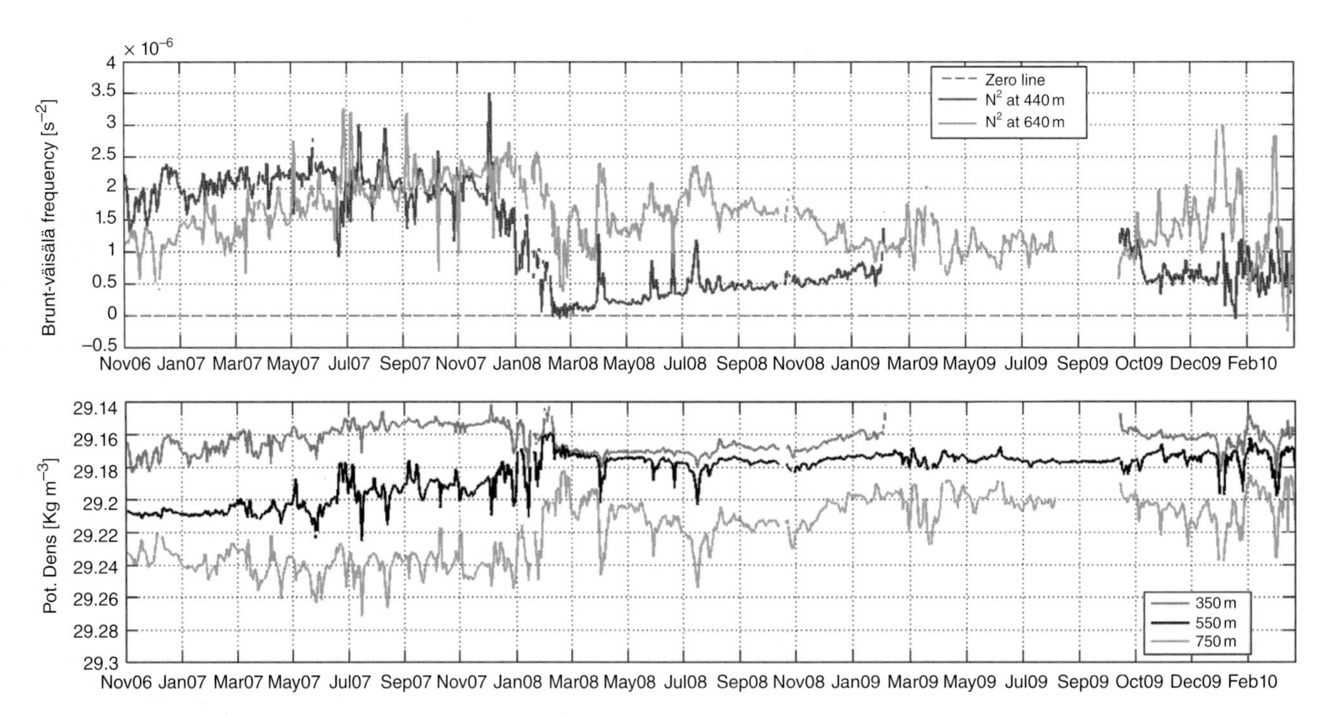

Figure 9.6. Stability of the water column calculated in the layers 350–550 m and 550–750 m (upper panel). The lower panel shows the time series of the potential density at 350 m, 550 m, and 750 m. For color detail, please see color plate section.

17–19 February 2008. These are similar to those reported by *Marshall and Schott* [1999] as typical for convection in the Gulf of Lion. Using the obtained values, a theoretical convective depth was calculated setting $B_0 = 2.5 \times 10^{-7} m^2 s^{-3}$, dt = 2 days and $N = 0.15 \times 10^{-6} s^{-1}$. The result gives values around 750 m, which is in agreement with the experimental observations. Note that, using dt = 3 or $N = 0.1 \times 10^{-6} s^{-1}$, the theoretical convective depth increases to 900 m. The temperature and salinity vertical gradient (and consequently the vertical density gradient) in the intermediate layer decreased progressively from 2008 onward. As a consequence, deep convection events observed in 2009 and 2010 required less energy to be generated.

From the analysis of observed vertical distribution of potential temperature (θ), salinity, potential density, and dissolved oxygen acquired along the Bari-Dubrovnik section during September 2007, January 2008, and February 2008, it is possible to infer the evolution of the water column leading eventually to DWF (Figure. 9.7). During September 2007 (Figure. 9.7a), a doming of the isopycnals due to the prevailing cyclonic circulation was evident along the section. Very probably, this doming remained without significant variations until November–December 2007. At the end of January 2008, the sea surface experienced a strong cooling episode due to sustained Bora wind that produced heat losses larger than 300 Wm^{-2} (see Figure. 9.3). This episode triggered an initial phase of DWF (Figure. 9.7b) extending down to 400–500-m depth. One month later, an even stronger DWF event was again triggered by vigorous Bora wind, which led to a heat loss larger than 800 Wm^{-2} (see Figure. 9.3). *In situ* data collected at the end of February 2008 show a surfacing of the picnocline: indeed, density values up to 29.16–29.17 $kg m^{-3}$ were reached at the sea surface (Figure. 9.7c). The associated deep convection was able to homogenize the water column down to 800-m depth. Observed values were $\theta = 13.69°C$, S = 38.75–38.76, and DO = 225–230 μM. As a result, the LIW core was completely eroded by vertical mixing.

Further information can be extracted by an analysis of the ADCP data obtained at the E2M3A. Winter 2007 was characterized by two major events with large vertical velocities at 300-m depth (Figure 9.8). However, in the second event only, a noticeable temperature decrease was recorded. During winter 2008, vertical velocities up to 3–4 $cm s^{-1}$ at 300-m depth, recorded between 17 and 19 February, were associated with a large temperature decrease. In this case, vertical convection, which reached 800-m depth, had a duration of about 24–36 hours. Similar vertical velocities were registered by the ADCP during February 2009 and were associated with a very large temperature decrease.

9.3.4. On the Relationship between Thermodynamic Forcing and Current Excitation: Identification of Eddylike Patterns

Mesoscale eddies can be important in influencing deep convection: cyclonic eddies can contribute to destabilizing the upper water column, thus exerting preconditioning for DWF, and anticyclonic eddies can push water toward deeper layers, thus simulating the effect of a convection phase [*Legg et al.*, 1998; *Lilly et al.*, 1999; *Straneo and Kawase*, 1999; *Gascard et al.*, 2002]. In the Adriatic Sea, large mesoscale activity has been demonstrated by both *in situ* and remote sensing measurements [*Burrage et al.*, 2009; *Ursella et al.*, 2011].

Current data collected in the upper layer (100–300 m) and close to the bottom (1180 m) of the southern Adriatic reveal a great number of rotational events, some of which seem to be associated with the passage of mesoscale eddies in the proximity of E2M3A. Eddy polarity was determined from the current velocity fields together with the density structure through the water column. Isopycnal fluctuations and temperature variability typically accompany the passage of eddies, so part of the variability observed in the temperature and salinity time series at E2M3A can be attributed to the passage of mesoscale eddies.

In particular, density derived from temperature and salinity measured at E2M3A indicates that eddies can extend down to 1000 m and even reach the bottom. Hence, vortices in the southern Adriatic could contribute to heat and salt transfer throughout the whole water column and they could be one of the major mechanisms of water exchange between intermediate and deep layers, contributing to reducing the partial stagnation of the deep layer, which is characteristic of this region. A comparison between density and current time series at E2M3A reveals a good correlation between sudden density variations and veering of the horizontal velocity field. Figure 9.9 shows the effect on the water column of a cyclonic eddy passage at E2M3A during the 2008 post-convection phase. At 300-m water depth, horizontal velocities reached 20 $cm s^{-1}$ during a cyclone passage. Figure 9.9 also shows that, during April 2008, the cyclonic eddy forced a major upward displacement of deep waters causing a sudden density increase throughout the water column (i.e., ~0.05 Kg m^{-3} at 750-m depth). This phenomenon was clearly visible for several days in the density field down to 1000-m depth. Note that the eddy possibly contributed to the transporting of waters belonging to the boundary current of the southern Adriatic gyre toward the center of the basin.

Most of the observed eddies preserved their rotational structure throughout their thickness, but in some of them this structure changed with depth. In the vortex whose

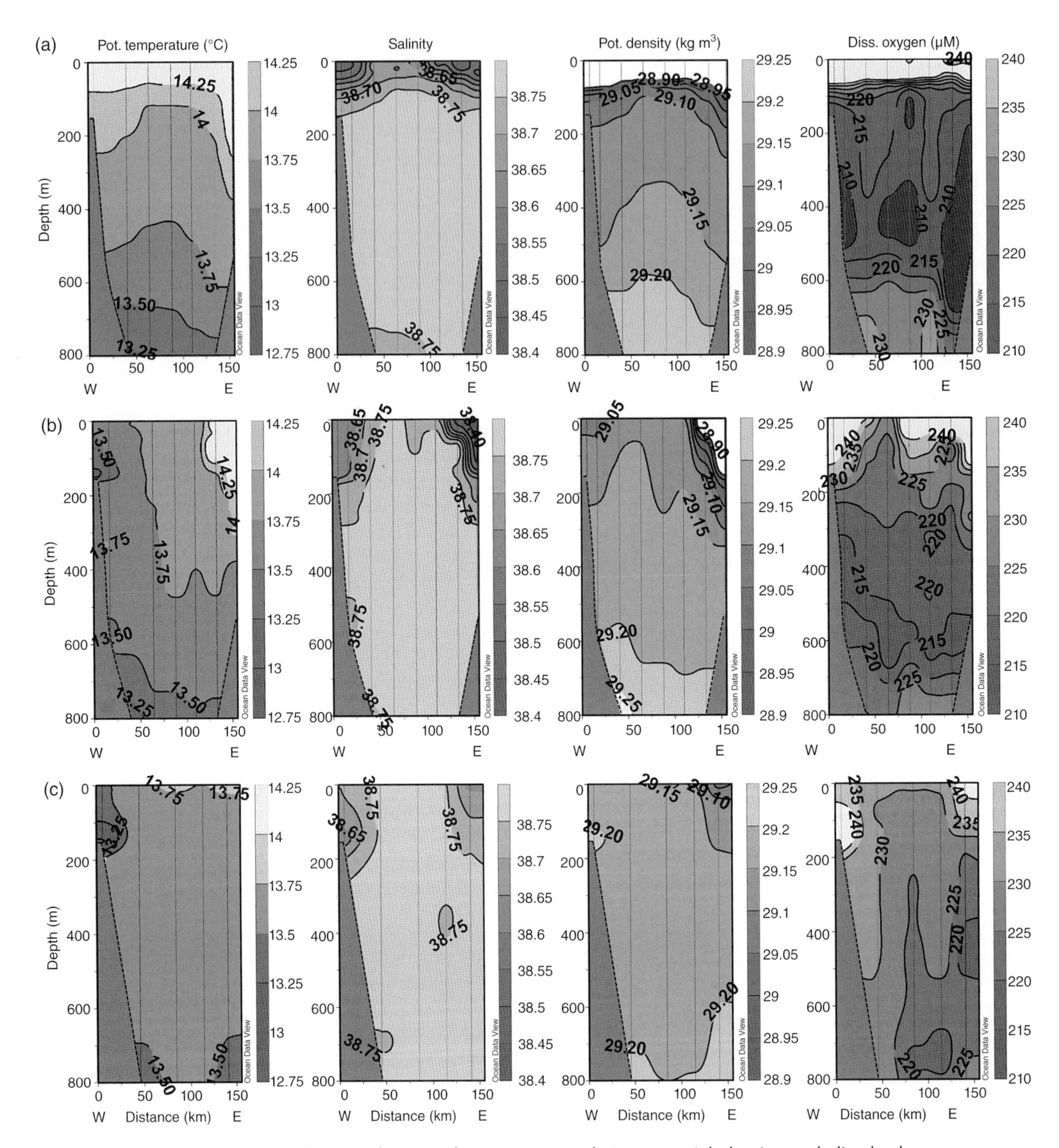

Figure 9.7. Vertical distribution of potential temperature, salinity, potential density, and dissolved oxygen acquired along the Bari-Dubrovnik section during the following periods: a) 14–15 September 2007, b) 31 January–2 February 2008, and c) 23–24 February 2008. The Italian shelf is on the left side of each panel.

evolution is depicted in Figure 9.9, for instance, the rotation of the horizontal velocity in the upper and intermediate layers was cyclonic but that of the bottom layer was anticyclonic. This phenomenon has often been observed in mesoscale and submesoscale eddies in convectively active regions (see, e.g., *Rubino et al.* [2007]). Note also that a cyclonic eddy followed the anticyclonic one, showing a possible dipole-like vortex structure close to the bottom (Figure 9.9c). In general, the analysis of the mooring time series points to an apparent dominance of cyclones in the

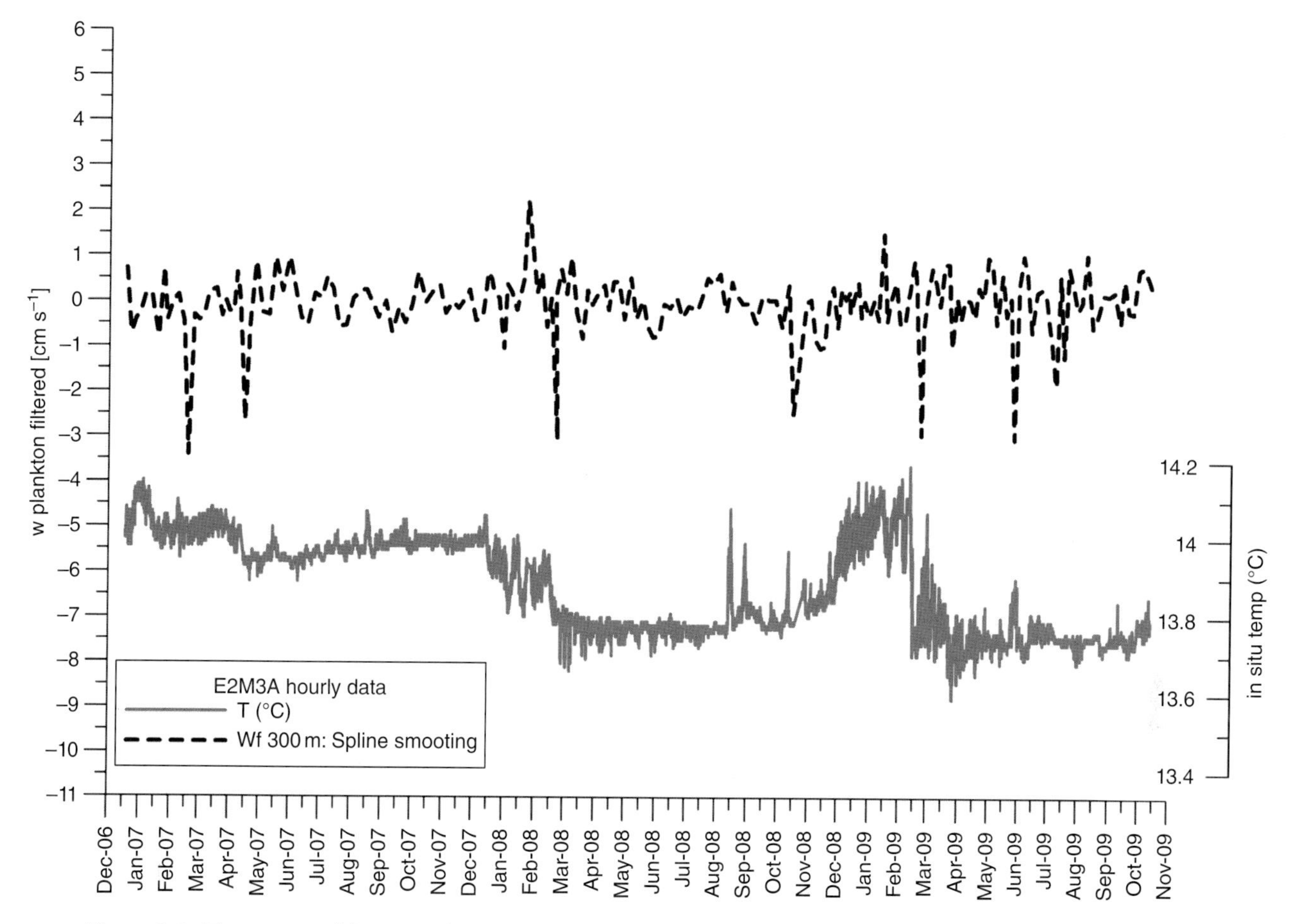

Figure 9.8. Time series of the vertical component w at 300-m depth (black dashed line), cleaned by DVM of zooplankton as measured at E2M3A. The *in situ* temperature time series recorded by the ADCP is also shown in the figure (grey line).

proximity of E2M3A. We counted approximately 19 passages of cyclones and 10 passages of anticyclones during the period 2006–2009 (2007, eight cyclones and four anticyclones; 2008, six cyclones and three anticyclones; 2009, five cyclones and three anticyclones).

An analysis of chlorophyll-*a* distribution from satellite images confirmed the presence of eddylike structures at the sea surface during the eddy events observed at E2M3A. In particular, Figure 9.10 shows the presence of a cyclonic eddy at E2M3A on 7 April 2008, and another three cyclonic vortices in the proximity of the mooring site. Typical diameters of the observed eddies in the southern Adriatic oscillated between 25 and 30 km.

9.4. CONCLUSIONS

In this paper, we analyzed high-frequency, almost continuous, temperature, salinity, current, and heat flux measurements collected in the period 2006–2010 at the deep-ocean observatory E2M3A, located in the Southern Adriatic Pit (maximum depth ~1250 m). This time series is the longest one available so far in the region. The data were complemented with measurements from CTD casts carried out in the study area during eight dedicated oceanographic cruises.

From our data, valuable information about mechanisms responsible for heat and salt exchanges among the surface, the intermediate, and the deep layers, and for local vertical motions, emerges. In general, the restratification period (May–November) was characterized by heat and salt increases in the intermediate layer, clearly evident down to 500–700-m depth. Winter 2007, a year with mild wintertime cooling, was characterized by weak convection, while in the winters 2008, 2009, and 2010, much stronger convection able to reach a considerable depth was observed.

In particular, the most severe episode of DWF was registered between 17 and 19 February 2008, when the water column appeared vertically mixed down to 800-m depth; however, our data do not show any evidence that convection can reach much deeper levels. These depths are usually renewed by waters coming from the northern Adriatic.

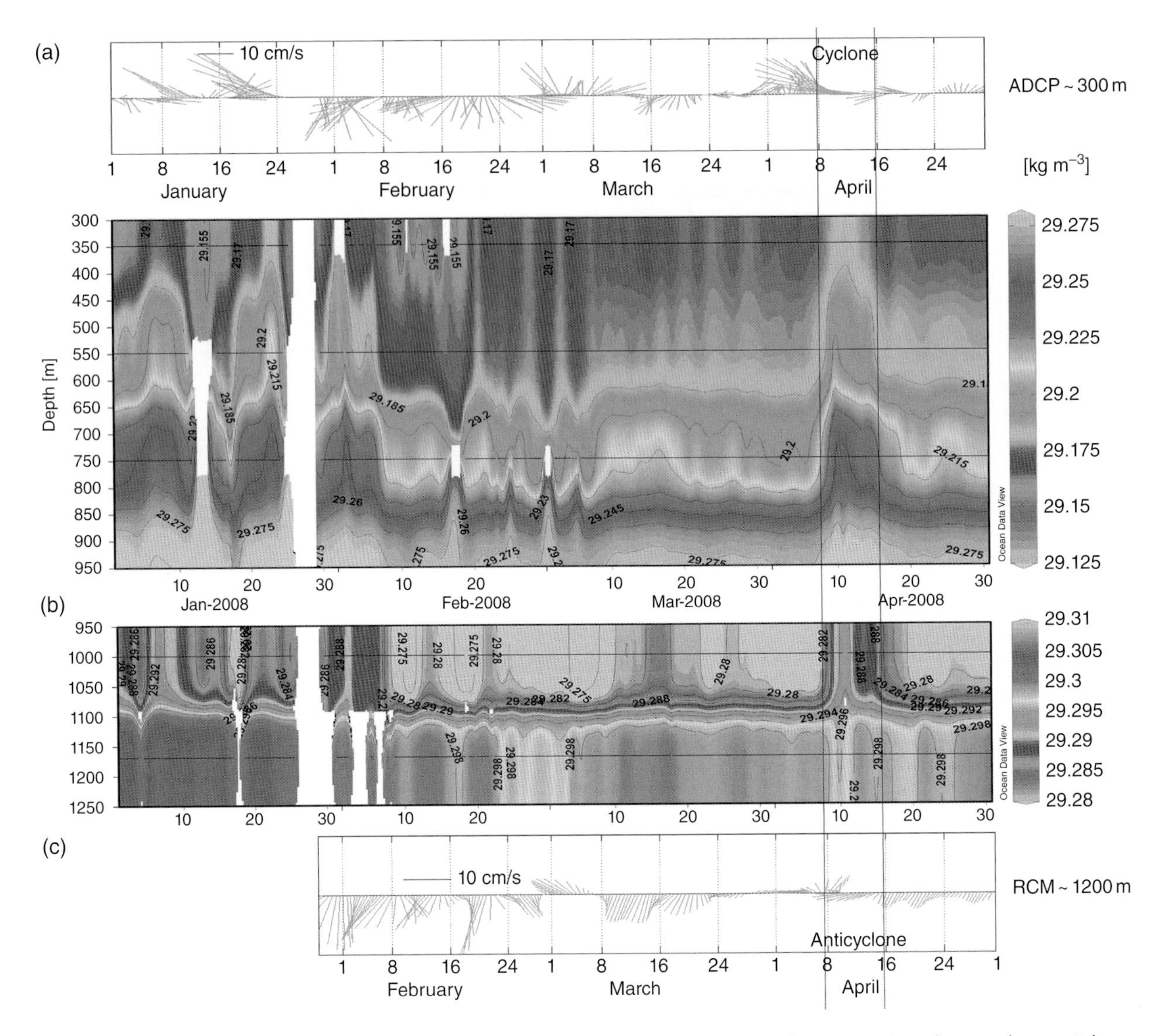

Figure 9.9. Stick diagram of horizontal currents at a) 300 m and at c) 1180 m; b) temporal evolution of potential density throughout the water column. The passage of a cyclonic eddy in the intermediate layer and the concurrent passage of an anticyclonic eddy in the bottom layer are marked by the black lines. The data were recorded between 1 January 2008 and 30 April 2008 at E2M3A. For color detail, please see color plate section.

Maximum temperature and salinity decreases in the intermediate layer, of about 0.4°C and 0.06, respectively, were recorded. Locally, intense mixing caused a heat and salt loss in the intermediate layer of $\sim 0.02 \times 10^{10}$ MJ m^{-2} and $\sim 0.003 \times 10^{7}$ kg m^{-2}, respectively.

During convective events, current velocities in the layer 100–300 m increased noticeably, reaching maximum horizontal values of 30–40 cm s^{-1} and vertical values of 3–4 cm s^{-1}. Besides convection, lateral advection plays an important role in driving the thermohaline variability in the area. Its activity appears to be linked to the periodic oscillation (alternately cyclonic and anticyclonic) of the Ionian surface and intermediate circulation [*Borzelli et al.*, 2009; *Gačić et al.*, 2010, 2011], which favors the intrusion of saltier or fresher waters into the Adriatic.

From our data, it can be evinced that, from the end of 2006 to 2007, a large quantity of water of Levantine origin (high salinity and temperature values) reached the intermediate layer of the southern Adriatic. From 2008 to 2010, however, a reduced influx of salty and warm waters took place. This fact, together with the strong episodes of DWF after 2007, contributed to generation of a fresher and colder Adriatic outflow, which will probably influence the bottom layer circulation of the Ionian basin in the near future.

The deep layer of the observation area experienced increasing heat and salt content. Positive trends of $\sim 0.01 \times 10^{10}$ MJ m^{-2} for heat content and of $\sim 0.0012 \times 10^{6}$ kg m^{-2} for salt content were calculated from our data. In particular, a continuous temperature and

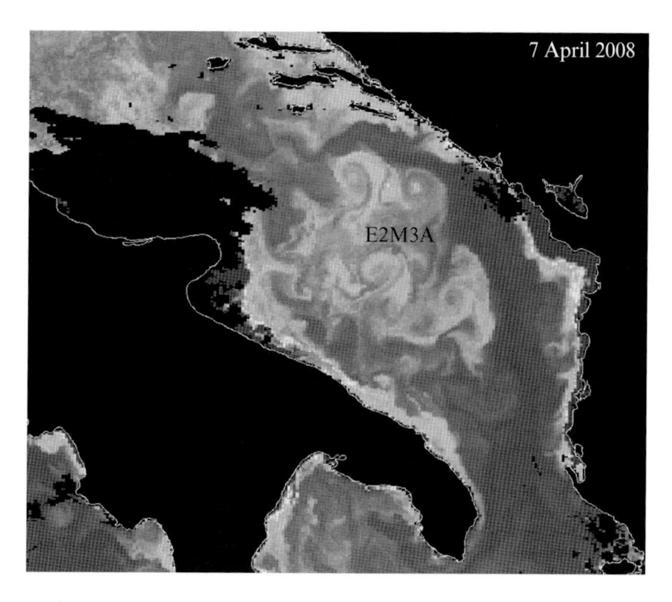

Figure 9.10. Moderate Resolution Imaging Spectroradiometer (MODIS) chlorophyll-*a* distribution in the Southern Adriatic Pit for 7 April 2008. The image has been processed at OGS with SeaWiFS Data Analysis System (Seadas). Data are courtesy of OceanColor database (freely available on http://oceancolor.gsfc.nasa.gov). For color detail, please see color plate section.

salinity increase of ~0.05°C y^{-1} and ~0.004 y^{-1} were observed in the bottom layer during the whole study period. These positive trends could be the result of the arrival of NAdDW formed before 2006 and characterized by relatively high temperature and salinity values. This NAdDW anomaly was, in turn, probably a consequence of the higher atmospheric temperatures over the generation region and of the larger inflow of waters of Levantine origin due to the larger-scale circulation variability observed in recent years.

Current measurements at E2M3A also show a rather frequent passage of mesoscale eddies in the area. These features, which often seem to occupy the whole water column, tend to produce sudden thermohaline changes lasting for as much as 10–15 days. Cyclonic eddies pump water from deeper layers toward the surface while anticyclonic eddies tend to push surface water toward deeper layers. From our data, cyclones seem to be more frequent in the proximity of the mooring site. They also seem to be more energetic than anticyclones: recorded horizontal currents were frequently higher than 20 cm s^{-1}. The effect of the eddy passage is considerable: they contribute to the restratification of the water column during the post-convection phase by exchanging buoyancy between the mixed patch and the surrounding waters, and they transfer heat, salt, and nutrients between the deep and the intermediate layers. They allow efficient communication between waters of the boundary of the southern Adriatic gyre with waters of its central part. They also seem to reduce the effect of stagnation, characteristic of the deepest part of the Southern Adriatic Pit. From a biological point of view, cyclonic eddies are able to raise nutrient concentrations in the surface layer, and thus they are important for phytoplankton development.

Long *in situ* time series are fundamental to understanding the low-frequency oceanic variability. From a comparison between E2M3A and data from other deep-site observatories located in the Mediterranean Sea, valuable information on the climate of the area could be obtained [*Schroeder et al.*, 2012]. Further analyses performed with the help of autonomous Lagrangian platforms (i.e., Argo floats, gliders, and drifters) could also provide additional information on the spatial variability in the southern Adriatic, as well as on the mesoscale structures and the evolution of new-formed deep waters.

Acknowledgments. We are grateful to all OGS technical staff (Fabio Brunetti, Paolo Mansutti, Alessandro Bubbi, Stefano Kückler, and Franco Arena) for their work on the E2M3A mooring, and to Rajesh Nair and Nevio Medeot for the instrument calibration. Corrado Fragiacomo processed the MODIS satellite image used in this paper. This study was supported by the European FP7 project EuroSITES.

REFERENCES

Bensi, M., and S. Kückler (2009), Progetti VECTOR e SESAME: metodologia di elaborazione dei dati ctd raccolti nelle crociere di febbraio e marzo 2008. Technical report 2009/100 OGA 17 OCE. IN-OGS, Trieste (Available from OGS, Borgo Grotta Gigante 42/c 34010 Sgonico [Ts], Italy, mbensi@ogs.trieste.it).

Bensi, M., A. Rubino, V. Cardin, D. Hainbucher, and I. Mancero-Mosquera (2012), Structure and variability of the abyssal water masses in the Ionian Sea in the period 2003–2010, submitted to *JGR-Oceans*.

Borzelli, G. L. E., M. Gačić, V. Cardin, and G. Civitarese (2009), Eastern Mediterranean transient and reversal of the Ionian Sea circulation, *Geophys. Res. Lett. (36)L15108, doi:10.1029/2009GL039261*.

Burrage, D. M., J. W. Book, and P. J. Martin (2009), Eddies and filaments of the western Adriatic current near Cape Gargano, analysis and prediction, *Journal of Marine Systems*, *78*, Supplement (0), S205–S226, doi:10.1016/j.jmarsys.2009.01.024.

Cardin, V., M. Bensi, and M. Pacciaroni (2011), Variability of water mass properties in the last two decades in the southern Adriatic Sea with emphasis on the period 2006–2009, *Cont. Shelf Res.*, *31*, 951–965. doi: 10.1016/j.csr.2011.03.002.

Cardin, V., M. Gačić (2003), Long-term heat flux variability and winter convection in the Adriatic Sea, *Journal of Geophysical Research*, *108*, C9, 8103,doi:10.1029/2002JC001645.

Cozzi, S., and M. Giani (2011), River water and nutrient discharges in the northern Adriatic Sea: Current importance and long term changes, *Continental Shelf Research*, 10.1016/j.csr.2011.08.010.

Dur, G., F. G. Schmitt, and S. Souissi (2007), Analysis of high frequency temperature time series in the Seine estuary from the Marel autonomous monitoring buoy, *Hydrobiologia*, *588*, 59–68, DOI 10.1007/s10750-007-0652-3.

Flagg, C. N., J. A. Vennersch, and R. C. Beardsley (1976), 1974 MIT New England Shelf Dynamics Experiment (March, 1974), Data Report, Part II: The Moored Array, *MIT Rpt.* 76–1.

Gačić, M., G. Civitarese, G. L. E. Borzelli, V. Kovačević, P.-M. Poulain, A. Theocharis, M. Menna, A. Catucci, and N. Zarokanellos (2011), On the relationship between the decadal oscillations of the northern Ionian Sea and the salinity distributions in the eastern Mediterranean, *J. Geophys. Res.*, *116*, 9 pp., doi:201110.1029/2011JC007280.

Gačić, M., G. L. E. Borzelli, G. Civitarese, V. Cardin, and S. Yari (2010), Can internal processes sustain reversals of the ocean upper circulation? The Ionian Sea example, *Geophys. Res. Lett.*, *37*(9), doi:10.1029/2010GL043216.

Gačić, M., P.-M. Poulain, M. Zore-Armanda, and V. Barale (2001), Overview *in Physical Oceanography of the Adriatic Sea: Past, Present and Future*, B. Cushman-Roisin, M. Gacic, P. Poulain, and A. Artegiani, ed., Kluwer Academic Publishers.

Gačić, M., S. Marullo, R. Santoleri, and A. Bergamasco (1997), Analysis of the seasonal and interannual variability of the sea surface temperature field in the Adriatic Sea from AVHRR data, *Journal of Geophysical Research*, *102*, 22937–22946.

Gascard, J.-C., A. J. Watson, M.-J. Messias, K. A. Olsson, T. Johannessen, and K. Simonsen (2002), Long-lived vortices as a mode of deep ventilation in the Greenland Sea, *Nature*, *416* (6880), 525–527, doi:10.1038/416525a.

Hainbucher, D., A. Rubino, and B. Klein, (2006), Water mass characteristics in the deep layers of the western Ionian basin observed during May 2003, *Geophys. Res. Lett.*, *33*, 4 pp.

Hecht, A., Z. Rosentroub, and J. Bishop (1985), Temporal and spatial variations of heat storage in the eastern Mediterranean, *Isr. J. Earth Sci.*, *34*, 51–64.

Hopkins, T. S. (1978), Physical processes in the Mediterranean basins, pp. 285–286, in B. Kjerfve, ed., *Estuarine Transport Processes*, University of South Carolina Press, Columbia, S.C.

Hopkins, T. S. (1985), Physics of the sea, pp. 116–117, in R. Margalef, ed., *Key Environments — Western Mediterranean*, Pergamon Press, Oxford.

Killworth, P. D. (1983), Deep convection in the World Ocean, *Rev. Geophys.*, *21*(1), 1–26, doi:10.1029/RG021i001p00001.

Lascaratos A., W. Roether, K. Nittis, and B. Klein (1999), Recent changes in deep water formation and spreading in the eastern Mediterranean Sea, *Progress in Oceanography*, *44*, 5–36. http://dx.doi.org/10.1016/S0079-6611(99)00019-1.

Legg, S., J. McWilliams, and J. Gao (1998), Localization of deep ocean convection by a mesoscale eddy, *Journal of Physical Oceanography*, *28*(5), 944–970.

Lilly, J. M., P. B. Rhines, F. Schott, K. Lavender, J. Lazier, U. Send, and E. D'Asaro (2003), Observations of the Labrador Sea eddy field, *Progress in Oceanography*, *59*(1), 75–176, doi:10.1016/j.pocean.2003.08.013.

Lilly, J. M., P. B. Rhines, M. Visbeck, R. Davis, J. R. N. Lazier, F. Schott, and D. Farmer (1999), Observing deep convection in the Labrador Sea during winter 1994/95, *Journal of Physical Oceanography*, *29*, 2065–2098.

Malanotte-Rizzoli, P., B. B. Manca, M. Ribera d'Alcala, A. Theocharis, A. Bergamasco, D. Bregant, G. Budillon, G. Civitarese, D. Georgopoulos, A. Michelato, E. Sansone, P. Scarazzato, and E. Souvermezoglou (1997), A synthesis of the Ionian Sea hydrography, circulation and water mass pathways during POEM-Phase I, *Progress in Oceanography 39*, 153–204.

Mantziafou, A., and A. Lascaratos (2008), Deep-water formation in the Adriatic Sea: Interannual simulations for the years 1979–1999, *Deep Sea Research Part I: Oceanographic Research Papers*, *55*, 1403–1427.

Marshall, J., and F. Schott (1999), Open-ocean convection: Observations theory, and models, *Rev. Geophys.*, *37* (1), 1–64.

Oddo P., N. Pinardi, and M. Zavatarelli (2005), A numerical study of the interannual variability of the Adriatic Sea circulation (1999–2002), *Science of the Total Environment*, *353*, 39–56.1.

Orlić, M., M. Kuzmić, and Z. Pasarić (1994), Response of the Adriatic Sea to the Bora and sirocco forcing, *Cont. Shelf Res.*, *14*, 91–116.

Roether, W., B. B. Manca, B. Klein, D. Bregant, D. Georgopoulos, V. Beitzel, V. Kovačević, and A. Luchetta (1996), Recent changes in eastern Mediterranean deep waters, *Science*, *271*, 333–335.

Rubino, A., A. Androssov, and S. Dotsenko (2007), Intrinsic dynamics and long-term evolution of a convectively generated oceanic vortex in the Greenland Sea, *Geophys. Res. Lett.*, *34*, L16607, doi: 10.1029/2007GL030634.

Rubino, A., and D. Hainbucher, (2007), A large abrupt change in the abyssal water masses of the eastern Mediterranean, *Geophys. Res. Lett.*, *34*, 5 pp.

Rubino, A., D. Romanenkov, D. Zanchettin, V. Cardin, D. Hainbucher, M. Bensi, A. Boldrin, L. Langone, S. Miserocchi, and M. Turchetto (2012), On the descent of dense water on a complex canyon system in the southern Adriatic basin, *Continental Shelf Research*, *44* (0), 20–29, doi:10.1016/j.csr.2010.11.009.

Schlitzer, R., W. Roether, H. Oster, H. Junghans, M. Hausmann, H. Johannsen, and A. Michelato (1991), Chlorofluoromethane and oxygen in the eastern Mediterranean, *Deep-Sea Res.*, *38*, 1531–1535.

Schott, F., M. Visbeck, and J. Fischer (1993), Observations of vertical currents and convection in the central Greenland Sea during the winter of 1988–1989, *J. Geophys. Res.*, *98*(C8), 14,401–14,421, doi:10.1029/93JC00658.

Schroeder, K., C. Millot, L. Bengara, S. Ben Ismail, M. Bensi, M. Borghini, G. Budillon, V. Cardin, L. Coppola, C. Curtil,, A. Drago, B. El Moumni, J. Font, J. L. Fuda, J. García-Lafuente, G. P. Gasparini, H. Kontoyiannis, D. Lefevre, P. Puig, P. Raimbault, G. Rougier, J. Salat, C. Sammari, J. C. Sánchez Garrido, A. Sanchez-Roman, S. Sparnocchia, C. Tamburini, I. Taupier-Letage, A. Theocharis, M. Vargas-Yáñez, and A. Vetrano (2012), Long-term monitoring programme of the hydrological variability in the Mediterranean Sea: A first overview of the HYDROCHANGES network, *Ocean Sci. Discuss.*, *9*, 1741–1812, doi:10.5194/osd-9-1741-2012, 2012.

Straneo, F., and M. Kawase (1999), Comparisons of localized convection due to localized forcing and to preconditioning, *Journal of Physical Oceanography*, *29*, 1, 55–68.

Supić, N., and I. Vilibić (2006), Dense water characteristics in the northern Adriatic in the 1967–2000 interval with respect to surface fluxes and Po River discharge rates, *Estuarine, Coastal and Shelf Science*, *66*, 580–593.

Ursella, L., V. Kovačević, and M. Gačić (2011), Footprints of mesoscale eddy passages in the Strait of Otranto (Adriatic Sea), *J. Geopys. Res.*, *116*, C04005, doi:10.1029/2010JC006633.

Vilibić, I., 2003, An analysis of dense water production on the north Adriatic shelf, *Estuarine, Coastal and Shelf Science*, *56*, 697–707.

Wüst, G. (1961), On the vertical circulation of the Mediterranean Sea, *J. Geophys. Res.*, *66*, 261–271.

INDEX

Note: Italicized page locators refer to figures; tables are noted with *t*.

The Mediterranean Sea: Temporal Variability and Spatial Patterns, Geophysical Monograph 202. First Edition.
Edited by Gian Luca Eusebi Borzelli, Miroslav Gačić, Piero Lionello, and Paola Malanotte-Rizzoli.